Device Modeling for Analog and RF CMOS Circuit Design

Device Modeling for Analog and RF CMOS Circuit Design

Trond Ytterdal
Norwegian University of Science and Technology

Yuhua Cheng
Skyworks Solutions Inc., USA

Tor A. Fjeldly
Norwegian University of Science and Technology

WILEY

Other Wiley Editorial Offices

John Wiley & Sons Inc., 111 River Street, Hoboken, NJ 07030, USA

Jossey-Bass, 989 Market Street, San Francisco, CA 94103-1741, USA

Wiley-VCH Verlag GmbH, Boschstr. 12, D-69469 Weinheim, Germany

John Wiley & Sons Australia Ltd, 33 Park Road, Milton, Queensland 4064, Australia

John Wiley & Sons (Asia) Pte Ltd, 2 Clementi Loop #02-01, Jin Xing Distripark, Singapore 129809

John Wiley & Sons Canada Ltd, 22 Worcester Road, Etobicoke, Ontario, Canada M9W 1L1

Wiley also publishes its books in a variety of electronic formats. Some content that appears in print may not be available in electronic books.

British Library Cataloguing in Publication Data

A catalogue record for this book is available from the British Library

ISBN 0-471-49869-6

Typeset in 10/12pt Times by Laserwords Private Limited, Chennai, India
Printed and bound in Great Britain by Antony Rowe Ltd, Chippenham, Wiltshire
This book is printed on acid-free paper responsibly manufactured from sustainable forestry in which at least two trees are planted for each one used for paper production.

Contents

Preface

We are fortunate to live in an age in which microelectronics still enjoy an accelerating growth in performance and complexity. Fortunate, since we are experiencing a remarkable progress in science, in communication technology, in our ability to acquire new knowledge, and in the many other wonderful amenities of modern society, all of which are permeated by and made possible by modern microelectronics. This exponential evolutionary trend, as described by Moore's Law, has now lasted for more than three decades, and is still on track, fueled by a seemingly unending demand for ever better performance and by fierce global competition.

A driving force behind this fantastic progress is the long-term commitment to a steady downscaling of MOSFET/CMOS technology needed to meet the requirements on speed, complexity, circuit density, and power consumption posed by the many advanced applications relying on this technology. The degree of scaling is measured in terms of the half-pitch size of the first-level interconnect in DRAM technology, also termed the "technology node" by the International Technology Roadmap for Semiconductors. At the time of the 2001 ITRS update, the technology node had reached 130 nm, while the smallest features, the MOSFET gate lengths, were a mere 65 nm. Within a decade, these numbers are expected to be close to 40 nm and 15 nm, respectively.

Very important issues in this development are the increasing levels of complexity of the fabrication process and the many subtle mechanisms that govern the properties of deep submicrometer FETs. These mechanisms, dictated by device physics, have to be described and implemented into circuit design tools to empower the circuit designers with the ability to fully utilize the potential of existing and future technologies.

Hence, circuit designers are faced with the relentless challenge of staying updated on the properties, potentials, and the limitations of the latest device technology and device models. This is especially true for designers of analog and radio frequency (RF) integrated circuits, where the sensitivity to the modeling details and the interplay between individual devices is more acute than for digital electronics. A deeper insight into these issues is therefore crucial for gaining the competitive edge needed to ensure first-time-right silicon and to reduce time-to-market for new products.

Existing textbooks on analog and RF CMOS circuit design traditionally lack a thorough treatment of the device modeling challenges outlined above. Our primary objectives with the present book is to bridge the gap between device modeling and analog circuit design by presenting the state-of-the-art MOSFET models that are available in analog and SPICE-type circuit simulators today, together with related modeling issues of importance to both circuit designers and students, now and in the future.

This book is intended as a main or supplementary text for senior and graduate-level courses in analog integrated circuit design, as well as a reference and a text for self or group studies by practicing design engineers. Especially in student design projects, we foresee that this book will be a valuable handbook as well as a reference, both on basic modeling issues and on specific MOSFET models encountered in circuit simulators. Likewise, practicing engineers can use the book to enhance their insight into the principles of MOSFET operation and modeling, thereby improving their design skills.

We assume that the reader already has a basic knowledge of common electronic devices and circuits, and fundamental concepts such as small-signal operation and equivalent circuits.

The book is organized into twelve chapters. In Chapter 1, the reader is introduced to the basic physics, the principles of operation, and the modeling of MOS structures and MOSFETs. This chapter also discusses many of the issues that are important in the modeling of modern-day MOSFETs. Chapter 2 walks the reader through the fabrication steps of modern MOSFET and CMOS technology. In Chapter 3, the special concerns and the challenges of accurate modeling of MOSFETs operating at radio frequencies are discussed. Chapter 4 deals with modeling of noise in MOSFETs. Distortion analysis, discussed in Chapter 5, is of special concern for analog MOSFET circuit design. In Chapters 6, 7, and 8, we present the state-of-the-art MOSFET models that are commonly used by the analog design community today. The models covered are BSIM4, EKV, MOS Model 9 and MOSA1. These chapters are written in a reference style to provide quick lookup when the book is used like a handbook. Chapters 9 and 10 are devoted to the modeling of other devices that are of importance in typical analog CMOS circuits, such as bipolar transistors (Chapter 9) and passive devices, including resistors, capacitors, and inductors (Chapter 10). The remaining two chapters deal with essential industry-related issues of circuit design. Chapter 11 discusses the important topic of modeling of process variations and device mismatch effects and Chapter 12 deals with the quality assurance of the device models used by the design houses.

The book is accompanied by two software application tools, AIM-Spice and MOSCalc. AIM-Spice is a version of SPICE with standard SPICE parameters, very familiar to many electrical engineers and electrical engineering students. Running under the Microsoft Windows family of operating systems, it takes full advantage of the available graphics user interface. The AIM-Spice software will run on all PCs equipped with Windows 95, 98, ME, NT 4, 2000, or XP. In addition to all the models included into Berkeley SPICE (Version 3e.1), AIM-Spice incorporates BSIM4, EKV, and MOSA1, which were covered in Chapters 6, 7, and 8. A limited version of AIM-Spice can be downloaded from *www.aimspice.com*. The second tool, MOSCalc, is a Web-based calculator for rapid estimates of MOSFET large- and small-signal parameters. The designer enters the gate length and width, and a range of biasing voltages and/or the transistor currents, whereupon quantities such as gate overdrive voltage, effective threshold voltage, drain-source saturation voltage, all terminal currents, transconductance, channel conductance, and all small signal intrinsic capacitances are calculated. MOSCalc is available at *ngl.fysel.ntnu.no*.

These dedicated software tools allow students to solve real engineering problems, which brings semiconductor device physics and modeling home to the user at a very practical level, bridging the gap between theory and practice. AIM-Spice and MOSCalc can be used routinely by practicing engineers during the design phase of analog integrated circuits.

We are grateful to the following colleagues for their suggestions and/or for reviewing portions of this book: Matthias Bucher and Bjørnar Hernes. We would also like to express our appreciation to the staff at Wiley, UK, and in particular to Kathryn Sharples, for making possible the timely production of the book.

Finally, we would like to thank our families for their great support, patience, and understanding provided throughout the period of writing.

1

MOSFET Device Physics and Operation

1.1 INTRODUCTION

A field effect transistor (FET) operates as a conducting semiconductor channel with two ohmic contacts – the *source* and the *drain* – where the number of charge carriers in the channel is controlled by a third contact – the *gate*. In the vertical direction, the gate-channel-substrate structure (gate junction) can be regarded as an orthogonal two-terminal device, which is either a MOS structure or a reverse-biased rectifying device that controls the mobile charge in the channel by capacitive coupling (field effect). Examples of FETs based on these principles are metal-oxide-semiconductor FET (MOSFET), junction FET (JFET), metal-semiconductor FET (MESFET), and heterostructure FET (HFETs). In all cases, the stationary gate-channel impedance is very large at normal operating conditions. The basic FET structure is shown schematically in Figure 1.1.

The most important FET is the MOSFET. In a silicon MOSFET, the gate contact is separated from the channel by an insulating silicon dioxide (SiO_2) layer. The charge carriers of the conducting channel constitute an inversion charge, that is, electrons in the case of a *p*-type substrate (*n*-channel device) or holes in the case of an *n*-type substrate (*p*-channel device), induced in the semiconductor at the silicon-insulator interface by the voltage applied to the gate electrode. The electrons enter and exit the channel at n^+ source and drain contacts in the case of an *n*-channel MOSFET, and at p^+ contacts in the case of a *p*-channel MOSFET.

MOSFETs are used both as discrete devices and as active elements in digital and analog monolithic integrated circuits (ICs). In recent years, the device feature size of such circuits has been scaled down into the deep submicrometer range. Presently, the 0.13-μm technology node for complementary MOSFET (CMOS) is used for very large scale ICs (VLSIs) and, within a few years, sub-0.1-μm technology will be available, with a commensurate increase in speed and in integration scale. Hundreds of millions of transistors on a single chip are used in microprocessors and in memory ICs today.

CMOS technology combines both *n*-channel and *p*-channel MOSFETs to provide very low power consumption along with high speed. New silicon-on-insulator (SOI) technology may help achieve three-dimensional integration, that is, packing of devices into many

Device Modeling for Analog and RF CMOS Circuit Design. T. Ytterdal, Y. Cheng and T. A. Fjeldly
© 2003 John Wiley & Sons, Ltd ISBN: 0-471-49869-6

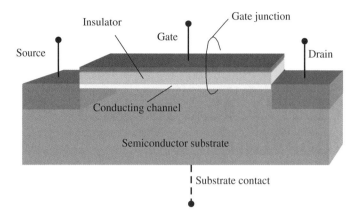

Figure 1.1   Schematic illustration of a generic field effect transistor. This device can be viewed as a combination of two orthogonal two-terminal devices

layers, with a dramatic increase in integration density. New improved device structures and the combination of bipolar and field effect technologies (BiCMOS) may lead to further advances, yet unforeseen. One of the rapidly growing areas of CMOS is in analog circuits, spanning a variety of applications from audio circuits operating at the kilohertz (kHz) range to modern wireless applications operating at gigahertz (GHz) frequencies.

1.2 THE MOS CAPACITOR

To understand the MOSFET, we first have to analyze the MOS capacitor, which constitutes the important gate-channel-substrate structure of the MOSFET. The MOS capacitor is a two-terminal semiconductor device of practical interest in its own right. As indicated in Figure 1.2, it consists of a metal contact separated from the semiconductor by a dielectric insulator. An additional ohmic contact is provided at the semiconductor substrate. Almost universally, the MOS structure utilizes doped silicon as the substrate and its native oxide, silicon dioxide, as the insulator. In the silicon–silicon dioxide system, the density of surface states at the oxide–semiconductor interface is very low compared to the typical channel carrier density in a MOSFET. Also, the insulating quality of the oxide is quite good.

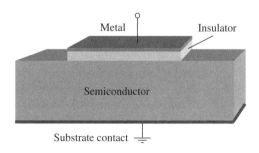

Figure 1.2   Schematic view of a MOS capacitor

We assume that the insulator layer has infinite resistance, preventing any charge carrier transport across the dielectric layer when a bias voltage is applied between the metal and the semiconductor. Instead, the applied voltage will induce charges and counter charges in the metal and in the interface layer of the semiconductor, similar to what we expect in the metal plates of a conventional parallel plate capacitor. However, in the MOS capacitor we may use the applied voltage to control the type of interface charge we induce in the semiconductor – majority carriers, minority carriers, and depletion charge.

Indeed, the ability to induce and modulate a conducting sheet of minority carriers at the semiconductor–oxide interface is the basis for the operation of the MOSFET.

1.2.1 Interface Charge

The induced interface charge in the MOS capacitor is closely linked to the shape of the electron energy bands of the semiconductor near the interface. At zero applied voltage, the bending of the energy bands is ideally determined by the difference in the work functions of the metal and the semiconductor. This band bending changes with the applied bias and the bands become flat when we apply the so-called flat-band voltage given by

$$V_{FB} = (\Phi_m - \Phi_s)/q = (\Phi_m - X_s - E_c + E_F)/q, \tag{1.1}$$

where Φ_m and Φ_s are the work functions of the metal and the semiconductor, respectively, X_s is the electron affinity for the semiconductor, E_c is the energy of the conduction band edge, and E_F is the Fermi level at zero applied voltage. The various energies involved are indicated in Figure 1.3, where we show typical band diagrams of a MOS capacitor at zero bias, and with the voltage $V = V_{FB}$ applied to the metal contact relative to the semiconductor–oxide interface. (Note that in real devices, the flat-band voltage may be

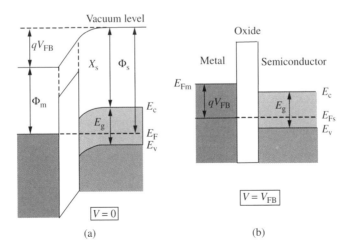

Figure 1.3 Band diagrams of MOS capacitor (a) at zero bias and (b) with an applied voltage equal to the flat-band voltage. The flat-band voltage is negative in this example

affected by surface states at the semiconductor–oxide interface and by fixed charges in the insulator layer.)

At stationary conditions, no net current flows in the direction perpendicular to the interface owing to the very high resistance of the insulator layer (however, this does not apply to very thin oxides of a few nanometers, where tunneling becomes important, see Section 1.5). Hence, the Fermi level will remain constant inside the semiconductor, independent of the biasing conditions. However, between the semiconductor and the metal contact, the Fermi level is shifted by $E_{\text{Fm}} - E_{\text{Fs}} = qV$ (see Figure 1.3(b)). Hence, we have a quasi-equilibrium situation in which the semiconductor can be treated as if in thermal equilibrium.

A MOS structure with a p-type semiconductor will enter the *accumulation* regime of operation when the voltage applied between the metal and the semiconductor is more negative than the flat-band voltage ($V_{\text{FB}} < 0$ in Figure 1.3). In the opposite case, when $V > V_{\text{FB}}$, the semiconductor–oxide interface first becomes depleted of holes and we enter the so-called *depletion* regime. By increasing the applied voltage, the band bending becomes so large that the energy difference between the Fermi level and the bottom of the conduction band at the insulator–semiconductor interface becomes smaller than that between the Fermi level and the top of the valence band. This is the case indicated for $V = 0\,\text{V}$ in Figure 1.3(a). Carrier statistics tells us that the electron concentration then will exceed the hole concentration near the interface and we enter the *inversion* regime. At still larger applied voltage, we finally arrive at a situation in which the electron volume concentration at the interface exceeds the doping density in the semiconductor. This is the strong inversion case in which we have a significant conducting sheet of inversion charge at the interface.

The symbol ψ is used to signify the potential in the semiconductor measured relative to the potential at a position x deep inside the semiconductor. Note that ψ becomes positive when the bands bend down, as in the example of a p-type semiconductor shown in Figure 1.4. From equilibrium electron statistics, we find that the intrinsic Fermi level E_i in the bulk corresponds to an energy separation $q\varphi_b$ from the actual Fermi level E_F of the doped semiconductor,

$$\varphi_b = V_{\text{th}} \ln \left(\frac{N_a}{n_i} \right), \tag{1.2}$$

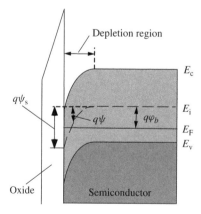

Figure 1.4   Band diagram for MOS capacitor in weak inversion ($\varphi_b < \psi_s < 2\varphi_b$)

where V_{th} is the thermal voltage, N_a is the shallow acceptor density in the p-type semicon-ductor and n_i is the intrinsic carrier density of silicon. According to the usual definition, strong inversion is reached when the total band bending equals $2q\varphi_b$, corresponding to the surface potential $\psi_s = 2\varphi_b$. Values of the surface potential such that $0 < \psi_s < 2\varphi_b$ corre-spond to the depletion and the weak inversion regimes, $\psi_s = 0$ is the flat-band condition, and $\psi_s < 0$ corresponds to the accumulation mode.

The surface concentrations of holes and electrons are expressed in terms of the surface potential as follows using equilibrium statistics,

$$p_s = N_a \exp(-\psi_s/V_{th}), \tag{1.3}$$

$$n_s = n_i^2/p_s = n_{po} \exp(\psi_s/V_{th}), \tag{1.4}$$

where $n_{po} = n_i^2/N_a$ is the equilibrium concentration of the minority carriers (electrons) in the bulk.

The potential distribution $\psi(x)$ in the semiconductor can be determined from a solution of the one-dimensional Poisson's equation:

$$\frac{d^2\psi(x)}{dx^2} = -\frac{\rho(x)}{\varepsilon_s}, \tag{1.5}$$

where ε_s is the semiconductor permittivity, and the space charge density $\rho(x)$ is given by

$$\rho(x) = q(p - n - N_a). \tag{1.6}$$

The position-dependent hole and electron concentrations may be expressed as

$$p = N_a \exp(-\psi/V_{th}), \tag{1.7}$$

$$n = n_{po} \exp(\psi/V_{th}). \tag{1.8}$$

Note that deep inside the semiconductor, we have $\psi(\infty) = 0$.

In general, the above equations do not have an analytical solution for $\psi(x)$. How-ever, the following expression can be derived for the electric field F_s at the insula-tor–semiconductor interface, in terms of the surface potential (see, e.g., Fjeldly *et al.* 1998),

$$F_s = \sqrt{2}\frac{V_{th}}{L_{Dp}} f\left(\frac{\psi_s}{V_{th}}\right), \tag{1.9}$$

where the function f is defined by

$$f(u) = \pm\sqrt{[\exp(-u) + u - 1] + \frac{n_{po}}{N_a}[\exp(u) - u - 1]}, \tag{1.10}$$

and

$$L_{Dp} = \sqrt{\frac{\varepsilon_s V_{th}}{q N_a}} \tag{1.11}$$

is called the *Debye length*. In (1.10), a positive sign should be chosen for a positive ψ_s and a negative sign corresponds to a negative ψ_s.

Using Gauss' law, we can relate the total charge Q_s per unit area (carrier charge and depletion charge) in the semiconductor to the surface electric field by

$$Q_s = -\varepsilon_s F_s. \tag{1.12}$$

At the flat-band condition ($V = V_{FB}$), the surface charge is equal to zero. In accumulation ($V < V_{FB}$), the surface charge is positive, and in depletion and inversion ($V > V_{FB}$), the surface charge is negative. In accumulation (when $|\psi_s|$ exceeds a few times V_{th}) and in strong inversion, the mobile sheet charge density is proportional to $\exp[|\psi_s|/(2V_{th})])$. In depletion and weak inversion, the depletion charge is dominant and its sheet density varies as $\psi_s^{1/2}$. Figure 1.5 shows $|Q_s|$ versus ψ_s for p-type silicon with a doping density of 10^{16}/cm^3.

In order to relate the semiconductor surface potential to the applied voltage V, we have to investigate how this voltage is divided between the insulator and the semiconductor. Using the condition of continuity of the electric flux density at the semiconductor–insulator interface, we find

$$\varepsilon_s F_s = \varepsilon_i F_i, \tag{1.13}$$

where ε_i is the permittivity of the oxide layer and F_i is the constant electric field in the insulator (assuming no space charge). Hence, with an insulator thickness d_i, the voltage drop across the insulator becomes $F_i d_i$. Accounting for the flat-band voltage, the applied voltage can be written as

$$V = V_{FB} + \psi_s + \varepsilon_s F_s/c_i, \tag{1.14}$$

where $c_i = \varepsilon_i/d_i$ is the insulator capacitance per unit area.

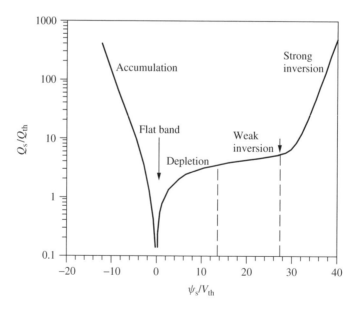

Figure 1.5   Normalized total semiconductor charge per unit area versus normalized surface potential for p-type Si with $N_a = 10^{16}$/cm^3. $Q_{th} = (2\varepsilon_s q N_a V_{th})^{1/2} \approx 9.3 \times 10^{-9}$ C/cm^2 and $V_{th} \approx 0.026$ V at $T = 300$ K. The arrows indicate flat-band condition and onset of strong inversion

1.2.2 Threshold Voltage

The threshold voltage $V = V_T$, corresponding to the onset of the strong inversion, is one of the most important parameters characterizing metal-insulator-semiconductor devices. As discussed above, strong inversion occurs when the surface potential ψ_s becomes equal to $2\varphi_b$. For this surface potential, the charge of the free carriers induced at the insulator–semiconductor interface is still small compared to the charge in the depletion layer, which is given by

$$Q_{dT} = -qN_a d_{dT} = -\sqrt{4\varepsilon_s q N_a \varphi_b}, \tag{1.15}$$

where $d_{dT} = (4\varepsilon_s \varphi_b / q N_a)^{1/2}$ is the width of the depletion layer at threshold. Accordingly, the electric field at the semiconductor–insulator interface becomes

$$F_{sT} = -Q_{dT}/\varepsilon_s = \sqrt{4q N_a \varphi_b / \varepsilon_s}. \tag{1.16}$$

Hence, substituting the threshold values of ψ_s and F_s in (1.14), we obtain the following expression for the threshold voltage:

$$V_T = V_{FB} + 2\varphi_b + \sqrt{4\varepsilon_s q N_a \varphi_b}/c_i. \tag{1.17}$$

Figure 1.6 shows typical calculated dependencies of V_T on doping level and dielectric thickness.

For the MOS structure shown in Figure 1.2, the application of a bulk bias V_B is simply equivalent to changing the applied voltage from V to $V - V_B$. Hence, the threshold

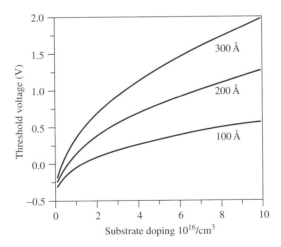

Figure 1.6 Dependence of MOS threshold voltage on the substrate doping level for different thicknesses of the dielectric layer. Parameters used in calculation: energy gap, 1.12 eV; effective density of states in the conduction band, $3.22 \times 10^{25}/m^3$; effective density of states in the valence band, $1.83 \times 10^{25}/m^3$; semiconductor permittivity, 1.05×10^{-10} F/m; insulator permittivity, 3.45×10^{-11} F/m; flat-band voltage, -1 V; temperature: 300 K. Reproduced from Lee K., Shur M., Fjeldly T. A., and Ytterdal T. (1993) *Semiconductor Device Modeling for VLSI*, Prentice Hall, Englewood Cliffs, NJ

referred to the ground potential is simply shifted by V_B. However, the situation will be different in a MOSFET where the conducting layer of mobile electrons may be maintained at some constant potential. Assuming that the inversion layer is grounded, V_B biases the effective junction between the inversion layer and the substrate, changing the amount of charge in the depletion layer. In this case, the threshold voltage becomes

$$V_T = V_{FB} + 2\varphi_b + \sqrt{2\varepsilon_s q N_a(2\varphi_b - V_B)}/c_i. \tag{1.18}$$

Note that the threshold voltage may also be affected by so-called fast surface states at the semiconductor–oxide interface and by fixed charges in the insulator layer. However, this is not a significant concern with modern day fabrication technology.

As discussed above, the threshold voltage separates the subthreshold regime, where the mobile carrier charge increases exponentially with increasing applied voltage, from the above-threshold regime, where the mobile carrier charge is linearly dependent on the applied voltage. However, there is no clear point of transition between the two regimes, so different definitions and experimental techniques have been used to determine V_T. Sometimes (1.17) and (1.18) are taken to indicate the onset of so-called moderate inversion, while the onset of strong inversion is defined to be a few thermal voltages higher.

1.2.3 MOS Capacitance

In a MOS capacitor, the metal contact and the neutral region in the doped semiconductor substrate are separated by the insulator layer, the channel, and the depletion region. Hence, the capacitance C_{mos} of the MOS structure can be represented as a series connection of the insulator capacitance $C_i = S\varepsilon_i/d_i$, where S is the area of the MOS capacitor, and the capacitance of the active semiconductor layer C_s,

$$C_{mos} = \frac{C_i C_s}{C_i + C_s}. \tag{1.19}$$

The semiconductor capacitance can be calculated as

$$C_s = S \left| \frac{dQ_s}{d\psi_s} \right|, \tag{1.20}$$

where Q_s is the total charge density per unit area in the semiconductor and ψ_s is the surface potential. Using (1.9) to (1.12) for Q_s and performing the differentiation, we obtain

$$C_s = \frac{C_{so}}{\sqrt{2}f(\psi_s/V_{th})} \left\{ 1 - \exp\left(-\frac{\psi_s}{V_{th}}\right) + \frac{n_{po}}{N_a}\left[\exp\left(\frac{\psi_s}{V_{th}}\right) - 1\right] \right\}. \tag{1.21}$$

Here, $C_{so} = S\varepsilon_s/L_{Dp}$ is the semiconductor capacitance at the flat-band condition (i.e., for $\psi_s = 0$) and L_{Dp} is the Debye length given by (1.11). Equation (1.14) describes the relationship between the surface potential and the applied bias.

The semiconductor capacitance can formally be represented as the sum of two capacitances – a depletion layer capacitance C_d and a free carrier capacitance C_{fc}. C_{fc} together with a series resistance R_{GR} describes the delay caused by the generation/recombination mechanisms in the buildup and removal of inversion charge in response to changes in the bias voltage (see following text). The depletion layer capacitance is given by

$$C_d = S\varepsilon_s/d_d, \tag{1.22}$$

where

$$d_d = \sqrt{\frac{2\varepsilon_s \psi_s}{q N_a}} \tag{1.23}$$

is the depletion layer width. In strong inversion, a change in the applied voltage will primarily affect the minority carrier charge at the interface, owing to the strong dependence of this charge on the surface potential. This means that the depletion width reaches a maximum value with no significant further increase in the depletion charge. This maximum depletion width d_{dT} can be determined from (1.23) by applying the threshold condition, $\psi_s = 2\varphi_b$. The corresponding minimum value of the depletion capacitance is $C_{dT} = S\varepsilon_s/d_{dT}$.

The free carrier contribution to the semiconductor capacitance can be formally expressed as

$$C_{fc} = C_s - C_d. \tag{1.24}$$

As indicated, the variation in the minority carrier charge at the interface comes from the processes of generation and recombination mechanisms, with the creation and removal of electron–hole pairs. Once an electron–hole pair is generated, the majority carrier (a hole in p-type material and an electron in n-type material) is swept from the space charge region into the substrate by the electric field of this region. The minority carrier is swept in the opposite direction toward the semiconductor–insulator interface. The variation in minority carrier charge at the semiconductor–insulator interface therefore proceeds at a rate limited by the time constants associated with the generation/recombination processes. This finite rate represents a delay, which may be represented electrically in terms of an RC product consisting of the capacitance C_{fc} and the resistance R_{GR}, as reflected in the equivalent circuit of the MOS structure shown in Figure 1.7. The capacitance C_{fc} becomes important in the inversion regime, especially in strong inversion where the mobile charge is important. The resistance R_s in the equivalent circuit is the series resistance of the neutral semiconductor layer and the contacts.

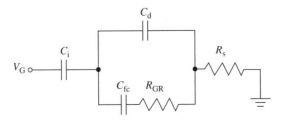

Figure 1.7 Equivalent circuit of the MOS capacitor. Reproduced from Shur M. (1990) *Physics of Semiconductor Devices*, Prentice Hall, Englewood Cliffs, NJ

This equivalent circuit is clearly frequency-dependent. In the low-frequency limit, we can neglect the effects of R_{GR} and R_s to obtain (using $C_s = C_d + C_{fc}$)

$$C_{mos}^o = \frac{C_s C_i}{C_s + C_i}.$$ (1.25)

In strong inversion, we have $C_s \gg C_i$, which gives

$$C_{mos}^o \approx C_i$$ (1.26)

at low frequencies.

In the high-frequency limit, the time constant of the generation/recombination mechanism will be much longer than the signal period ($R_{GR}C_{fc} \gg 1/f$) and C_d effectively shunts the lower branch of the parallel section of the equivalent in Figure 1.7. Hence, the high-frequency, strong inversion capacitance of the equivalent circuit becomes

$$C_{mos}^\infty = \frac{C_{dT} C_i}{C_{dT} + C_i}.$$ (1.27)

The calculated dependence of C_{mos} on the applied voltage for different frequencies is shown in Figure 1.8. For applied voltages well below threshold, the device is in accumulation and C_{mos} equals C_i. As the voltage approaches threshold, the semiconductor passes the flat-band condition where C_{mos} has the value C_{FB}, and then enters the depletion and the weak inversion regimes where the depletion width increases and the capacitance value drops steadily until it reaches the minimum value at threshold given by (1.27). The calculated curves clearly demonstrate how the MOS capacitance in the strong inversion regime depends on the frequency, with a value of C_{mos}^∞ at high frequencies to C_i at low frequencies.

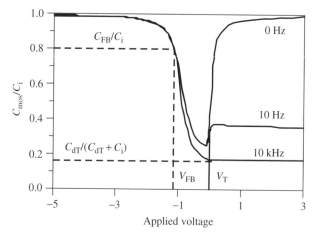

Figure 1.8   Calculated dependence of C_{mos} on the applied voltage for different frequencies. Parameters used: insulator thickness, 2×10^{-8} m; semiconductor doping density, $10^{15}/cm^3$; generation time, 10^{-8} s. Reproduced from Shur M. (1990) *Physics of Semiconductor Devices*, Prentice Hall, Englewood Cliffs, NJ

We note that in a MOSFET, where the highly doped source and drain regions act as reservoirs of minority carriers for the inversion layer, the time constant $R_{GR}C_{fc}$ must be substituted by a much smaller time constant corresponding to the time needed for transporting carriers from these reservoirs in and out of the MOSFET gate area. Consequently, high-frequency strong inversion MOSFET gate-channel $C-V$ characteristics will resemble the zero frequency MOS characteristic.

Since the low-frequency MOS capacitance in the strong inversion is close to C_i, the induced inversion charge per unit area can be approximated by

$$qn_s \approx c_i(V - V_T). \qquad (1.28)$$

This equation serves as the basis of a simple charge control model (SCCM) allowing us to calculate MOSFET current–voltage characteristics in strong inversion.

From measured MOS $C-V$ characteristics, we can easily determine important parameters of the MOS structure, including the gate insulator thickness, the semiconductor substrate doping density, and the flat-band voltage. The maximum measured capacitance C_{max} (capacitance C_i in Figure 1.7) yields the insulator thickness

$$d_i \approx S\varepsilon_i/C_{max}. \qquad (1.29)$$

The minimum measured capacitance C_{min} (at high frequency) allows us to find the doping concentration in the semiconductor substrate. First, we determine the depletion capacitance in the strong inversion regime using (1.27),

$$1/C_{min} = 1/C_{dT} + 1/C_i. \qquad (1.30)$$

From C_{dT} we obtain the thickness of the depletion region at threshold as

$$d_{dT} = S\varepsilon_s/C_{dT}. \qquad (1.31)$$

Then we calculate the doping density N_a using (1.23) with $\psi_s = 2\varphi_b$ and (1.2) for φ_b. This results in the following transcendental equation for N_a:

$$N_a = \frac{4\varepsilon_s V_{th}}{q d_{dT}^2} \ln\left(\frac{N_a}{n_i}\right). \qquad (1.32)$$

This equation can easily be solved by iteration or by approximate analytical techniques. Once d_i and N_a have been obtained, the device capacitance C_{FB} under flat-band conditions can be determined using $C_s = C_{so}$ ((1.21) at flat-band condition) in combination with (1.19):

$$C_{FB} = \frac{C_{so}C_i}{C_{so} + C_i} = \frac{S\varepsilon_s\varepsilon_i}{\varepsilon_s d_i + \varepsilon_i L_{Dp}}. \qquad (1.33)$$

The flat-band voltage V_{FB} is simply equal to the applied voltage corresponding to this value of the device capacitance.

We note that the above characterization technique applies to ideal MOS structures. Different nonideal effects, such as geometrical effects, nonuniform doping in the substrate, interface states, and mobile charges in the oxide may influence the $C-V$ characteristics of the MOS capacitor.

1.2.4 MOS Charge Control Model

Well above threshold, the charge density of the mobile carriers in the inversion layer can be calculated using the parallel plate charge control model of (1.28). This model gives an adequate description for the strong inversion regime of the MOS capacitor, but fails for applied voltages near and below threshold (i.e., in the weak inversion and depletion regimes). Several expressions have been proposed for a unified charge control model (UCCM) that covers all the regimes of operation, including the following (see Byun *et al.* 1990):

$$V - V_T = q(n_s - n_o)/c_a + \eta V_{th} \ln\left(\frac{n_s}{n_o}\right), \tag{1.34}$$

where $c_a \approx c_i$ is approximately the insulator capacitance per unit area (with a small correction for the finite vertical extent of the inversion channel, see Lee *et al.* (1993)), $n_o = n_s(V = V_T)$ is the density of minority carriers per unit area at threshold, and η is the so-called subthreshold ideality factor, also known as the subthreshold swing parameter. The ideality factor accounts for the subthreshold division of the applied voltage between the gate insulator and the depletion layer, and $1/\eta$ represents the fraction of this voltage that contributes to the interface potential. A simplified analysis gives

$$\eta = 1 + C_d/C_i, \tag{1.35}$$

$$n_o = \eta V_{th} c_a/2q. \tag{1.36}$$

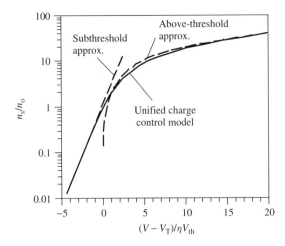

Figure 1.9 Comparison of various charge control expression for the MOS capacitor. Equation (1.38) is a close approximation to (1.34), while the above- and below-threshold approximations are given by (1.28) and (1.37), respectively. Reproduced from Fjeldly T. A., Ytterdal T., and Shur M. (1998) *Introduction to Device Modeling and Circuit Simulation*, John Wiley & Sons, New York

In the subthreshold regime, (1.34) approaches the limit

$$n_s = n_o \exp\left(\frac{V - V_T}{\eta V_{th}}\right).$$ (1.37)

We note that (1.34) does not have an exact analytical solution for the inversion charge in terms of the applied voltage. However, for many purposes, the following approximate solution may be suitable:

$$n_s = 2n_o \ln\left[1 + \frac{1}{2}\exp\left(\frac{V - V_T}{\eta V_{th}}\right)\right].$$ (1.38)

This expression reproduces the correct limiting behavior both in strong inversion and in the subthreshold regime, although it deviates slightly from (1.34) near threshold. The various charge control expressions of the MOS capacitor are compared in Figure 1.9.

1.3 BASIC MOSFET OPERATION

In the MOSFET, an inversion layer at the semiconductor–oxide interface acts as a conducting channel. For example, in an n-channel MOSFET, the substrate is p-type silicon and the inversion charge consists of electrons that form a conducting channel between the n^+ ohmic source and the drain contacts. At DC conditions, the depletion regions and the neutral substrate provide isolation between devices fabricated on the same substrate. A schematic view of the n-channel MOSFET is shown in Figure 1.10.

As described above for the MOS capacitor, inversion charge can be induced in the channel by applying a suitable gate voltage relative to other terminals. The onset of strong inversion is defined in terms of a threshold voltage V_T being applied to the gate electrode relative to the other terminals. In order to assure that the induced inversion channel extends all the way from source to drain, it is essential that the MOSFET gate structure either overlaps slightly or aligns with the edges of these contacts (the latter is achieved by a self-aligned process). Self-alignment is preferable since it minimizes the parasitic gate-source and gate-drain capacitances.

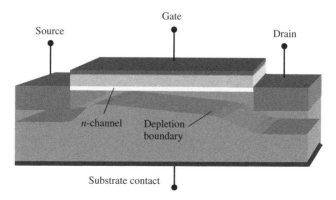

Figure 1.10   Schematic view of an n-channel MOSFET with conducting channel and depletion region

When a drain-source bias V_{DS} is applied to an n-channel MOSFET in the above-threshold conducting state, electrons move in the channel inversion layer from source to drain. A change in the gate-source voltage V_{GS} alters the electron sheet density in the channel, modulating the channel conductance and the device current. For $V_{GS} > V_T$ in an n-channel device, the application of a positive V_{DS} gives a steady voltage increase from source to drain along the channel that causes a corresponding reduction in the local gate-channel bias V_{GX} (here X signifies a position x within the channel). This reduction is greatest near drain where V_{GX} equals the gate-drain bias V_{GD}.

Somewhat simplistically, we may say that when $V_{GD} = V_T$, the channel reaches threshold at the drain and the density of inversion charge vanishes at this point. This is the so-called pinch-off condition, which leads to a saturation of the drain current I_{ds}. The corresponding drain-source voltage, $V_{DS} = V_{SAT}$, is called the *saturation voltage*. Since $V_{GD} = V_{GS} - V_{DS}$, we find that $V_{SAT} = V_{GS} - V_T$. (This is actually a result of the SCCM, which is discussed in more detail in Section 1.4.1.)

When $V_{DS} > V_{SAT}$, the pinched-off region near drain expands only slightly in the direction of the source, leaving the remaining inversion channel intact. The point of transition between the two regions, $x = x_p$, is characterized by $V_{XS}(x_p) \approx V_{SAT}$, where $V_{XS}(x_p)$ is the channel voltage relative to source at the transition point. Hence, the drain current in saturation remains approximately constant, given by the voltage drop V_{SAT} across the part of the channel that remain in inversion. The voltage $V_{DS} - V_{SAT}$ across the pinched-off region creates a strong electric field, which efficiently transports the electrons from the strongly inverted region to the drain.

Typical current–voltage characteristics of a long-channel MOSFET, where pinch-off is the predominant saturation mechanism, are shown in Figure 1.11. However, with shorter MOSFET gate lengths, typically in the submicrometer range, velocity saturation will occur in the channel near drain at lower V_{DS} than that causing pinch-off. This leads to more evenly spaced saturation characteristics than those shown in this figure, more in

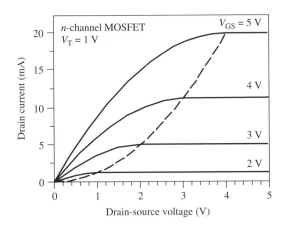

Figure 1.11   Current–voltage characteristics of an n-channel MOSFET with current saturation caused by pinch-off (long-channel case). The intersections with the dotted line indicate the onset of saturation for each characteristic. The threshold voltage is assumed to be $V_T = 1$ V. Reproduced from Fjeldly T. A., Ytterdal T., and Shur M. (1998) *Introduction to Device Modeling and Circuit Simulation*, John Wiley & Sons, New York

agreement with those observed for modern devices. Also, phenomena such as a finite channel conductance in saturation, a drain bias–induced shift in the threshold voltage, and an increased subthreshold current are important consequences of shorter gate lengths (see Section 1.5).

1.4 BASIC MOSFET MODELING

Analytical or semianalytical MOSFET models are usually based on the so-called gradual channel approximation (GCA). Contrary to the situation in the ideal two-terminal MOS device, where the charge density profile is determined from a one-dimensional Poisson's equation (see Section 1.2), the MOSFET generally poses a two-dimensional electrostatic problem. The reason is that the geometric effects and the application of a drain-source bias create a lateral electric field component in the channel, perpendicular to the vertical field associated with the ideal gate structure. The GCA states that, under certain conditions, the electrostatic problem of the gate region can be expressed in terms of two coupled one-dimensional equations – a Poisson's equation for determining the vertical charge density profile under the gate and a charge transport equation for the channel. This allows us to determine self-consistently both the channel potential and the charge profile at any position along the gate. A direct inspection of the two-dimensional Poisson's equation for the channel region shows that the GCA is valid if we can assume that the electric field gradient in the lateral direction of the channel is much less than that in the vertical direction perpendicular to the channel (Lee *et al.* 1993).

Typically, we find that the GCA is valid for long-channel MOSFETs, where the ratio between the gate length and the vertical distance of the space charge region from the gate electrode, the so-called aspect ratio, is large. However, if the MOSFET is biased in saturation, the GCA always becomes invalid near drain as a result of the large lateral field gradient that develops in this region. In Figure 1.12, this is schematically illustrated for a MOSFET in saturation.

Next, we will discuss three relatively simple MOSFET models, the simple charge control model, the Meyer model, and the velocity saturation model. These models, with extensions, can be identified with the models denoted as MOSFET Level 1, Level 2, and Level 3 in SPICE.

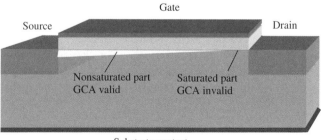

Figure 1.12 Schematic representation of a MOSFET in saturation, where the channel is divided into a nonsaturated region where the GCA is valid and a saturated region where the GCA is invalid

We should note that the analysis that follows is based on idealized device structures. Especially in modern MOSFET/CMOS technology, optimized for high-speed and low-power applications, the devices are more complex. Additional oxide and doping regions are used for the purpose of controlling the threshold voltage and to avoid deleterious effects of high electric fields and so-called short- and narrow-channel phenomena associated with the steady downscaling device dimensions. These effects will be discussed more in Section 1.5 and in later chapters.

1.4.1 Simple Charge Control Model

Consider an n-channel MOSFET operating in the above-threshold regime, with a gate voltage that is sufficiently high to cause inversion in the entire length of the channel at zero drain-source bias. We assume a long-channel device, implying that GCA is applicable and that the carrier mobility can be taken to be constant (no velocity saturation). As a first approximation, we can describe the mobile inversion charge by a simple extension of the parallel plate expression (1.28), taking into account the potential variation $V(x)$ along the channel, that is,

$$qn_s(x) \approx c_i[V_{GT} - V(x)], \qquad (1.39)$$

where $V_{GT} \equiv V_{GS} - V_T$. This simple charge control expression implies that the variation of the depletion layer charge along the channel, which depends on $V(x)$, is negligible. Furthermore, since the expression relies on GCA, it is only applicable for the nonsaturated part of the channel. Saturation sets in when the conducting channel is pinched-off at the drain side, that is, for $n_s(x = L) \geq 0$. Using the pinch-off condition and $V(x = L) = V_{DS}$ in (1.39), we obtain the following expression for the saturation drain voltage in the SCCM:

$$V_{SAT} = V_{GT}. \qquad (1.40)$$

The threshold voltage in this model is given by (1.18), where we have accounted for the substrate bias V_{BS} relative to the source. We note that this expression is only valid for negative or slightly positive values of V_{BS}, when the junction between the source contact and the p-substrate is either reverse-biased or slightly forward-biased. For high V_{BS}, a significant leakage current will take place.

Figure 1.13 shows an example of calculated dependences of the threshold voltage on substrate bias for different values of gate insulator thickness. As can be seen from this figure and from (1.18), the threshold voltage decreases with decreasing insulator thickness and is quite sensitive to the substrate bias. This so-called body effect is essential for device characterization and in threshold voltage engineering. For real devices, it is important to be able to carefully adjust the threshold voltage to match specific application requirements.

Equation (1.18) also shows that V_T can be adjusted by changing the doping or by using different gate metals (including heavily doped polysilicon). As discussed in Section 1.2, the gate metal affects the flat-band voltage through the work-function difference between the metal and the semiconductor. Threshold voltage adjustment by means of doping is often performed with an additional ion implantation through the gate oxide.

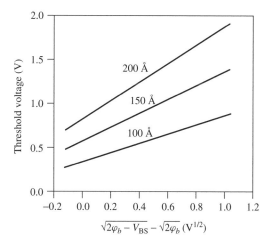

Figure 1.13 Body plot, the dependence of the threshold voltage on substrate bias in MOSFETs with different insulator thicknesses. Parameters used in the calculation: flat-band voltage $-1\,\mathrm{V}$, substrate doping density $10^{22}/\mathrm{m}^3$, temperature $300\,\mathrm{K}$. The slope of the plots are given in terms of the body-effect parameter $\gamma = (2\varepsilon_s q N_a)^{1/2}/c_i$. Reproduced from Fjeldly T. A., Ytterdal T., and Shur M. (1998) *Introduction to Device Modeling and Circuit Simulation*, John Wiley & Sons, New York

Assuming a constant electron mobility μ_n, the electron velocity can be written as $v_n = -\mu_n\, dV/dx$. Neglecting the diffusion current, which is important only near threshold and in the subthreshold regime, the absolute value of the drain current can be written as

$$I_{ds} = W\mu_n q n_s F, \tag{1.41}$$

where $F = |dV/dx|$ is the magnitude of the electric field in the channel and W is the channel width. Integrating this expression over the gate length and using the fact that I_{ds} is independent of position x, we obtain the following expression for the current–voltage characteristics:

$$I_{ds} = \frac{W\mu_n c_i}{L} \times \begin{cases} (V_{GT} - V_{DS}/2)V_{DS}, & \text{for } V_{DS} \le V_{SAT} = V_{GT} \\ V_{GT}^2/2, & \text{for } V_{DS} > V_{SAT} \end{cases}. \tag{1.42}$$

As implied above, the pinch-off condition implies a vanishing carrier concentration at the drain side of the channel. Hence, at a first glance, one might think that the drain current should also vanish. However, instead the saturation drain current I_{dsat} is determined by the resistance of nonsaturated part of the channel and the current across it. In fact, this channel resistance changes very little when V_{DS} increases beyond V_{SAT}, since the pinch-off point x_p moves only slightly away from the drain, leaving the nonsaturated part of the channel almost intact. Moreover, the voltage at the pinch-off point will always be approximately V_{SAT} since the threshold condition at x_p is determined by $V_G - V(x_p) = V_T$, or $V(x_p) = V_{GT} = V_{SAT}$. Hence, since the resistance of the nonsaturated part is constant and the voltage across it is constant, I_{dsat} will also remain constant. Therefore, the saturation current I_{SAT} is determined by substituting $V_{DS} = V_{SAT}$ from (1.40) into the nonsaturation expression in (1.42). In reality, of course, the electron concentration never vanishes, nor does the electric field become infinite. This is simply a consequence of the breakdown of GCA near drain in saturation, pointing to the need for a more accurate and detailed analysis of the saturation regime.

The MOSFET current–voltage characteristics shown in Figure 1.11 were calculated using this simple charge control model.

Important device parameters are the channel conductance,

$$g_d = \left. \frac{\partial I_d}{\partial V_{DS}} \right|_{V_{GS}} = \begin{cases} \beta(V_{GT} - V_{DS}), & \text{for } V_{DS} \leq V_{SAT} \\ 0, & \text{for } V_{DS} > V_{SAT} \end{cases}, \tag{1.43}$$

and the transconductance,

$$g_m = \left. \frac{\partial I_d}{\partial V_{GS}} \right|_{V_{DS}} = \begin{cases} \beta V_{DS}, & \text{for } V_{DS} \leq V_{SAT} \\ \beta V_{GT}, & \text{for } V_{DS} > V_{SAT} \end{cases}, \tag{1.44}$$

where $\beta = W \mu_n c_i / L$ is called the *transconductance parameter*. As can be seen from these expressions, high values of channel conductance and transconductance are obtained for large electron mobilities, large gate insulator capacitances (i.e., thin gate insulator layers), and large gate width to length ratios.

The SCCM was developed at a time when the MOSFET gate lengths were typically tens of micrometers long, justifying some of the above approximations. With today's deep submicron technology, however, the SCCM is clearly not applicable. We therefore introduce two additional models that include significant improvements. In the first of these, the Meyer model, the lateral variation of the depletion charge in the channel is taken into account. In the second, the velocity saturation model (VSM), we introduce the effects of saturation in the carrier velocity. The former is important at realistic levels of substrate doping, and the latter is important because of the high electric fields generated in short-channel devices. Additional effects of small dimensions and high electric fields will be discussed in Section 1.5.

1.4.2 The Meyer Model

The total induced charge q_s per unit area in the semiconductor of an n-channel MOSFET, including both inversion and depletion charges, can be expressed in terms of Gauss's law as follows, assuming that the source and the semiconductor substrate are both connected to ground (see Section 1.2),

$$q_s = -c_i[V_{GS} - V_{FB} - 2\varphi_b - V(x)]. \tag{1.45}$$

Here, the content of the bracket expresses the voltage drop across the insulator layer. The induced sheet charge density includes both the inversion charge density $q_i = -q n_s$ and the depletion charge density q_d, that is, $q_s = q_i + q_d$. Using (1.15) and including the added channel-substrate bias caused by the channel voltage, the depletion charge per unit area can be expressed as

$$q_d = -q N_a d_d = -\sqrt{2\varepsilon_s q N_a [2\varphi_b + V(x)]}, \tag{1.46}$$

where d_d is the local depletion layer width at position x. Hence, the inversion sheet charge density becomes

$$q_i = -qn_s = -c_i[V_{GS} - V_{FB} - 2\varphi_b - V(x)] + \sqrt{2\varepsilon_s q N_a[2\varphi_b + V(x)]}. \tag{1.47}$$

A constant electron mobility is also assumed in the Meyer model. Hence, the nonsaturated drain current can again be obtained by substituting the expression for n_s in

$$I_{ds} = W\mu_n q n_s(x) F(x) \tag{1.48}$$

to give (Meyer 1971)

$$I_{ds} = \frac{W\mu_n c_i}{L} \left\{ \left(V_{GS} - V_{FB} - 2\varphi_b - \frac{V_{DS}}{2} \right) V_{DS} \right. \\ \left. - \frac{2\sqrt{2\varepsilon_s q N_a}}{3c_i} [(V_{DS} + 2\varphi_b)^{3/2} - (2\varphi_b)^{3/2}] \right\}. \tag{1.49}$$

The saturation voltage is obtained using the pinch-off condition $n_s = 0$,

$$V_{SAT} = V_{GS} - 2\varphi_b - V_{FB} + \frac{\varepsilon_s q N_a}{c_i^2} \left[1 - \sqrt{1 + \frac{2c_i^2(V_{GS} - V_{FB})}{\varepsilon_s q N_a}} \right]. \tag{1.50}$$

At low doping levels, we see that V_{SAT} approaches V_{GT}, which is the result found for the simple charge control model.

1.4.3 Velocity Saturation Model

The linear velocity-field relationship (constant mobility) used in the above MOSFET models works reasonably well for long-channel devices. However, the implicit notion of a diverging carrier velocity as we approach pinch-off is, of course, unphysical. Instead, current saturation is better described in terms of a saturation of the carrier drift velocity when the electric field near drain becomes sufficiently high. The following two-piece model is a simple, first approximation to a realistic velocity-field relationship:

$$v(F) = \begin{cases} \mu_n F & \text{for } F < F_s \\ v_s & \text{for } F \geq F_s \end{cases}, \tag{1.51}$$

where $F = |dV(x)/dx|$ is the magnitude of lateral electrical field in the channel, v_s is the saturation velocity, and $F_s = v_s/\mu_n$ is the saturation field. In this description, current saturation in FETs occurs when the field at the drain side of the gate reaches the saturation field. A somewhat more precise expression, which is particularly useful for n-channel MOSFETs, is the so-called Sodini model (Sodini et $al.$ 1984),

$$v(F) = \begin{cases} \dfrac{\mu_n F}{1 + F/2F_s} & \text{for } F < 2F_s \\ v_s & \text{for } F \geq 2F_s \end{cases}. \tag{1.52}$$

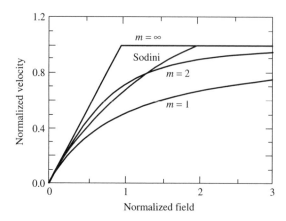

Figure 1.14 Velocity-field relationships for charge carriers in silicon MOSFETs. The electric field and the velocity are normalized to F_s and v_s, respectively. Two of the curves are calculated from (1.53) using $m = 1$ for holes and $m = 2$ for electrons. The curve marked $m = \infty$ corresponds to the linear two-piece model in (1.51). The Sodini model (1.52) is also shown

Even more realistic velocity-field relationships for MOSFETs are obtained from

$$v(F) = \frac{\mu F}{[1 + (F/F_s)^m]^{1/m}},\tag{1.53}$$

where $m = 2$ and $m = 1$ are reasonable choices for n-channel and p-channel MOSFETs, respectively. The two-piece model in (1.51) corresponds to $m = \infty$ in (1.53). Figure 1.14 shows different velocity-field models for electrons and holes in silicon MOSFETs.

Using the simple velocity-field relationship of (1.51), current–voltage characteristics can easily be derived from either the SCCM or the Meyer model, since the form of the nonsaturated parts of the characteristics will be the same as before (see (1.42) and (1.49)). However, the saturation voltage will now be identical to the drain-source voltage that initiates velocity saturation at the drain side of the channel. In terms of (1.51), this occurs when $F(L) = F_s$. Hence, using this condition in combination with the SCCM, we obtain the following expressions for the drain current and the saturation voltage:

$$I_{ds} = \frac{W \mu_n c_i}{L} \times \begin{cases} V_{GT} V_{DS} - V_{DS}^2/2, & \text{for } V_{DS} \leq V_{SAT} \\ (V_{GT} - V_{SAT}) V_L, & \text{for } V_{DS} > V_{SAT} \end{cases},\tag{1.54}$$

$$V_{SAT} = V_{GT} - V_L \left[\sqrt{1 + (V_{GT}/V_L)^2} - 1 \right],\tag{1.55}$$

where $V_L = F_s L = L v_s / \mu_n$. The Meyer VSM leads to a much more complicated relationship for V_{SAT}.

For large values of V_L such that $V_L \gg V_{GT}$, the square root terms in (1.55) may be expanded into a Taylor series, yielding the previous long-channel result for the SCCM without velocity saturation. Assuming, as an example, that $V_{GT} = 3$ V, $\mu_n = 0.08\,\text{m}^2/\text{Vs}$, and $v_s = 1 \times 10^5$ m/s, we find that velocity saturation effects may be neglected for $L \gg 2.4\,\mu\text{m}$. Hence, velocity saturation is certainly important in modern MOSFETs with gate lengths typically in the deep submicrometer range.

In the opposite limiting case, when $V_L \ll V_{GT}$, we obtain $V_{SAT} \approx V_L$ and $I_{dsat} \approx \beta V_L V_{GT}$. Since I_{dsat} is proportional to V_{GT}^2 in long-channel devices and proportional to V_{GT} in short-channel devices, we can use this difference to identify the presence of short-channel effects on the basis of measured device characteristics.

1.4.4 Capacitance Models

For the simulation of dynamic events in MOSFET circuits, we also have to account for variations in the stored charges of the devices. In a MOSFET, we have stored charges in the gate electrode, in the conducting channel, and in the depletion layers. Somewhat simplified, the variation in the stored charges can be expressed through different capacitance elements, as indicated in Figure 1.15.

We distinguish between the so-called parasitic capacitive elements and the capacitive elements of the intrinsic transistor. The parasitics include the overlap capacitances between the gate electrode and the highly doped source and drain regions (C_{os} and C_{od}), the junction capacitances between the substrate and the source and drain regions (C_{js} and C_{jd}), and the capacitances between the metal electrodes of the source, the drain, and the gate.

The semiconductor charges of the intrinsic gate region of the MOSFET are divided between the mobile inversion charge and the depletion charge, as indicated in Figure 1.15. In addition, these charges are nonuniformly distributed along the channel when drain-source bias is applied. Hence, the capacitive coupling between the gate electrode and the semiconductor is also distributed, making the channel act as an RC transmission line. In practice, however, because of the short gate lengths and limited bandwidths of FETs, the distributed capacitance of the intrinsic device is usually very well represented in terms of a lumped capacitance model, with capacitive elements between the various intrinsic device terminals.

An accurate modeling of the intrinsic device capacitances still requires an analysis of how the inversion charge and the depletion charge are distributed between source, drain, and substrate for different terminal bias voltages. As discussed by Ward and Dutton (1978), such an analysis leads to a set of charge-conserving and nonreciprocal capacitances between the different intrinsic terminals (nonreciprocity means $C_{ij} \neq C_{ji}$, where i and j denote source, drain, gate, or substrate).

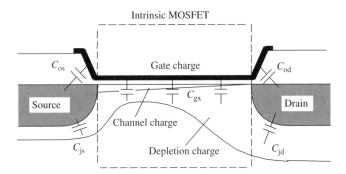

Figure 1.15 Intrinsic and parasitic capacitive elements of the MOSFET. Reproduced from Fjeldly T. A., Ytterdal T., and Shur M. (1998) *Introduction to Device Modeling and Circuit Simulation*, John Wiley & Sons, New York

In a simplified and straightforward analysis by Meyer (1971) based on the SCCM, a set of reciprocal capacitances ($C_{ij} = C_{ji}$) were obtained as derivatives of the total gate charge with respect to the various terminal voltages. Although charge conservation is not strictly enforced in this case, since the Meyer capacitances represent only a subset of the Ward–Dutton capacitances, the resulting errors in circuit simulations are usually small, except in some cases of transient analyzes of certain demanding circuits. Here, we first consider Meyer's capacitance model for the long-channel case, but return with modifications of this model and comments on charge-conserving capacitance models in Section 1.5.3.

In Meyer's capacitance model, the distributed intrinsic MOSFET capacitance can be split into the following three lumped capacitances between the intrinsic terminals:

$$C_{GS} = \left.\frac{\partial Q_G}{\partial V_{GS}}\right|_{V_{GD}, V_{GB}}, \quad C_{GD} = \left.\frac{\partial Q_G}{\partial V_{GD}}\right|_{V_{GS}, V_{GB}}, \quad C_{GB} = \left.\frac{\partial Q_G}{\partial V_{GB}}\right|_{V_{GS}, V_{GD}}, \quad (1.56)$$

where Q_G is the total intrinsic gate charge. The intrinsic MOSFET equivalent circuit corresponding to this model is shown in Figure 1.16.

In general, the gate charge reflects both the inversion charge and the depletion charge and can therefore be written as $Q_G = Q_{Gi} + Q_{Gd}$. However, in the SCCM for the drain current, the depletion charge is ignored in strong inversion, except for its influence on the threshold voltage (see (1.18)). Likewise, in the Meyer capacitance model, the gate-source capacitance C_{GS} and the gate-drain capacitance C_{GD} can be assumed to be dominated by the inversion charge. Here, we include gate-substrate capacitance C_{GB} in the subthreshold regime, where the depletion charge is dominant.

The contribution of the inversion charge to the gate charge is determined by integrating the sheet charge density given by (1.39), over the gate area, that is,

$$Q_{Gi} = W c_i \int_0^L [V_{GT} - V(x)] \, dx. \quad (1.57)$$

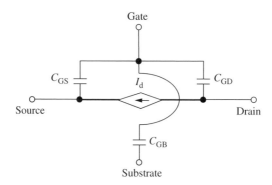

Figure 1.16 Large-signal equivalent circuit of intrinsic MOSFET based on Meyer's capacitance model. Reproduced from Fjeldly T. A., Ytterdal T., and Shur M. (1998) *Introduction to Device Modeling and Circuit Simulation*, John Wiley & Sons, New York

From (1.41), we notice that $dx = W\mu_n c_i (V_{GT} - V)\,dV/I_{ds}$, which allows us to make a change of integration variable from x to V in (1.57). Hence, we obtain for the nonsaturated regime

$$Q_{Gi} = \frac{W\mu_n C_i^2}{LI_{ds}} \int_0^{V_{DS}} (V_{GT} - V)^2\,dV = \frac{2}{3}C_i \frac{(V_{GS} - V_T)^3 - (V_{GD} - V_T)^3}{(V_{GS} - V_T)^2 - (V_{GD} - V_T)^2}, \qquad (1.58)$$

where C_i is the total gate oxide capacitance and where we expressed I_{ds} using (1.42) and replaced V_{DS} by $V_{GS} - V_{GD}$ everywhere.

Using the above relationships, the following strong inversion, long-channel Meyer capacitances are obtained:

$$C_{GS} = \frac{2}{3}C_i \left[1 - \left(\frac{V_{GT} - V_{DS}}{2V_{GT} - V_{DS}} \right)^2 \right], \qquad (1.59)$$

$$C_{GD} = \frac{2}{3}C_i \left[1 - \left(\frac{V_{GT}}{2V_{GT} - V_{DS}} \right)^2 \right], \qquad (1.60)$$

$$C_{GB} = 0. \qquad (1.61)$$

We recall that $V_{SAT} = V_{GT}$ is the saturation voltage in the SCCM. The capacitances at saturation are found by replacing $V_{DS} = V_{SAT}$ in the above expressions, that is,

$$C_{GSs} = \frac{2}{3}C_i, \quad C_{GDs} = C_{GBs} = 0. \qquad (1.62)$$

This result indicates that in saturation, a small change in the applied drain-source voltage does not contribute to the gate or the channel charge, since the channel is pinched off. Instead, the entire channel charge is assigned to the source terminal, giving a maximum value of the capacitance C_{GS}. Normalized dependencies of the Meyer capacitances C_{GS} and C_{GD} on bias conditions are shown in Figure 1.17.

In the subthreshold regime, the inversion charge becomes negligible compared to the depletion charge, and the MOSFET gate-substrate capacitance will be the same as that of a MOS capacitor in depletion, with a series connection of the gate oxide capacitance C_i and the depletion capacitance C_d (see (1.19) to (1.23)). According to the discussion in Section 1.2, the applied gate-substrate voltage V_{GB} can be subdivided as follows:

$$V_{GB} = V_{FB} + \psi_s - q_{dep}/C_i, \qquad (1.63)$$

where V_{FB} is the flat-band voltage, ψ_s is the potential across the semiconductor depletion layer (i.e., the surface potential relative to the substrate interior), and $-q_{dep}/c_i$ is the voltage drop across the oxide. In the depletion approximation, the depletion charge per unit area q_{dep} is related to ψ_s by $q_{dep} = -\gamma c_i \psi_s^{1/2}$ where $\gamma = (2\varepsilon_s q N_a)^{1/2}/c_i$ is the body-effect parameter. Using this relationship to substitute for ψ_s in (1.63), we find

$$Q_{Gd} = -WL q_{dep} = \gamma C_i \left(\sqrt{\gamma^2/4 + V_{GB} - V_{FB}} - \gamma/2 \right), \qquad (1.64)$$

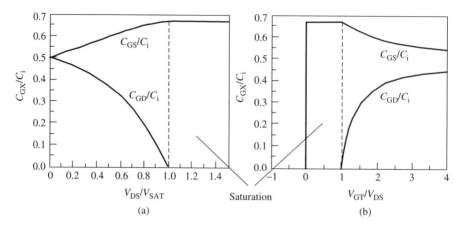

Figure 1.17 Normalized strong inversion Meyer capacitances according to (1.59) to (1.62) versus (a) drain-source bias and (b) gate-source bias. Note that $V_{SAT} = V_{GT}$ in this model. Reproduced from Fjeldly T. A., Ytterdal T., and Shur M. (1998) *Introduction to Device Modeling and Circuit Simulation*, John Wiley & Sons, New York

from which we obtain the following subthreshold capacitances:

$$C_{GB} = \frac{C_i}{\sqrt{1 + 4(V_{GB} - V_{FB})/\gamma^2}}, \quad C_{GS} = C_{GD} = 0. \qquad (1.65)$$

We note that (1.65) gives $C_{GB} = C_i$ at the flat-band condition, which is different from the flat-band capacitance of (1.33). This discrepancy arises from neglecting the effects of the free carriers in the subthreshold regime in the present simplified treatment. For the same reason, we observe the presence of discontinuities in the Meyer capacitances at threshold. Discontinuities in the derivatives of the Meyer capacitances occur at the onset of saturation as a result of additional approximations. Such discontinuities should be avoided in the device models since they give rise to increased simulation time and conversion problems in circuit simulators. These issues will be discussed further in Section 1.5.

In the MOSFET VSM, the above-threshold capacitance expressions derived on the basis of the SCCM are still valid in the nonsaturated regime $V_{DS} \leq V_{SAT}$. The capacitance values at the saturation point are found by replacing V_{DS} in (1.59) and (1.60) by V_{SAT} from (1.55), yielding

$$C_{GSs} = \frac{2}{3} C_i \left[1 - \left(\frac{V_{SAT}}{2V_L} \right)^2 \right], \qquad (1.66)$$

$$C_{GDs} = \frac{2}{3} C_i \left[1 - \left(1 - \frac{V_{SAT}}{2V_L} \right)^2 \right]. \qquad (1.67)$$

However, well into saturation, the intrinsic gate charge will change very little with increasing V_{DS}, similar to what takes place in the case of saturation by pinch-off (see preceding text). Hence, the real capacitances have to approach the same limiting values in saturation as the Meyer capacitances, that is, $C_{GS}/C_i \to 2/3$ and $C_{GD}/C_i \to 0$. In fact, since the behavior of C_{GS} and C_{GD} in the VSM and in the SCCM coincide for $V_{DS} < V_{SAT}$ and have the same asymptotic values in saturation, the Meyer capacitance model offers a reasonable approximation for the MOSFET capacitances also in short-channel devices. This suggests a separate "saturation" voltage for the capacitances close to the long-channel pinch-off voltage ($\approx V_{GT}$), which is larger than V_{SAT} associated with the onset of velocity saturation.

1.4.5 Comparison of Basic MOSFET Models

The I–V characteristics shown in Figure 1.18 were calculated using the three basic MOSFET models discussed above – the simple charge control model (SCCM), the Meyer I–V model (MM), and the velocity saturation model (VSM). The same set of MOSFET parameters were used in all cases. We note that all models coincide at small drain-source voltages. However, in saturation, SCCM always gives the highest current. This is a direct consequence of omitting velocity saturation and spatial variation in the depletion charge in SCCM, resulting in an overestimation of both carrier velocity and inversion charge. The characteristics for VSM and MM clearly demonstrate how inclusion of velocity saturation and distribution of depletion charge, respectively, affect the saturation current.

The intrinsic capacitances according to Section 1.4.4 are shown in Figure 1.19. Meyer's capacitance model can be used in conjunction with all the MOSFET models illustrated in Figure 1.18 (SCCM, MM and VSM). In the present device example, we note that velocity saturation and depletion charge may be quite important. Therefore, we emphasize that SCCM is usually applicable only for long-channel, low-doped devices, while MM applies to long-channel devices with an arbitrary doping level. VSM gives a reasonable description of short-channel devices, although important short-channel effects such as channel-length modulation and drain-induced barrier lowering (DIBL) are still unaccounted for in these

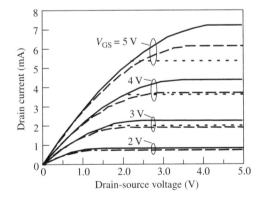

Figure 1.18 Comparison of I–V characteristics obtained for a given set of MOSFET parameters using the three basic MOSFET models: simple charge control model (solid curves), Meyer's I–V model (dashed curves), and velocity saturation model (dotted curves). The MOSFET device parameters are $L = 2\,\mu m$, $W = 20\,\mu m$, $d_i = 300\,\text{Å}$; $\mu_n = 0.06\,\text{m}^2/\text{Vs}$, $v_s = 10^5\,\text{m/s}$; $N_a = 10^{22}/\text{m}^3$, $V_T = 0.43\,\text{V}$; $V_{FB} = -0.75\,\text{V}$; $\varepsilon_i = 3.45 \times 10^{-11}\,\text{F/m}$; $\varepsilon_s = 1.05 \times 10^{-10}\,\text{F/m}$; $n_i = 1.05 \times 10^{16}/\text{m}^3$. Reproduced from Fjeldly T. A., Ytterdal T., and Shur M. (1998) *Introduction to Device Modeling and Circuit Simulation*, John Wiley & Sons, New York

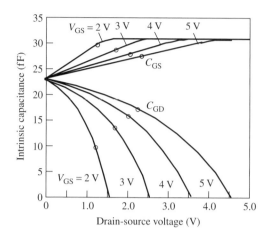

Figure 1.19 Intrinsic MOSFET C–V characteristics for the same devices as in Figure 1.18, obtained from the Meyer capacitance model. The circles indicate the onset of saturation according to (1.66) and (1.67). Reproduced from Fjeldly T. A., Ytterdal T., and Shur M. (1998) *Introduction to Device Modeling and Circuit Simulation*, John Wiley & Sons, New York

models. Likewise, we have ignored certain high-field effects (avalanche breakdown), and advanced MOSFET designs. Some of these issues will be discussed in Section 1.5 and in later chapters of this book.

1.4.6 Basic Small-signal Model

So far, we have considered large-signal MOSFET models, which are suitable for digital electronics and for determining the operating point in small-signal applications. The small-signal regime is, of course, a very important mode of operation of MOSFETs as well as for other active devices. Typically, the AC signal amplitudes are so small relative to the DC values of the operating point that a linear relationship can be assumed between an incoming signal and its response. Normally, if sufficiently accurate large-signal models are available, the AC designers will use such large-signal models also for small-signal applications, since this mode is readily available in circuit simulators such as SPICE. However, in cases when suitable large-signal models are unavailable or when simple hand calculations are needed, it is convenient to use a dedicated small-signal MOSFET model based on a linearized network.

Figure 1.20 shows an intrinsic, common-source, small-signal model for MOSFETs. The model is generalized to include inputs at both the gate and the substrate terminal, and the response is observed at the drain (Fonstad 1994). The network elements are obtained as first derivatives of current–voltage and charge–voltage characteristics, resulting in fixed small-signal conductances, transconductances, and capacitances for a given operating point.

To build a more complete model, some of the extrinsic parasitics may be added, including the gate overlap capacitances and the source and drain junction capacitances, shown in Figure 1.15, and the source and drain series resistances. At very high frequencies, in the radio frequency (RF) range, the junction capacitances become very important since

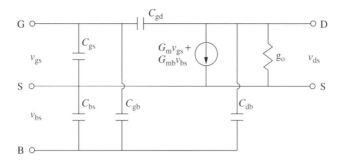

Figure 1.20   Basic small-signal equivalent circuit of an intrinsic, common-source MOSFET. Reproduced from Fonstad C. G. (1994) *Microelectronic Devices and Circuits*, McGraw-Hill, New York

they couple efficiently to the MOSFET substrate. Other important parasitics in this range are the gate resistance and the series inductances associated with the conducting paths. RF CMOS modeling will be discussed in more detail in Chapter 3 of this book.

1.5 ADVANCED MOSFET MODELING

The rapid evolution of semiconductor electronics technology is fueled by a never-ending demand for better performance, combined with a fierce global competition. For silicon CMOS technology, this evolution is often measured in generations of three years – the time it takes for manufactured memory capacity on a chip to be increased by a factor of 4 and for logic circuit density to increase by a factor of between 2 and 3. Technologically, this long-term trend is made possible by a steady downscaling of CMOS feature size by about a factor of 2 per two generations.

At present, CMOS in high volume manufacturing has progressed to the 130-nm technology node. The technology node, used as a measure of the technology scaling, typically signifies the half-pitch size of the first-level interconnect in dynamic RAM (DRAM) technology, while the smallest features, the MOSFET gate lengths, are presently at 65 nm. Following the evolutionary trend, the technology node is expected to decrease below 100 nm within a few years, as indicated in Figure 1.21. Simultaneously, the performance of CMOS ICs rises steeply, packing several 100 million transistors on a chip and operating with clock rates well into the gigahertz range.

Very important issues in this development are the increasing levels of complexity of the fabrication process and the many subtle mechanisms that govern the properties of deep submicrometer FETs. These mechanisms, dictated by the device physics, have to be described and implemented into process modeling and circuit design tools, to empower the circuit designers with abilities to fully utilize the potential of existing and future technologies.

The downscaling of FETs tends to augment important nonideal phenomena, most of which have to be incorporated into any viable device model for use in circuit simulation and device design. These include the so-called short-channel effects, which tend to weaken the gate control over the channel charge. Among the manifestations of short-channel phenomena are serious leakage currents associated with punch-through and threshold voltage shifts resulting from increasing influence of the source and drain contacts over the intrinsic

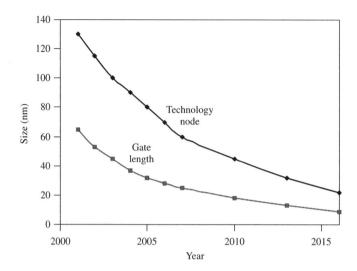

Figure 1.21 Projected CMOS scaling according to International Technology Roadmap for Semiconductors. Reproduced from *ITRS – International Technology Roadmap for Semiconductor*, Semiconductor Industry Assoc., Austin, TX (2001)

channel and depletion charges. The drain-source bias induces an additional lowering of the injection barrier near the source, giving rise to further shifts in the threshold voltage. The latter also causes an increased output conductance in saturation. The loss of gate control may be interpreted as resulting from an improper collective scaling of dimensions, doping levels, and voltages in the device, since an ideal scaling scheme is difficult to enforce in practice.

Gate leakage is another deleterious effect that occurs in radically downscaled devices with gate oxide thicknesses of one to two nanometers. This leakage is the result of quantum-mechanical tunneling, an effect that actually poses a fundamental limitation for further MOSFET scaling within the next few decades.

In addition to these "new" phenomena, well-known effects from earlier FET generations become magnified at short gate lengths owing to enhanced electric fields associated with improper scaling of voltages. Examples are channel-length modulation (CLM), bias dependence of the field effect mobility, and phenomena related to hot electron–induced impact ionization near drain.

The above mechanisms also have important consequences for the modeling of MOSFETs. All the presently accepted MOSFET models used by industry, including the latest BSIM models (Berkeley short-channel IGFET models), are, in effect, based on the GCA. As discussed in Section 1.4, the GCA allows a separation of the model development into two coupled equations, one describing the local vertical field and charge distribution by means of a one-dimensional Poisson's equation and another describing the lateral charge transport in the channel. In improperly scaled devices, this description becomes seriously flawed since the electrostatic problem of the gate region truly becomes a two-dimensional one, with lateral and vertical fields and field gradients of similar magnitudes. The consequence is that the GCA-based models have to be augmented by numerous empirical and semiempirical "fixes" to maintain the required accuracy. This has resulted in a plethora of device parameters, counting in the hundreds for the latest BSIM models.

1.5.1 Modeling Approach

For any FET, the threshold gate voltage V_T is a key parameter. It separates the *on-* (above-threshold) and the *off-* (subthreshold) states of operation. As indicated in Figure 1.22, the average potential energy of the channel electrons in the *off*-state is high relative to those of the source, creating an effective barrier against electron transport from source to drain. In the *on*-state, this barrier is significantly lowered, promoting a high population of free electrons in the channel region. For long-channel devices, with gate lengths of several micrometers and with high power supply voltages, the behavior in the transition region near threshold is not important in digital applications. However, for MOSFETs with deep submicrometer feature size and reduced power supply voltages (such as in low-power operation), the transition region becomes increasingly important, and the distinction between *on-* and *off*-states becomes blurred. Accordingly, a precise modeling of all regimes of device operation, including the near-threshold regime, is needed for short-channel devices, both for digital and high-frequency analog applications.

In the basic MOSFET models considered in Section 1.4, the subthreshold regime is simply considered an *off*-state of the device, ideally blocking all drain current (although the SPICE implementations of some of these models include descriptions of this regime). In practice, however, there will always be some leakage current in the *off*-state owing to a finite amount of mobile charge in the channel and a finite rate of carrier injection from the source to the channel.

This effect is enhanced in modern day downscaled MOSFETs owing to short-channel phenomena such as drain-induced barrier lowering. DIBL is a mechanism whereby the application of a drain-source bias causes a lowering of the source-channel junction barrier. In a long-channel device biased in the subthreshold regime, the applied drain-source voltage drop will be confined to the channel-drain depletion zone. The remaining part of the channel is essentially at a constant potential (flat energy bands), where diffusion is the primary mode of charge transport. However, in a short-channel device the effect of the applied drain-source voltage will be distributed over the length of the channel, giving rise to a shift of the conduction band edge near the source end of the channel, as illustrated in Figure 1.23. Such a shift represents an effective lowering of the injection barrier between the source and the channel. Since the dominant injection mechanism is thermionic emission, this barrier lowering translates into a significant increase of the injected current. This phenomenon can be described in terms of a shift in the threshold voltage (see, e.g., Fjeldly and Shur 1993). Well above threshold, the injection barrier is much reduced, and the DIBL effect eventually disappears.

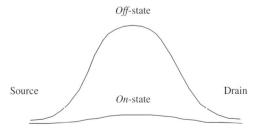

Figure 1.22 Schematic conduction band profile through the channel region of a short-channel MOSFET in the *on*-state and the *off*-state

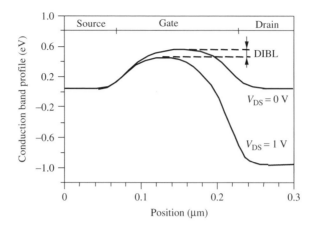

Figure 1.23 Conduction band profile at the semiconductor–oxide interface of a short n-channel MOSFET with and without drain bias. The figure indicates the origin of DIBL. Reproduced from Fjeldly T. A., Ytterdal T., and Shur M. (1998) *Introduction to Device Modeling and Circuit Simulation*, John Wiley & Sons, New York

The magnitude of the subthreshold current is obviously very important since it has consequences for the power supply voltages and the logic levels needed to achieve a satisfactory *off*-state in digital operations. Hence, it affects the power dissipation in logic circuits. Likewise, the holding time in dynamic memory circuits is affected by the level of subthreshold current.

To correctly model the subthreshold operation of MOSFETs, we need a charge control model for this regime. Also, to avoid convergence problems when using the model in circuit simulators, it is preferable to use a UCCM that covers both the above- and below-threshold regimes with continuous expressions. One such model is a generalization of the UCCM that was introduced in Section 1.2.4 for the purpose of accurately describing the inversion charge density in MOS structures (Lee *et al.* 1993),

$$V_{GT} - \alpha V_F(x) = \eta V_{th} \ln \left(\frac{n_s(x)}{n_o} \right) + a[n_s(x) - n_o]. \tag{1.68}$$

Here, V_F is the quasi-Fermi potential in the channel measured relative to the Fermi potential at the source and α is a constant with a value close to unity called the *bulk effect parameter*. We note that in strong inversion, $V_F(x)$ can be replaced by the channel potential $V(x)$ and the linear term in $n_s(x)$ will dominate on the right-hand side, signifying that charge transport in the channel will be drift current. Below threshold, the logarithmic term dominates on the right-hand side and the charge transport is primarily by diffusion.

Although (1.68) does not have an analytical solution with respect to n_s, we can use a generalized version of the approximate analytical expression introduced for the MOS capacitor in (1.38),

$$n_s = 2n_o \ln \left[1 + \frac{1}{2} \exp \left(\frac{V_{GT} - \alpha V_F}{\eta V_{th}} \right) \right] \tag{1.69}$$

This and related models have since been successfully applied to various FETs including MOSFETs, MESFETs, HFETs, poly-Si thin film transistors (TFTs), and a-Si TFTs (see

Fjeldly *et al.* 1998). The unified MOSFET model based on the UCCM expression in (1.68) is discussed in Chapter 8 (see also Shur *et al.* 1992). More elaborate MOSFET models such as the BSIM4 and EKV models are discussed in Chapters 6 and 7, respectively. They include a number of advanced features relating to small feature sizes and scaling of device dimensions.

BSIM4 is presently the most advanced MOSFET model supplied with Berkeley SPICE, and has been adopted in most commercial simulators. Although the BSIM models are characterized by a large number of SPICE parameters (in the hundreds), they have gained a wide popularity for use in professional circuit simulation and design, and have been accepted as an industry standard in the United States.

In Section 1.5.2, we consider more closely some of the advanced features included in modern MOSFET models, such as velocity saturation, gate bias–dependent mobility, impact ionization, drain and source series resistances (extrinsic modeling), channel-length modulation, and DIBL. In Section 1.5.3, we continue the discussion of the MOSFET capacitances from Section 1.4.4 and present a unified and charge-conserving description of the intrinsic capacitance–voltage characteristics.

1.5.2 Nonideal Effects

1.5.2.1 High-field effects

Channel-length modulation   When the drain-source bias of a FET approaches the drain saturation voltage, a region of high electric field forms near the drain and the electron velocity in this region saturates (in long devices, we instead have pinch-off where n_s becomes very small near drain). In saturation, the length ΔL of the high-field region expands in the direction of the source with increasing V_{DS}, and the MOSFET behaves as if the effective channel length has been reduced by ΔL. This phenomenon is called channel-length modulation (CLM). The following simplified expression links V_{DS} to the length of the saturated region (see Lee *et al.* 1993):

$$V_{DS} = V_p + V_\alpha \left[\exp\left(\frac{\Delta L}{l} \right) - 1 \right] \qquad (1.70)$$

where V_p, V_α, and l are parameters related to the electron saturation velocity, the field effect mobility, and the drain conductance in the saturation regime. In fact, V_p is the potential at the point of saturation in the channel, which is usually approximated by the saturation voltage V_{SAT}. Good agreement has been obtained between the potential profile described by (1.70) and that obtained from a two-dimensional simulation for the saturated region of an n-channel MOSFET.

The CLM effect manifests itself as a finite output conductance in saturation, which tends to remain constant over a wide range of drain biases. The output conductance also increases steadily with increasing gate bias. This observation suggests an even simpler model than that in (1.70) for describing CLM, where the basic expression for the drain current is simply multiplied by the first-order term $(1 + \lambda V_{DS})$. In this case, the CLM parameter λ can easily be extracted from the output conductance in the saturation regime, well above threshold. This first-order approximation is implemented in several

FET models used in circuit simulators, while expressions similar to (1.70) are used in the BSIM models.

Hot-carrier effects   Hot-carrier effects are among the main concerns when shrinking FET dimensions into the deep submicrometer regime. Reducing the channel length while retaining high power supply levels, known as constant voltage scaling, results in increased electric field strengths in the channel, causing acceleration and heating of the charge carriers.

Some of the manifestations of hot electrons on device operation are breakdown and substrate current caused by impact ionization, creation of interface states, gate current resulting from hot-electron emission across the interface barrier, oxide charges owing to tunneling of charge carriers into oxide states, and photocurrents caused by electron–hole recombination with emission of photons (see following text).

The substrate current resulting from electron–hole pair generation may overload substrate-bias generators, introduce snapback breakdown, cause CMOS latch-up, and generate a significant increase in the subthreshold drain current. A complete model for the substrate current is too complex for use in circuit level simulation. Instead, the following, approximate, analytical expression is widely used:

$$I_{\text{substr}} = I_{\text{ds}} \frac{A_i}{B_i} (V_{\text{DS}} - V_{\text{SAT}}) \exp \left(-\frac{l_{\text{d}} B_i}{V_{\text{DS}} - V_{\text{SAT}}} \right), \qquad (1.71)$$

where I_{ds} is the channel current, A_i and B_i are the ionization constants, V_{SAT} is the saturation voltage, and l_{d} is the effective ionization length. This expression is also applicable in the subthreshold regime by using $V_{\text{SAT}} = 0$ (Iñiguez and Fjeldly 1997).

In FETs fabricated on an insulating layer, such as silicon-on-insulator (SOI) MOSFETs, impact ionization may give rise to a charging of the transistor body, causing a shift in the threshold voltage. This effect results in an increased drain current in saturation (floating body effect). Related mechanisms are also observed in amorphous TFTs (Wang *et al.* 2000) and polysilicon TFTs (Iñiguez *et al.* 1999).

At sufficiently high drain bias, we have impact ionization and avalanche breakdown in all types of FETs. In MOSFETs, a substantial amount of the majority carriers created by impact ionization near drain will flow toward source and forward-bias the source–substrate junction, causing injection of minority carriers into the substrate. This effect can be modeled in terms of conduction in a parasitic bipolar transistor, as described by Sze (1981). In MESFETs, the breakdown usually takes place in the high-field depletion extension toward the drain.

Electron trapping in the oxide and generation of interface traps caused by hot-electron emission induce degradation of the MOSFET channel near drain in conventional MOSFETs or cause changes in the parasitic drain resistances in low-doped drain (LDD) MOSFET (Ytterdal *et al.* 1995). Reduced current drive capability and transconductance degradation are manifestations of interface traps in *n*-MOSFET characteristics. Reduced current also leads to circuit speed degradation, such that the circuits may fail to meet speed specifications after aging.

Photon emission and subsequent absorption in a different location of the device may cause unwanted photocurrent, which, for example, may degrade the performance of memory circuits.

Temperature dependence and self-heating   Since electronic devices and circuits have to operate in different environments, including a wide range of temperatures, it is imperative to establish reliable models for such eventualities. Heat generated from power dissipation in an integrated circuit chip can be considerable, and the associated temperature rise must be accounted for both in device and circuit design. In conventional silicon substrates, the thermal conductivity is relatively high such that a well-designed chip placed on a good heat sink may achieve a reasonably uniform and tolerable operating temperature. However, such design becomes increasingly difficult as the device dimensions are scaled down and power dissipation increases. The thermal behavior of MOSFETs has been extensively studied in the past, and the temperature dependencies of major model parameters have been incorporated in SPICE models.

Circuits fabricated on substrates that are poor heat conductors, such as GaAs and silicon dioxide, are more susceptible to a significant self-heating effect (SHE). In thin film SOI CMOS, the buried SiO_2 layer inhibits an effective heat dissipation, and the self-heating manifests itself as a reduced drain current and even as a negative differential conductance at high power inputs. Hence, for a reliable design of SOI circuits, accurate and self-consistent device models that account for SHE are needed for use in circuit simulation.

The influence of SHE on the electrical characteristics of SOI MOSFETs can be evaluated using a two-dimensional device simulator incorporating heat flow or by combining a temperature rise model with an $I–V$ expression through an iteration procedure. But the effect can also be described in terms of a temperature-dependent model for the device's $I–V$ characteristics combined with the following simplified relationship between temperature rise and power dissipation:

$$T - T_{\rm o} = R_{\rm th} I_{\rm d} V_{\rm ds}. \tag{1.72}$$

Here T is the actual temperature, $T_{\rm o}$ is the ambient (substrate) temperature, and $R_{\rm th}$ is a thermal resistance that contains information on thermal conductivity and geometry. The equations can be solved self-consistently, either numerically or analytically (see Cheng and Fjeldly 1996). Once the temperature dependence of the device parameters are established, the same procedure can also be used for describing self-heating in other types of devices, such as amorphous TFTs (Wang *et al.* 2000), GaAs MESFETs and HFETs.

Gate bias–dependent mobility   In submicron MOSFETs, scaling dictates that the gate dielectric must be made very thin. In a sub-0.1-μm device, the gate dielectric may be as thin as a few nanometers. With a gate-source voltage of 1 V, this corresponds to a transverse electric field of nearly 500 kV/cm. In this case, electrons are confined to a very narrow region at the silicon–silicon dioxide interface, and their motion in the direction perpendicular to the gate oxide is quantized. This close proximity of the carriers to the interface enhances the scattering rate by surface nonuniformities, drastically reducing the field effect mobility in comparison to that of bulk silicon.

To a first-order approximation, the following simple expression accurately describes the dependence of the field effect mobility on the gate bias (Park *et al.* 1991)

$$\mu_n = \mu_{on} - \kappa_n (V_{\rm GS} + V_{\rm T}). \tag{1.73}$$

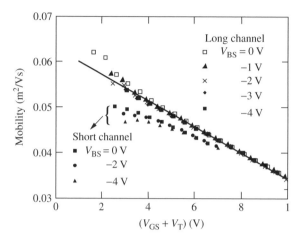

Figure 1.24   Electron mobility versus $V_{GS} + V_T$ for a long-channel ($L = 20\,\mu m$) and a short-channel ($L = 1\,\mu m$) NMOS for different values of substrate bias. The solid line corresponds to the linear approximation used in (1.73). Reproduced from Park C. K. *et al.* (1991) A unified charge control model for long channel n-MOSFETs, *IEEE Trans. Electron Devices*, **ED-38**, 399–406

The experimental MOSFET mobility data in Figure 1.24 shows that this expression can be applied with the same set of parameters for different values of substrate bias. The parameter values are fairly close even for devices with quite different gate lengths. All in all, this leads to a reduction in the number of parameters needed for accurate modeling of the MOSFET characteristics.

More complete expression for the MOSFET field effect mobility, which takes both temperature variations and scaling into account, are used in the BSIM models (see Chapter 6).

1.5.2.2 Short-channel effects

Aspect ratio   To first order, FET dimensions are scaled by preserving the device aspect ratio, that is, the ratio between the gate length and the active vertical dimension of the device. In MOSFETs, the vertical dimension accounts for the oxide thickness d_i, the source and drain junction depths r_j, and the source and drain junction depletion depths W_s and W_d. A low aspect ratio is synonymous with short-channel behavior. The following empirical relationship indicates the transition from long-channel to short-channel behavior (Brews *et al.* 1980):

$$L < L_{min}(\mu m) = 0.4[r_j(\mu m)d_i(\text{Å})(W_d + W_s)^2(\mu m^2)]^{1/3} \qquad (1.74)$$

When $L < L_{min}$, the MOSFET threshold voltage V_T will be affected in several ways as a result of reduced gate control. First, the *depletion* charges near source and drain are under the shared control of these contacts and the gate. In a short-channel device, the shared charge will constitute a relatively large fraction of the total gate depletion charge, giving rise to an increasingly large shift in V_T with decreasing L. Also, the shared depletion charge near drain expands with increasing drain-source bias, resulting in an additional V_{DS}-dependent shift in V_T (DIBL effect, see following text).

Drain-induced barrier lowering   The threshold voltage is a measure of the strength of the barrier against carrier injection from source to channel. In the short-channel regime ($L < L_{\min}$), this barrier may be significantly modified by the application of a drain bias, as was schematically depicted in Figure 1.23. In n-channel FETs, this drain-induced barrier lowering (DIBL) translates into a lowering of the threshold voltage (n-channel MOSFET) and a concomitant rise in the subthreshold current with increasing V_{DS}. The combined scaling and DIBL effect on the threshold voltage may be expressed as follows:

$$V_T(L) = V_{To}(L) - \sigma(L)V_{DS} \qquad (1.75)$$

where $V_{To}(L)$ describes the scaling of V_T at zero drain bias resulting from charge sharing and $\sigma(L)$ is the channel-length-dependent DIBL parameter. In the long-channel case, where $L > L_{\min}$, V_T should become independent of L and V_{DS}. This behavior can be modeled by letting both $V_{To}(L)$ and $\sigma(L)$ scale approximately as $\exp(-L/L_{\min})$. In BSIM, somewhat more detailed scaling functions and also a dependence on substrate bias are used (see Chapter 6).

In Figure 1.25(a), we show experimental data of V_T versus V_D for two n-channel MOSFETs with short gate lengths. A good agreement with the linear relationship of (1.75) is indicated. Also, the exponential scaling for V_T versus L is confirmed by experiments, except for a deviation at the shortest gate lengths, as shown in Figure 1.25(b) (Fjeldly and Shur 1993).

As stated above, DIBL vanishes well above threshold. For modeling purposes, we therefore adopt the following empirical expression for σ (Lee *et al.* 1993):

$$\sigma = \frac{\sigma_0}{1 + \exp\left(\dfrac{V_{gto} - V_{\sigma t}}{V_\sigma}\right)} \qquad (1.76)$$

where V_{gto} is the gate voltage overdrive at zero drain-source bias and the parameters $V_{\sigma t}$ and V_σ determine the voltage and the width of the DIBL fade-out, respectively. We note that $\sigma \to \sigma_0$ for $V_{gto} < V_{\sigma t}$ and $\sigma \to 0$ for $V_{gto} > V_{\sigma t}$.

The DIBL effect can be accounted for in our I–V models by adjusting the threshold voltage according to (1.75) in the expressions for the saturation current and the linear channel conductance. Likewise, the UCCM expression in (1.68) is modified as follows:

$$V_{GTo} + \sigma V_{DS} - \alpha V_F \approx \eta V_{th} \ln\left(\frac{n_s}{n_o}\right) + a(n_s - n_o) \qquad (1.77)$$

where V_{GTo} is the intrinsic threshold voltage overdrive at zero drain-source bias.

A related effect of device miniaturization is observed in narrow-channel FETs, where charges associated with the extension of the gate depletion regions beyond the nominal width of the gate may become a significant fraction of the total gate depletion charge. In this case, a one-dimensional analysis will underestimate the total depletion charge and give a wrong prediction of the threshold voltage. In practice, the threshold voltage increases (n-MOSFET) as the channel width is reduced. A common method of modeling this effect is to add an additional term in the threshold voltage expression containing a $1/W$ term, where W is the effective width of the gate.

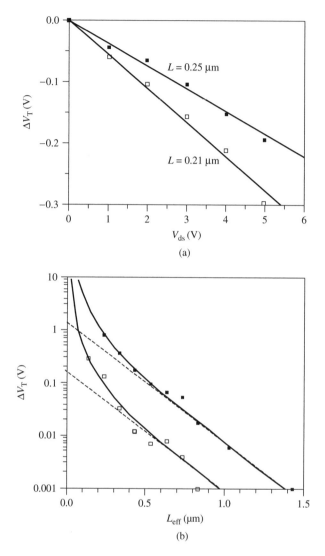

Figure 1.25 DIBL effect: (a) experimental threshold voltage shift versus drain-source voltage for two *n*-MOSFETs with different gate lengths and (b) experimental threshold voltage shifts versus gate length compared with exponential scaling. Reproduced from Fjeldly T. A. and Shur M. (1993) Threshold voltage modeling and the subthreshold regime of operation of short-channel MOSFETs, *IEEE Trans. Electron Devices*, **TED-40**, 137–145

1.5.2.3 Gate leakage and effective oxide thickness

The basic properties and the integrity of the silicon dioxide gate dielectric has been essential for the success of the silicon MOSFETs. However, as the CMOS technology node (half-pitch size in DRAMs) of MOSFETs in large-scale integration moves into the sub-100-nm range, this very success factor harbors one of the most difficult issues facing a continued evolution along the trend described by Moore's law. The reason for this lies in the nonyielding rules of device scaling combined with the well-known quantum-mechanical phenomenon of tunneling. In fact, to derive sufficient advantage of sub-100 nm technology, the gate oxide thickness has to be scaled down to just a couple of nanometers or less, corresponding to only a handful of atomic layers (see ITRS 2001). At such small dimensions, the tunneling leakage current through the oxide from the gate to the semiconductor becomes significant enough to add noticeably to the power consumption and to interfere with the device operation.

An additional problem arises from the long-term reliability of such ultrathin dielectric films (Stathis 2002). These problems grow rapidly with further scaling, ultimately with completely debilitating consequences. The limits for viable scaling have recently been predicted to be at a technology node of 50 nm (gate length of 25 nm) and a silicon dioxide thickness of 1 nm (Wu *et al.* 2002).

A temporary solution to this impasse is to replace the silicon dioxide with materials that have much larger dielectric constants, so-called high-k insulators. This way, the same scaling advantage can be derived using a correspondingly thicker insulator with reduced tunneling current. Many such materials are presently being investigated, but it is hard to find candidates that can match the chemical and electrical properties of silicon dioxide and its excellent interface with silicon. If this development meets with success, the end of the present evolutionary trend in MOSFET/CMOS technology may be extended for yet another decade, bringing the technology node to about 20 nm (see ITRS 2001). Within this time frame, alternative MOSFET architectures with improved short-channel properties will also have to be developed for large-scale integration, including Vertical, FinFET, and planar double gate structures.

Another problem related to the thin dielectrics in MOSFETs is the relative importance of the inversion layer thickness in the semiconductor and the depletion layer thickness when using polysilicon gate electrodes. The former is a result of the lack of vertical confinement of the carriers, especially electrons, owing to the quantum-mechanical uncertainty and exclusion principles in combination with the finite steepness of the semiconductor band bending at the interface. In terms of device performance, these layers add to the effective oxide thickness d_{eff}, thereby reducing the gate's field effect coupling to the channel. The two layers may contribute a few tenths of a nanometer each to d_{eff}, which is quite significant for radically scaled MOSFETs. Development of suitable metal gate electrodes will alleviate some of the problem.

1.5.3 Unified MOSFET $C-V$ Model

1.5.3.1 Unified Meyer C–V model

In order to develop unified expressions for the intrinsic MOSFET capacitances, we return to the Meyer capacitances discussed in Section 1.4. The Meyer large-signal equivalent is shown in Figure 1.16 and the gate-source and gate-drain capacitances are given by (1.59) and (1.60), respectively. The only device-specific part of these equations is the gate-channel capacitance C_{ch} at zero drain bias ($V_F = 0$), which can be derived from the UCCM expression of (1.69), that is,

$$C_{ch} = WLq\frac{dn_s}{dV_{GT}} \approx C_i\left[1 + 2\exp\left(-\frac{V_{GT}}{\eta V_{th}}\right)\right]^{-1}. \qquad (1.78)$$

Well above or below threshold, this expression has the familiar asymptotic forms

$$C_{\text{ch}} \approx C_{\text{i}}, \tag{1.79}$$

$$C_{\text{ch}} \approx \frac{C_{\text{i}}}{2} \exp\left(\frac{V_{\text{GT}}}{\eta V_{\text{th}}}\right), \tag{1.80}$$

respectively. Figure 1.26 shows C_{ch} versus V_{GT} in a linear and a semilogarithmic plot. From UCCM and from (1.78), we find that $C_{\text{ch}} = C_{\text{i}}/3$ at threshold, which may serve as a convenient and straightforward way of determining the threshold voltage from experimental C_{ch} versus V_{GS} curves.

In the subthreshold regime, the gate-substrate capacitance C_{GB} of (1.65) is the dominant Meyer capacitance in MOSFETs. Above threshold, C_{GB} vanishes in the ideal long-channel case. A unified version of C_{GB} that includes a gradual phase-out above threshold can be modeled as follows:

$$C_{\text{GB}} = \frac{C_{\text{i}}/(1 + n_{\text{s}}/n_{\text{o}})}{\sqrt{1 + 4(V_{\text{GS}} - V_{\text{BS}} - V_{\text{FB}})/\gamma^2}}. \tag{1.81}$$

Here n_s is the unified electron density given by UCCM in (1.68) or its approximate solution (1.69). Equation (1.81) utilizes the fact that the increasing density of inversion charge above threshold gradually shields the substrate from the influence of the gate electrode. A typical plot of C_{GB} versus V_{GT} is shown in Figure 1.27.

Using this unified gate-channel capacitance in conjunction with Meyer's capacitance model, we obtain the following continuous expressions for the intrinsic gate-source capacitance C_{GS} and the gate-drain capacitance C_{GD}, valid for all regimes of operation:

$$C_{\text{GS}} = \frac{2}{3}C_{\text{ch}}\left[1 - \left(\frac{V_{\text{GTe}} - V_{\text{DSe}}}{2V_{\text{GTe}} - V_{\text{DSe}}}\right)^2\right], \tag{1.82}$$

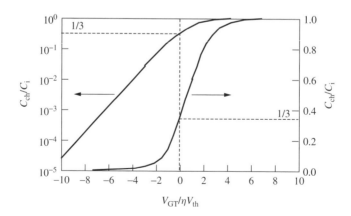

Figure 1.26 Normalized channel capacitance versus $V_{\text{GT}}/\eta V_{\text{th}}$ according to (1.78) in a linear plot (right) and a semilog plot (left). The condition $C_{\text{ch}}/C_{\text{i}} = 1/3$ at threshold is indicated. Reproduced from Fjeldly T. A., Ytterdal T., and Shur M. (1998) *Introduction to Device Modeling and Circuit Simulation*, John Wiley & Sons, New York

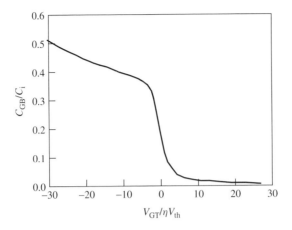

Figure 1.27 Normalized and unified Meyer-type gate-substrate capacitance versus $V_{GT}/\eta V_{th}$ according to (1.81) for $V_{BS} = 0$. Typical values for an *n*-channel MOSFET with a polysilicon gate were used: $V_T = 0.7 \, \text{V}$, $V_{FB} = -1 \, \text{V}$, $\gamma = 1 \, \text{V}^{1/2}$, and $\eta = 1.33$. Reproduced from Fjeldly T. A., Ytterdal T., and Shur M. (1998) *Introduction to Device Modeling and Circuit Simulation*, John Wiley & Sons, New York

$$C_{GD} = \frac{2}{3}C_{ch}\left[1 - \left(\frac{V_{GTe}}{2V_{GTe} - V_{DSe}}\right)^2\right]. \tag{1.83}$$

Here, V_{DSe} is an effective intrinsic drain-source voltage that is equal to V_{DS} for $V_{DS} < V_{GTe}$ and is equal to V_{GTe} for $V_{DS} > V_{GTe}$. V_{GTe} is the effective gate voltage overdrive, which equals V_{GT} above threshold and is of the order of the thermal voltage in the subthreshold regime. The following expression is used to model this behavior:

$$V_{GTe} = V_{th}\left[1 + \frac{V_{GT}}{2V_{th}} + \sqrt{\delta^2 + \left(\frac{V_{GT}}{2V_{th}} - 1\right)^2}\right], \tag{1.84}$$

where δ determines the width of the transition region at threshold ($V_{GT} = 0$). Typically, $\delta = 3$ is a good choice.

A smooth transition between the nonsaturated and the saturated regimes is assured by using the following type of interpolation expression for effective intrinsic drain-source voltage:

$$V_{DSe} = \frac{1}{2}\left[V_{DS} + V_{GTe} - \sqrt{V_\delta^2 + (V_{DS} - V_{GTe})^2}\right], \tag{1.85}$$

where V_δ is a constant voltage that determines the width of the transition region. This parameter may be treated as an adjustable parameter to be extracted from experiments. V_{GTe} is needed to assure a smooth transition between the correct limiting *I–V* and *C–V* expressions above and below threshold.

A comparison of the normalized dependencies of C_{GS} and C_{GD} on V_{DS} is shown in Figure 1.28 for $V_\delta/V_{GTe} = 0$, corresponding to the nonunified Meyer capacitances, and for

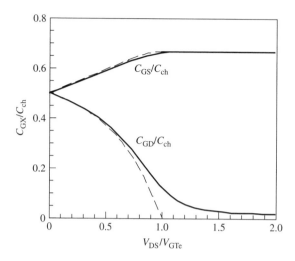

Figure 1.28 Normalized and nonunified Meyer capacitances according to (1.59) and (1.60) (dashed lines) and unified Meyer capacitances according to (1.82) and (1.83) (solid lines), using a transition width parameter $V_\delta = 0.2\, V_{GTe}$. Reproduced from Fjeldly T. A., Ytterdal T., and Shur M. (1998) *Introduction to Device Modeling and Circuit Simulation*, John Wiley & Sons, New York

a more realistic value of $V_\delta/V_{GTe} = 0.2$. On the basis of the discussion in Section 1.4.4, we can conclude that the present unified version of the Meyer capacitances is applicable also for short-channel devices. Still more flexible expressions for the capacitances are obtained by substituting V_{GT} by χV_{GT} in (1.82) and (1.83), where χ is an adjustable parameter close to unity.

1.5.3.2 Ward–Dutton model

As we discussed in Section 1.4, an accurate modeling of the intrinsic capacitances associated with the gate region of FETs requires an analysis of the charge distribution in the channel versus the terminal bias voltages. Normally, the problem is simplified by assigning the distributed charges to the various "intrinsic" terminals. Hence, the mobile charge Q_I of a MOSFET is divided into a source charge $Q_S = F_p Q_I$ and a drain charge $Q_D = (1 - F_p)Q_I$, where F_p is a partitioning factor. The depletion charge Q_B under the gate is assigned to the MOSFET substrate terminal. The total gate charge Q_G is the negative sum of these charges, that is, $Q_G = -Q_I - Q_B = -Q_S - Q_D - Q_B$. Note that by assigning the charges this way, charge conservation is always assured.

The net current flowing into terminal X can now be written as

$$I_X = \frac{dQ_X}{dt} = \sum_Y \frac{\partial Q_X}{\partial V_Y}\frac{\partial V_Y}{\partial t} = \sum_Y \chi_{XY} C_{XY} \frac{\partial V_Y}{\partial t}, \qquad (1.86)$$

where the indices X and Y run over the terminals G, S, D, and B. In this expression, we have introduced a set of intrinsic capacitance elements C_{XY}, the so-called transcapacitances, defined by

$$C_{XY} = \chi_{XY}\frac{\partial Q_X}{\partial V_Y} \quad \text{where} \quad \chi_{XY} = \begin{cases} -1 & \text{for } X \neq Y \\ 1 & \text{for } X = Y \end{cases}. \qquad (1.87)$$

These are equivalent to the charge-based nonreciprocal capacitances introduced by Ward and Dutton (1978) and by Ward (1981). The term *nonreciprocal* means that we have $C_{XY} \neq C_{YX}$ when $X \neq Y$. The elements C_{XX} are called self-capacitances. C_{XY} contain information on how much the charge Q_X assigned to terminal X changes by a small variation in the voltage V_Y at terminal Y. To illustrate why C_{XY} may be different from C_{YX}, assume a MOSFET in saturation. Then the gate charge changes very little when the drain voltage is perturbed since the inversion charge is very little affected, making C_{GD} small. However, if V_G is perturbed, the inversion charge changes significantly and so does Q_D, making C_{DG} large.

For the four-terminal MOSFET, the Ward–Dutton description leads to a total of 16 transcapacitances. This set of 16 elements can be organized as follows in a 4×4 matrix, a so-called indefinite admittance matrix:

$$\mathbf{C} = \begin{bmatrix} C_{GG} & C_{GS} & C_{GD} & C_{GB} \\ C_{SG} & C_{SS} & C_{SD} & C_{SB} \\ C_{DG} & C_{DS} & C_{DD} & C_{DB} \\ C_{BG} & C_{BS} & C_{BD} & C_{BB} \end{bmatrix}. \tag{1.88}$$

Here, the elements in each column and each row must sum to zero owing to the constraints imposed by charge conservation (which is equivalent to obeying Kirchhoff's current law) and for the matrix to be reference independent, respectively (see Arora 1993). This means that some of the transconductances will be negative, and of the 16 MOSFET elements, only 9 are independent. The complete MOSFET large-signal equivalent circuit, including the 16 transcapacitances, is shown in Figure 1.29. This compares with the simple Meyer model in Figure 1.16, which comprises 3 capacitances.

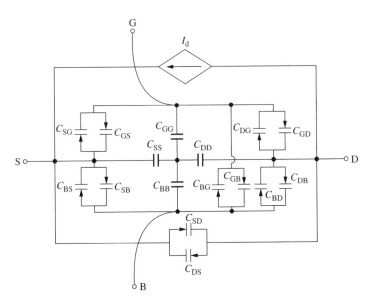

Figure 1.29 Intrinsic large-signal MOSFET equivalent circuit including a complete set of nonreciprocal and charge-conserving transcapacitances. The transcapacitances C_{XY} are defined in the text. Reproduced from Fjeldly T. A., Ytterdal T., and Shur M. (1998) *Introduction to Device Modeling and Circuit Simulation*, John Wiley & Sons, New York

We note that in three-terminal FETs, such as HFETs, MESFETs, and TFTs, we have a total of 9 transcapacitances, of which 4 are independent (Nawaz and Fjeldly 1997). The equivalent circuit for this case is obtained from Figure 1.29 by removing the substrate terminal B and all elements connected to it. Also, the 4×4 matrix in (1.88) reduces to a 3×3 matrix.

As explained in Section 1.4.4, the simplified C–V model by Meyer is obtained by taking derivatives of the total gate charge with respect to the various terminal voltages. The Meyer capacitances can be viewed as a subset of the Ward–Dutton capacitances. Although charge conservation is not assured in the Meyer model, the resulting errors in circuit simulations are usually small, but can in some cases lead to serious errors. The unified transcapacitances needed for the complete Ward–Dutton model can be obtained along the same lines as described for C_{GS} and C_{DS}. The accuracy of the model depends on the quality of the charge and current models used and on the partitioning of the inversion charge between the source and the drain terminal.

1.5.3.3 Non-quasi-static modeling

For very high-frequency operation of the MOSFET, comparable to the inverse carrier transport time of the channel (non-quasi-static (NQS) regime), we have to consider the temporal relaxation of the inversion and depletion charges. Most of the MOSFET models used in SPICE are based on the quasi-static assumption (QSA), in which an instantaneous charging of the inversion layer is assumed. Hence, circuit simulations will fail to accurately predict the performance of high-speed circuits.

The channel of a MOSFET is analogous to a bias-dependent distributed RC network as indicated schematically in Figure 1.30. In QSA, the distributed gate-channel capacitance is instead lumped into discrete capacitances between the gate and source and drain nodes, ignoring the finite charging time arising from the RC product associated with the channel resistance and the gate-channel capacitance.

The inclusion of the so-called Elmore equivalent circuit shown in Figure 1.31 can be viewed as a first step toward an NQS model. Using this equivalent circuit, the channel charge buildup is modeled with reasonable accuracy because the lowest frequency pole of the original RC network is retained. The Elmore resistance R_{Elmore} is calculated from the channel resistance in strong inversion as

$$R_{Elmore} \approx \frac{L_{eff}^2}{e \mu_{eff} Q_{ch}}. \tag{1.89}$$

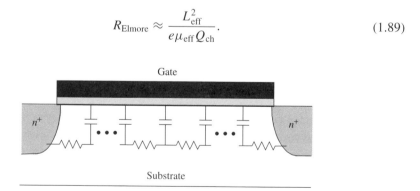

Figure 1.30 Equivalent RC network representing the MOSFET channel

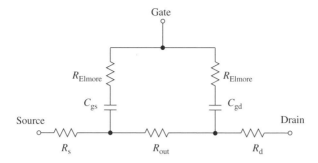

Figure 1.31   Elmore non-quasi-static equivalent circuit

where e is the Elmore constant with a theoretical value close to 5 and Q_{ch} is the total charge in the channel. This formulation is only valid above threshold where the drift current dominates.

To obtain a unified expression, including the subthreshold diffusion current, a relaxation time–based approach is adapted. The overall relaxation time for channel charging and discharging can be written as a combination of the contributions due to drift and diffusion as follows:

$$\frac{1}{\tau} = \frac{1}{\tau_{drift}} + \frac{1}{\tau_{diff}}, \tag{1.90}$$

where

$$\tau_{drift} = R_{Elmore} C_i, \tag{1.91}$$

$$\tau_{diff} = \frac{q(L_{eff}/4)^2}{\mu_{eff} k_B T}. \tag{1.92}$$

On the basis of this relaxation time concept, the NQS effect can be implemented in the SPICE MOSFET model using the subcircuit shown in Figure 1.32. The variable Q_{def} is an additional node created to keep track of the amount of deficit or surplus channel charge needed to achieve equilibrium. Q_{def} will decay exponentially into the channel with a bias-dependent NQS relaxation time τ, and the terminal currents can be written as

$$I_d = I_d(dc) + X_d \frac{Q_{def}}{\tau}, \tag{1.93}$$

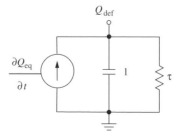

Figure 1.32   Non-quasi-static subcircuit implementation in MOSFET SPICE models. The RC time constant τ is determined by the resistance and capacitance values chosen

$$I_g = -\frac{Q_{\text{def}}}{\tau}. \tag{1.94}$$

Here $X_d = 1 - F_p$ and $X_s = F_p$, where F_p is the charge partitioning factor introduced in Section 1.5.3.2.

REFERENCES

Arora N. (1993) *MOSFET Models for VLSI Circuit Simulation. Theory and Practices*, Springer-Verlag, Wien.

Brews J. R., Fichtner W., Nicollian E. H., and Sze S. M. (1980) Generalized guide for MOSFET miniaturization, *IEEE Electron Device Lett.*, **EDL-1**, 2.

Byun Y., Lee K., and Shur M. (1990) Unified charge control model and subthreshold current in heterostructure field effect transistors, *IEEE Electron Device Lett.*, **EDL-11**, 50–53; (see erratum (1990) *IEEE Electron Device Lett.*, **EDL-11**, 273).

Cheng Y. and Fjeldly T. A. (1996) Unified physical I–V model including self-heating effect for fully depleted SOI/MOSFET's, *IEEE Trans. Electron Devices*, **ED-43**, 1291–1296.

Fjeldly T. A. and Shur M. (1993) Threshold voltage modeling and the subthreshold regime of operation of short-channel MOSFETs, *IEEE Trans. Electron Devices*, **TED-40**, 137–145.

Fjeldly T. A., Ytterdal T., and Shur M. (1998) *Introduction to Device Modeling and Circuit Simulation*, John Wiley & Sons, New York.

Fonstad C. G. (1994) *Microelectronic Devices and Circuits*, McGraw-Hill, New York.

Iñiguez B. and Fjeldly T. A. (1997) Unified substrate current model for MOSFETs, *Solid-State Electron.*, **41**, 87–94.

Iñiguez B., Xu Z., Fjeldly T. A., and Shur M. S. (1999) Unified model for short-channel poly-Si TFTs, *Solid-State Electron.*, **43**, 1821–1831.

ITRS – International Technology Roadmap for Semiconductor, Semiconductor Industry Assoc., Austin, TX (2001).

Lee K., Shur M., Fjeldly T. A., and Ytterdal T. (1993) *Semiconductor Device Modeling for VLSI*, Prentice Hall, Englewood Cliffs, NJ.

Meyer J. E. (1971) MOS models and circuit simulation, *RCA Rev.*, **32**, 42–63.

Nawaz M. and Fjeldly T. A. (1997) A new charge conserving capacitance model for GaAs MESFETs, *IEEE Trans. Electron Devices*, **ED-44**, 1813–1821.

Park C. K. *et al.* (1991) A unified charge control model for long channel n-MOSFETs, *IEEE Trans. Electron Devices*, **ED-38**, 399–406.

Shur M. (1990) *Physics of Semiconductor Devices*, Prentice Hall, Englewood Cliffs, NJ.

Shur M., Fjeldly T. A., Ytterdal T., and Lee K. (1992) Unified MOSFET model, *Solid-State Electron.*, **35**, 1795–1802.

Sodini C. G., Ko P. K., and Moll J. L. (1984) The effect of high fields on MOS device and circuit simulation, *IEEE Trans. Electron Devices*, **ED-31**, 1386–1393.

Stathis J. H. (2002) Reliability limits of the gate insulator in CMOS technology, *IBM J. Res. Dev.*, **46**(2/3), 265–286.

Sze S. M. (1981) *Physics of Semiconductor Devices*, Second Edition, John Wiley & Sons, New York.

Wang L. *et al.* (2000) Self-heating and kink effects in a-Si:H thin film transistors, *IEEE Trans. Electron Devices*, **ED-47**, 387–397.

Ward D. E. and Dutton R. W. (1978) A charge-oriented model for MOS transistor capacitances, *IEEE J. Solid-State Circuits*, **SC-13**, 703–708.

Ward D. E. (1981) *Charge Based Modeling of Capacitance in MOS Transistors*, Ph.D. thesis, Stanford University, Stanford, CA.

Wu E. Y. *et al.* (2002) CMOS scaling beyond the 100-nm node with silicon-dioxide-based gate dielectrics, *IBM J. Res. Dev.*, **46**(2/3), 287–298.

Ytterdal T., Kim S. H., Lee K., and Fjeldly T. A. (1995) A new approach for modeling of current degradation in hot-electron damaged LDD NMOSFETs, *IEEE Trans. Electron Devices*, **ED-42**, 362–364 (1995).

2
MOSFET Fabrication

2.1 INTRODUCTION

Semiconductor devices have long been used in electronics since the late nineteenth century. The galena crystal detector, invented in 1907, was widely used to build crystal radio sets. However, the idea of placing multiple electronic devices on the same substrate material came only after the late 1950s. In 1959, the first integrated circuit (IC) was constructed, which started a new era of modern semiconductor manufacturing. In less than 50 years, the IC technology, represented primarily by the complementary-metal-oxide-semiconductor (CMOS) process, has gone through the periods from producing very simple chips containing a few bipolar or MOS components to fabricating ultra-large-scale-integrated (ULSI) CMOS circuits with very high device densities from millions of transistors a chip for some circuits such as microprocessors to more than several billions of transistors a chip for some circuits such as memories. As predicted by Moore's law created in the early 1970s, the number of transistors per chip for a microprocessor has continued to double approximately every 18 to 24 months. Taking Intel's processors as an example, the number of transistors on a chip has increased more than 3200 times, from 2300 on the 4004 microprocessor in 1971 to 7.5 million on the Pentium II processor in 1996, and to 55 million on the Pentium 4 processor in 2001. At the same time, the minimum dimension of the transistors has reduced from about $20\,\mu m$ in 1960 to $0.35\,\mu m$ in 1996, and more rapidly recently, to $0.13\,\mu m$ in 2001, resulting in an amazing improvement in both speed and cost of the circuits.

The development of IC technology was driven mainly by the digital circuit (microprocessor and memory) market. Recently, however, CMOS technology has been extensively used in the analog circuit design because of the low cost of fabrication and compatibility of integrating both analog and digital circuits on the same chip, which improves the overall performance and reliability and may also reduce the cost of packaging. It has been the dominant technology to fabricate digital ICs and will be the mainstream technology for analog and mixed-signal applications. Currently, circuit designers are even exploring emerging pure CMOS approaches – integrating digital blocks, analog and radio-frequency (RF) circuits on a single chip to implement the so-called mixed-signal (MS) or system-on-chip (SOC) solutions.

Device Modeling for Analog and RF CMOS Circuit Design. T. Ytterdal, Y. Cheng and T. A. Fjeldly
© 2003 John Wiley & Sons, Ltd ISBN: 0-471-49869-6

In today's IC industry, much of the design efforts, including layout generators, device models, and technology files, have been automated in the design tools provided by either foundries or design automation vendors. However, a basic understanding of semiconductor devices and fabrication processes is essential to optimize the circuits, especially analog/RF circuits. This chapter provides a brief overview of the CMOS process. We first discuss the major process steps in CMOS fabrication. Then we will go though a typical digital process flow to understand the MOSFET structures and the concepts of MOSFET fabrication. Finally, additional fabrication steps for other components, mainly passive devices, in an analog/RF process will be discussed.

2.2 TYPICAL PLANAR DIGITAL CMOS PROCESS FLOW

The polysilicon gate CMOS process has been widely used for IC fabrication. A MOSFET process flow in a baseline polysilicon gate CMOS fabrication is described in the flowchart in Figure 2.1. The mask operations are illustrated in the figure.

Starting material

In this book, we only discuss CMOS technology that is fabricated from silicon, a very common and widely distributed element on earth. The mineral quartz consists entirely of silicon dioxide, also called silica. Ordinary sand is composed mainly of tiny grains of quartz and is therefore also mostly silica. Despite the abundance of its compounds, elemental silicon does not exist naturally. The element can be artificially produced by heating silica and carbon in an electric furnace. The carbon unites with the oxygen contained in the silica, leaving molten silicon. As it is solidified, it will be in a polycrystalline structure, that is, there is no regular crystal structure throughout the block of the material but simply small areas of crystals at different orientations to neighboring crystal areas. Impurities and disordering of the metallurgical-grade polysilicon make it unsuitable for semiconductor manufacture as a substrate material. The polysilicon can be refined in a purification process to produce an extremely pure semiconductor grade material. Once the material has been purified, it can be further processed into single crystal bars by using the so-called Czochralaki method, in which the purified material is completely molten and the seed crystal is dipped into the surface of the melt and slowly withdrawn and rotated. The speed of pull and the rate of cooling will determine the diameter of the final rod of the material. Dopant material can be introduced into the melt in the required ratio. Since ICs are formed upon the surface of a silicon crystal within a limited depth ($<10\,\mu$m), the crystal bar is customarily sliced into numerous thin circular pieces called *wafers*. The larger the wafer, the more ICs it can have, and so the lower the fabrication cost. Most modern processes currently employ 200-mm ($8''$) wafers. However, process lines for 300-mm ($12''$) wafers have been announced to operate in 2002 by the semiconductor foundry industry.

Although a certain amount of dopant material can be introduced into the material during the crystallization process, there is a limit on the doping level that can be introduced if a consistent dopant concentration is to be maintained throughout the material. To obtain the highly doped regions required by some of the active devices, a further crystal growth process called *epitaxy* is performed to provide a thin epitaxial region on the top of the "native" wafer. A continuous crystal structure has to be maintained, so the resulting wafer is still a single crystal throughout. Epitaxy allows the formation of buried layers. The formation of an n^+ buried layer is one of the key steps in most bipolar and BICMOS processes. CMOS ICs are normally fabricated on a p-type (100) substrate doped with boron. To provide a better immunity against CMOS latch-up, the substrate is usually doped as high as possible, limited by solid solubility, to minimize the substrate resistivity. In principle, this kind of p-type wafer can be used directly for fabrication. However, a lightly doped p-type epitaxial layer is usually formed to maintain the latch-up immunity but have more precise control of the electrical properties of the substrate material and hence control of the MOSFET electrical characteristics.

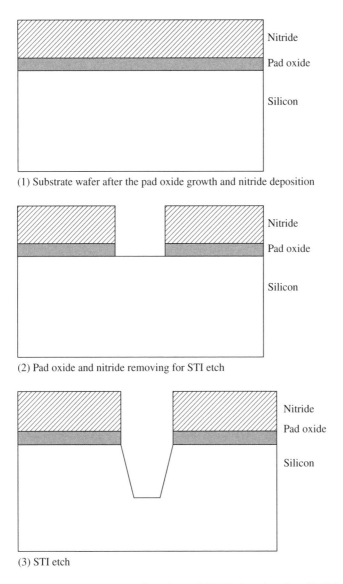

(1) Substrate wafer after the pad oxide growth and nitride deposition

(2) Pad oxide and nitride removing for STI etch

(3) STI etch

Figure 2.1   Illustration of the process flow for MOSFETs in a baseline CMOS technology

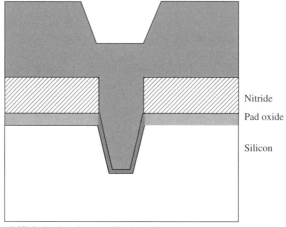

Nitride

Pad oxide

Silicon

(4) High-density plasma oxide deposition

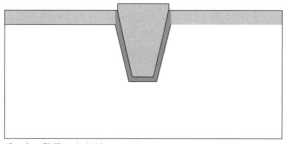

(5) After CMP and nitride removing

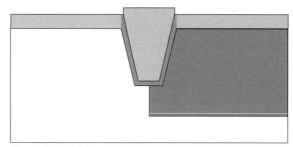

(6) *n*-well formation

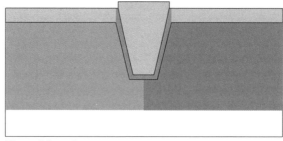

(7) *p*-well formation

Figure 2.1   (*continued*)

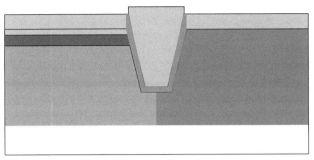

(8) n-channel V_t adjusting and punch-through implantations

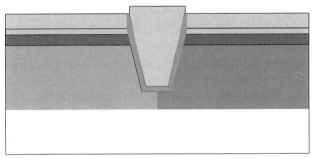

(9) p-channel V_t adjusting and punch-through implantation

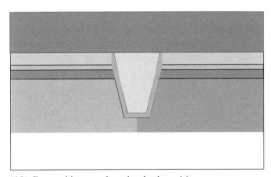

(10) Gate oxide growth and poly deposition

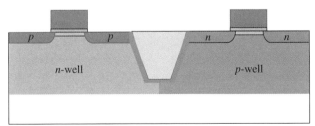

(11) Poly-gate formation for n/pFET and LDD implantation for
     both nFET and pFET

Figure 2.1   (*continued*)

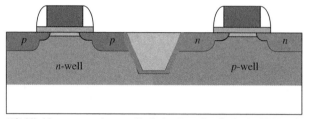

(12) Nitride spacer and source/drain implantation for both *n*FET and *p*FET

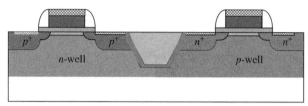

(13) Source/drain salicidation

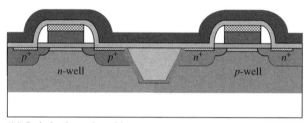

(14) Isolation layer deposition

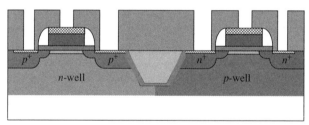

(15) Contact etch

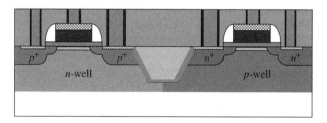

(16) Metallization for source/drain and gates

Figure 2.1   (*continued*)

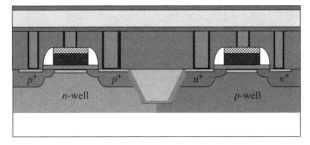

(17) Metal 1 layer deposition

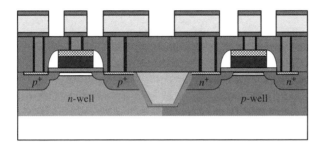

(18) Metal 1 etch

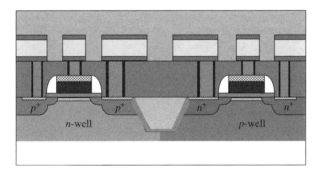

(19) Interlayer isolation dielectric deposition

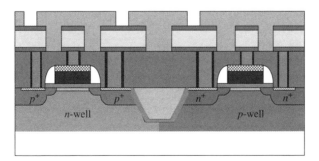

(20) Via etch

Figure 2.1   (*continued*)

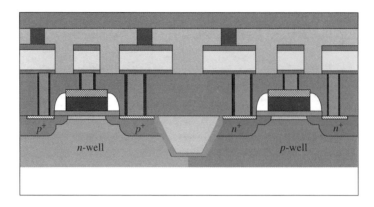

(21) Metal 2 deposition

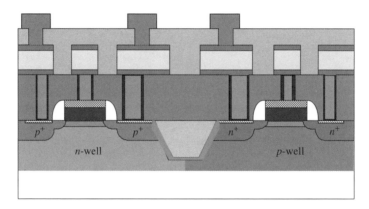

(22) Metal 2 etch

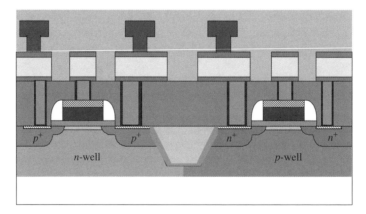

(23) Higher-level interlayer dielectric deposition

Figure 2.1 (*continued*)

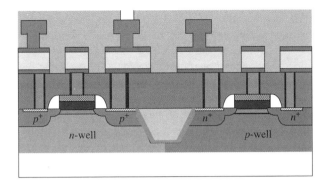

(24) Higher-level via etch

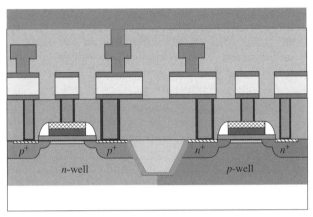

(25) High-level metal deposition

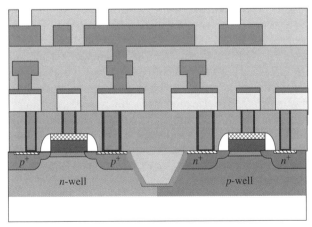

(26) Passivation layer deposition and pad open

Figure 2.1   (*continued*)

n-/p-well formation

The production of silicon wafers is only the first step in the fabrication of ICs. The construction of devices and circuits on the surface of the wafers in the planar process relies on processing to selectively deposit or remove the materials in some regions. Layout patterns, describing where areas of further doping, insulating layers, metal interconnections, and so on are located, need to be transferred onto the wafer. The technique of transferring the layer patterns onto the wafer is based on having layers of resistive material covering the surface, areas that can be selectively removed to expose the circuit below to doping, etching, or deposition of further layers. The process has many similarities to photographic techniques and is termed *photolithography*.

After the wafer has been thermally oxidized, a layer of photoresist that has been spun on to it is patterned using the *n*-well mask. Oxide-etch opens windows through which ion implantation deposits a controlled dose of phosphorus. A prolonged high-temperature drive creates a deep lightly doped *n*-type region called an *n*-well. The *n*-well for a typical 20-V CMOS process has a junction depth of about $5\,\mu$m. Thermal oxidation during the well drive covers the exposed silicon with a thin layer of pad oxide.

In an *n*-well CMOS process, *n*MOS transistors are formed in the epitaxial layer, and *p*MOS transistors reside in the well. The increased total dopant concentration caused by counterdoping the well slightly degrades the mobility of majority carriers within it. The *n*-well process therefore optimizes the performance of the *n*MOS transistor at the expense of the *p*MOS transistor. As a side effect, the *n*-well process also produces the grounded substrate favored by circuit designers.

A *p*-well CMOS process also exists. It uses an n^+ substrate, an *n*-epitaxial layer, and a *p*-well. *n*MOS transistors are formed in the *p*-well, and *p*MOS transistors are formed in the epitaxial layer/substrate. The *p*-well process optimizes the *p*MOS transistor at the expense of the *n*MOS transistors. Since the *n*-well process provides better *n*MOS transistors, which have better performance than *p*MOS transistors due to the higher carrier mobility, and the *p*-well process requires that the substrate is connected to the highest voltage supply instead of ground, which increases the design complexity in biasing the circuits, the *n*-well process is used more widely than the *p*-well process. Another advantage of the *n*-well process is that it is upwardly compatible with BICMOS technology, which has been used for high-frequency (HF) and high-speed applications.

In today's more advanced CMOS technologies ($0.18\,\mu$m and below), both *n*-well and *p*-well has been used to optimize device characteristics of both *n*FET and *p*FET[1]. In a twin-well process, the *p*-well formation does not require additional mask and uses a reverse mask for the *n*-well[2]. It has been proved that a twin-well process provides more benefits than a single-well process even though the complexity of the process has been increased. Actually, recently, a triple-well process has been reported for RF applications, which will be discussed later. In Figure 2.1, we use the twin-well process to illustrate the fabrication of CMOS technology.

Field region definition and channel stop implantation

Regions with thick oxide are defined outside the active regions to increase the threshold voltages in these regions and reduce parasitic capacitance between interconnects and the

---

[1] As we will discuss later, triple-well process exists also for RFCMOS process.
[2] In a dual-well process with device options such as native MOSFET, the *p*-well is a reverse mask of both *n*-well and optional devices.

underlying silicon. Local oxidation of silicon (LOCOS) or shallow trench isolation (STI) technique selectively grows thick oxide in the so-called field regions while leaving a thin pad oxide only over active regions. Taking the LOCOS process as an example, a patterned nitride layer is formed by first depositing nitride across the entire wafer, then defining the field region by using a specific mask, and finally removing the nitride over the field region selectively by using an etch step. The photomask used for this step is called an inverse moat (positive) mask because it consists of a color reverse of the moat regions, that is, the mask codes for areas where the moat is absent, not where it is present. The pad oxide between the silicon and the nitride layer in the field regions are critical because the conditions of nitride growth introduce mechanical stresses that can cause dislocations in the silicon lattice and the pad oxide provides mechanical compliance preventing strain produced by the nitride growth from damaging the underlying silicon.

Selective channel stop implantation underneath the field oxide usually requires to ensure that the threshold voltage in the field regions exceeds the operating voltages. p-substrate field regions receive a p-type channel stop implant, while n-well field regions receive an n-type channel stop implant. After the channel stop implantation, all photoresist will be stripped from the wafer in preparation for isolation processing.

Isolation processing

LOCOS used to be the most popular isolation technique for CMOS fabrication (Kooi 1991). Recently, however, STI is widely accepted as an effective approach to overcome the bird's beak encroachment (see, for example, Nandakumar *et al.* 1998) and the low isolation punch-through voltage of LOCOS-based isolation for subquarter micron devices (Tsai *et al.* 1989). CMOS devices and fabrication processes based on LOCOS have been very well discussed in many books. In the following, we show the processing and device structure with STI.

In spite of its superior isolation characteristics and extensibility, the STI process is much more complicated than LOCOS. In general, trench-etch, chemical vapor deposition (CVD) oxide filling, and planarization using a chemical mechanical polishing (CMP) are considered as key technologies for STI. STI requires a silicon-etch, so etch-damaged layers like crystallographic displacement and/or contamination of unwanted substances exist to a certain extent. Improper post-etching treatment can cause trap-assisted leakage current due to the damaged layers. To improve this, sidewall oxidation is used for crystallographic recovery of the silicon substrate. The roles of the sidewall oxidation include removal of plasma damage during trench-etch, active edge rounding for the suppression of parasitic channels, and reduction of interface traps for the decrease of junction leakage. Analogous to the gate oxidation process, in which a sacrificial oxidation step is adopted to remove damaged silicon formed during previous processes and, as a result, to improve the gate oxide quality, sidewall oxidation also requires an additional sacrificial oxidation step.

Threshold adjust and gate oxidation

Because of the existence of many different doping steps in the fabrication, the threshold voltage of a pFET without threshold adjust can be as high as $-2\,V$, which deviates from the range ($-0.7\,V \sim -0.4\,V$) in modern MOS transistors. So threshold adjust implantation is needed to move the threshold voltage to the desired targets.

Typically, two separate implants are used to adjust the threshold voltages (V_t) in nFETs and pFETs independently. After the wafer has been covered by photoresist, the V_t adjust mask is used to open windows over areas where MOSFETs will form. The boron adjust implant penetrates the dummy gate oxide, formed to eliminate the gate oxide integrity failures caused by nitride deposition, to dope the channel region underlying the dummy oxide. After the V_t adjust implantation, the dummy gate oxide is stripped away to reveal bare silicon in the active regions.

The real gate oxide will be formed by using oxygen at high temperature to ensure the quality of the $Si-SiO_2$ interface by minimizing the charges due to surface states and other traps. This oxidation process must be well controlled to obtain a very thin gate oxide, which is around 2.5 nm for advanced technology. The gate oxide will be the dielectric of the MOSFETs; it also covers the source/drain regions when implantations occur in these regions.

Polysilicon deposition and gate definition

Currently, CMOS technology uses polysilicon as the material for the gate electrodes. It is heavily doped to reduce its resistivity. The typical sheet resistance for a polysilicon gate ranges between 20 and 40 Ω/sq and can be reduced by a factor of 10 with a silicide process, and even more with a metal stack process. Although polysilicon gates do not conduct significant DC, switching signals at the gates do produce substantial AC, and low resistance polysilicon will improve the switching speed of the circuits. RF applications also prefer low gate resistance to reduce HF noise in the circuits. Furthermore, high doping concentration in the poly gate can help reduce/eliminate the so-called poly-depletion effect, which will influence both the DC and the AC characteristics of the device.

The deposited polysilicon layer must be patterned using the poly mask according to the designed geometry for the gates. Since the gate dimension is the most important parameter for a MOSFET, the definition and etching of poly gates are considered the most critical photolithographic steps in CMOS fabrication, especially since the rapid advance in VLSI manufacturing has brought the minimum device feature size and the spacing between devices below the wavelength of the light source. To accurately define the poly gates with critical feature dimensions, advanced compensation mechanisms are required that perturb either the shape via optical proximity correction (OPC) or the phase via phase-shifting masks (PSM) of the transmitting apertures in the reticle when preparing the masks used for fabrication.

Early CMOS technology used to adopt single poly gate (n^+ poly typically) for both nFETs and pFETs. In recent more advanced technologies, dual-gate (that is, different doped poly gates for nFETs and pFETs) technology has been widely used to improve the electrical performances in both nFETs and pFETs. In a dual-gate processing, a heavily doped n^+ poly layer is used as the gate electrodes for nFETs, while a heavily doped p^+ poly layer is the gate electrodes for pFETs. The introduction of the p^+ polysilicon gate in pFETs is to achieve surface channel operation, which offers the advantages of lower threshold voltage, superior short-channel effects, and subthreshold leakage compared to buried-channel pFETs. Also, shallower junction depth can be obtained as boron species are implanted in the self-aligned p^+ polysilicon gates.

Source/drain implantation

Source/drain implantation is another critical processing step determining the electrical performance of the device. n^+ source/drain implantation (NSDI) is needed for nFETs and p^+ source/drain implantation (PSDI) is needed for pFETs, followed by an annealing process to activate the dopants implanted in the source/drain regions. The polysilicon gates completed in the previous process steps are used as masks to self-align the source/drain implants for both nFETs and pFETs. Before NSDI starts, photoresist is applied to the wafer, followed by patterning using the NSDI mask. Shallow and heavily doped n^+ regions are then formed by implanting arsenic in the nFET S/D regions through the exposed gate oxide. The polysilicon gates block the S/D implants from the channel region underneath the gate and minimize the overlap capacitances between the gate and the source and between the gate and the drain. Once the NSDI is completed, the remaining photoresist is removed from the wafer. The PSDI begins with covering a photoresist layer patterned using the PSDI mask. Self-aligned shallow and heavily doped p^+ regions are formed by implanting boron in pFET S/D regions through the exposed gate oxide while keeping minimal overlap capacitances. After PSDI is completed, the photoresist is stripped from the wafer before the annealing process begins. The annealing process will activate the implanted dopants and slightly increase the oxide thickness over the source/drain regions. For CMOS technology, this annealing process is the final high-temperature step in the whole process. The junction depth is determined by the conditions of both the source/drain implantation and the annealing.

Contacts and metallization

When the source/drain region implantation is completed, a thick oxide layer is deposited as an insulation material between the active devices and the metal interconnects, so the metal interconnects can run over the field regions and the poly gates without influencing the device characteristics. Contact cuts are needed to open the thick oxide and form a good contact between the metal interconnect and the source/drain regions and the poly gates.

After the wafer is again coated with photoresist, the contacts in source/drain regions and in polysilicon gates are patterned using the contact mask. As the device sizes shrink, the contact sizes also shrink, which makes the formation of the contacts important to ensure the yield in the manufacturing.

Modern CMOS processes employ a metal silicidation technology to obtain good ohmic contacts in the gates and the source/drain regions, improving reliability by blocking any junction spiking. Metals with low sheet resistivity and contact resistance and high gate insulator reliability and heat stability are selected for use in silicided metallization. Since silicided polysilicon can reduce the gate resistance, it is suitable for digital IC applications. However, when dual-gate processing is implemented in CMOS technology, impurity inter-diffusion in the gate electrodes becomes an issue. Specific processing such as nitrogen ion implantation can be used to suppress this dopant interdiffusion in the gate electrodes. But this process requires a trade-off between dopant interdiffusion and gate depletion.

Heat resistance is another issue to be considered for salicide technology to be used for analog applications. Salicide technology is fully suitable for the conventional CMOS digital/ASIC ICs; however, metallization technology with high heat tolerance is needed because some process steps after salicide metallization need the high-temperature process to fabricate passive devices such as metal-insulator-metal (MIM) capacitors. Lower temperature process for MIM capacitors has been developed while advanced metal-stacked poly-gate structures are being developed. The metal-stacked gate structure can provide both low resistivity and high heat tolerance and is a potentially promising technology for analog applications.

Via and multilayer metal interconnects

After finishing the contact and the metallization, the basic transistors have been fabricated. In the next steps, the wafers need to complete the so-called back-end processing. Basically, it consists of multilayer metal interconnect formation, via etching and interlayer dielectric (ILD) deposition. The purpose of the multilayer interconnect with ILD is to isolate the metal layers from each other and to provide the various electrical connections needed on a chip by metal wires and via contacts. In the current 0.13-μm CMOS technology, the interconnects can extend up to eight metal layers. For digital processing, the back-end portion has become extremely critical in terms of the yield, reliability, and cost reduction.

Following the opening of the contact windows, metal 1, the first layer of metal interconnect, is deposited. It could be either aluminium or copper, depending on the technology node generation. Typically, for the 0.15-μm technology and older, aluminium is used for all interconnect layers (even though some foundries offer copper for the top two metal layers) and for a 0.13-μm technology and newer, copper is used for all interconnect layers. After the metal deposition, a lithography step will be processed with a mask to define all the needed metal connections, so the metal outside these defined areas will be selectively etched.

Other metal layers for multilevel interconnects are fabricated following the same procedures. Between adjacent metal layers, a dielectric material is deposited for isolation. To provide the signal paths between the metal layers, depending on the design of the circuit, contact holes are created by opening windows throughout the dielectric layers. The contact holes between the metal layers are called "vias" to distinguish them from the "contact" for the first level of metal to the active areas and to polysilicon.

In current advanced technologies such as 0.15 μm and newer, copper has been used for interconnect. Compared with aluminium, copper is more difficult to fabricate to achieve high-quality interconnects with reliable performance. New processing techniques such as metal slotting and via pattern as well as the metal dummy filling at each metal layer have been applied. A detailed discussion on advanced interconnect technology will be outside the scope of this book.

2.3 RF CMOS TECHNOLOGY

The logic CMOS technology discussed earlier includes various flavors of MOSFETs and some resistors such as polysilicon resistors, diffusion resistors, and *n*-well resistors. However, an RF CMOS technology should incorporate passive device options for high-frequency (HF) applications such as resistors, capacitors, and inductors in addition to active devices such as MOSFETs and bipolar junction transistors (BJTs). Also, the varactor is an important component in RF technology. So far, RF CMOS technology is developed on the basis of the available digital planar CMOS processing by adding necessary process steps and device structures for RF applications. For example, process steps to form deep *n*-well are added to reduce the substrate coupling and noise figures of MOSFETs. Additional implantation is adopted in a varactor to increase the tuning range and the quality factor. A much thicker top metal layer is introduced to increase the Q factor of the inductor devices, and so on. Even though most RF CMOS technologies are currently developed on the basis of digital CMOS process compatibility and cost saving, sufficient reason exists for developing a separate high-quality CMOS process for RF applications. The first problem is the lack of commonality between digital and analog device targets when optimizing the device performance in the process development. For digital CMOS, device parameters such as saturation current I_{dsat}, leakage current I_{off}, threshold voltage, and gate leakage are the most important ones to be optimized. But for analog applications, transconductance G_m, output conductance G_{DS}, and device-matching behavior are more important. For example, I_{off} (of great priority to the digital designer) is not highly prioritized by the analog designer. The second problem is that the desired analog process optimization strategies may be contradictory to traditional CMOS scaling. A clear example here is I_{on}/I_{off} versus G_m/G_{DS}. Traditional CMOS scaling methodologies incorporate halo (pocket) implants to control short-channel effects. However, halos have a detrimental effect on G_m/G_{DS} owing to the drain bias–induced modulation of the barrier created by the halo on the drain side of the device. Solutions such as using lateral work-function grading and asymmetric halos have been proposed to fix this problem. However, each of these approaches adds to the cost and complexity of the process and pushes the devices away from the baseline technology. Thus, a "special" RF CMOS technology with optimized device performance specifically for RF/analog applications could coexist with "regular" RF CMOS based on the baseline digital process (Woerlee *et al.* 2001). Designers could select different RF CMOS processes depending on their circuit applications. In some cases, an RF CMOS technology based on the baseline digital process could be used to meet the design specifications while the cost of a "special" RF process is much higher.

(1) Deep n-well options in an RF CMOS technology

The schematic cross section of an *n*-type MOSFET with the proposed deep *n*-well structure and its key process flow are shown in Figure 2.2 (Su *et al.* 2001). The process is based on a regular logic CMOS technology, with the addition of a deep *n*-well implantation and mask. To minimize the disturbance to the DC behavior of MOS transistors, high-energy ion implantation (I/I) steps, followed by postimplant annealing, were used to form the deep *n*-well. Specifically, I/I steps consist of a 2-MeV I/I with an arsenic dose of $2 \times 10^{13}\,\mathrm{cm}^2$ to form a deep enough *n*-well so as not to disturb the doping profile of the inner *p*-well and a 1-MeV I/I with a lower arsenic dose was used together with the regular *n*-well implant and a flat plateau of deep *n*-well. A postimplant annealing is

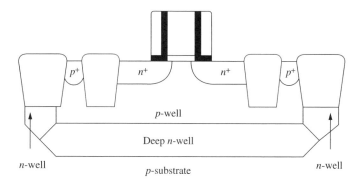

Figure 2.2   RF *n*FET with deep *n*-well option. Reproduced from Su J.-G. *et al.* (2001) Improving the RF performance of 0.18 μm CMOS with deep *n*-well implantation, *IEEE Electron Device Lett.*, **22**(10), 481–483

required. As a result, the resultant n-well (i.e., regular n-well and deep n-well combined) completely surrounds the p-well region for junction isolation.

(2) Varactor fabrication

Varactors (variable capacitors) are very important components in RF circuits such as voltage-controlled oscillators (VCOs). Historically, varactors in integrated technology were constructed by diodes. In a dual-well CMOS process, there exist at least four types of diodes: n^+/p-sub, n^+/p-well, n-well/p-sub, and p^+/n-well. However, p^+/n-well diodes are used more than others because they offer higher tuning range than n-well/p-sub junctions and provide better immunity against latch-up than n^+/p-sub and n^+/p-well junctions. In most RF CMOS technologies, both "free" varactors and high-quality varactors are offered. "Free" varactors use the p^+/n-well junction in the same way as that in MOSFETs and do not introduce any additional mask, but the tuning range and quality factor is lower than that in high-quality varactors which require one additional mask to adjust the implant to achieve an optimized doping profile for the p-/n-junction by considering together several figures of merit for a varactor such as tuning range, quality factor, breakdown voltage, leakage current, and so on. Figure 2.3 gives a cross section of a p^+/n-well diode varactor.

In addition to the diode varactors discussed above, other types of varactors have also been fabricated, such as MOS varactors. As in the case of diode varactors, both "free" and high-quality MOS varactors have been used in design. "Free" varactors are constructed by using MOS capacitors formed by an n^+ poly gate over an n-well. The designers can select different channel lengths and doping concentrations (different threshold voltage options offered in a specific process) to have a design trade-off between the tuning range and the quality factor. High-quality MOS varactors have been reported by modifying the "free" MOS varactors with added STI isolation between the channel (underneath the gate) and the n-well contact regions. With increased mask and process costs, the tuning range of this type of varactors can be increased without reducing the quality factor a lot. Figure 2.4 gives a cross section of a MOS varactor.

Design and fabrication of varactors with high tuning range and quality factor are needed to be considered in developing an RF CMOS process.

(3) MIM capacitor fabrication

Analog/mixed-signal processes use four major types of capacitors: polysilicon-insulator-polysilicon (PIP) capacitors, vertical metal-insulator-metal (VMIM) capacitors, Flux MIM

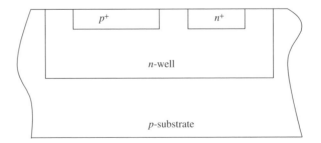

Figure 2.3   Cross section of a p^+/n-well diode varactor

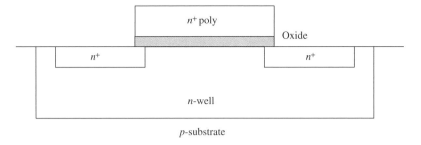

Figure 2.4 Cross section of a MOS varactor

(FMIM) capacitors, and MOS-style (depletion or accumulation) capacitors. Many older technologies used PIP capacitors, which are not suitable for RF applications in the giga-hertz range because of both the resistive losses in the plates and the contacts and because of the parasitic capacitance between the passive element and the lossy silicon substrate. Also, the poly in PIP capacitors is typically implanted at higher doses than the CMOS source-drain regions in order to minimize poly-depletion effects. This requires extra pro-cessing costs because of additional lithography layers that need to be added to support the implants. So far the most popular analog/mixed-signal capacitor is the VMIM (see Figure 2.5). VMIM capacitors have the inherent advantage that they are metal without any poly-depletion and poly-gate loss issues and, if implemented at the last metal layer, have the entire ILD stack between them and the substrate, so the parasitic capacitance is much smaller. The VMIM capacitors were widely used in the 0.18-μm and older aluminum-interconnect-based processes. In recent years, they have been implemented in commercial CMOS Cu-damascene processes. The excellent linearity with voltage and temperature illustrates the popularity of the device as an analog element. However, the issues about yield and reliability need still to be resolved for copper-interconnect-based processes.

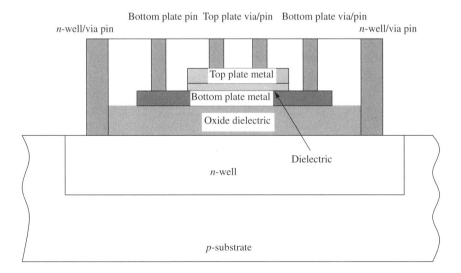

Figure 2.5 Cross section of a VMIM capacitor

One of the restrictions with MIM devices is that process technologies do not scale the vertical spacing in the back end nearly as fast as the lateral spacing. The reason is that digital circuit designs cannot tolerate large increases in the wiring capacitance from generation to generation. Lateral flux (finger) MIM capacitors solve this problem by using the lateral capacitance between the metal lines rather than the vertical capacitance between the different ILD layers. As a result, the capacitance is under design control and scales more effectively with the technology. Also, the FMIM capacitors do not need any extra mask to define the bottom and top plates, so the cost is lower. However, the matching property of the lateral FMIM is about one order (the actual numbers depend on the technology node) worse than that of a VMIM capacitor, so such FMIM capacitors may not be suitable for some analog applications that require precise matching behavior to the MIM capacitors. Figure 2.6 gives the cross section of an FMIM capacitor for a four-metal layer process.

Another of the limitations of the MIM device is the small capacitance per unit area due to the thickness of the insulator dielectric. Many designers take advantage of thin gate oxide processes to achieve high capacitance per unit area by using MOS capacitors. The disadvantage of using MOS capacitors is the high series resistance of a MOS capacitor due to the bottom plate that is formed with the doped channel/substrate. Also, the high gate leakage currents in modern devices with scaled oxides make MOS capacitors excessively leaky, which should be considered when using them for some leakage-sensitive designs such as for wireless applications.

(4) Resistor fabrication

Precision resistors are key passive elements in both digital and analog circuits (Ulrich *et al.* 2000). Different types of resistors exist in a CMOS process, such as *n* and *p* polysilicon resistors, *n* and *p* diffusion resistors, *n*-well resistors, and metal thin film

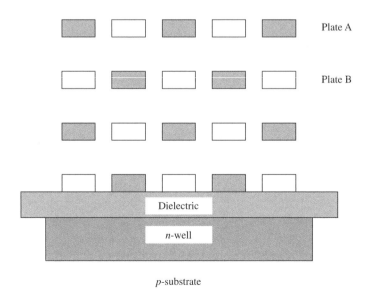

Figure 2.6   Cross section of a FMIM capacitor

resistors. Depending on the doping and the salicidation control, resistors of different resistance values can be fabricated. The simultaneous presence of both poly and metal resistors has added value in a CMOS process, because the metal resistors are at the top of the stack and the poly resistors at the bottom. The two widely separated locations allow designers to choose a resistor that minimizes parasitics for their particular circuit.

Typically, a CMOS process offers both low and high value polysilicon resistors. The low value resistors are formed by silicided polysilicon film. High value poly resistors are fabricated from unsilicided polysilicon by blocking the silicide formation from the polysilicon films. Since the resistance of polysilicon-silicided (polycide) resistors tends to be very low (5–15 Ω/sq) and the voltage coefficient tends to be relatively high (100–600 ppm/V), there is a strong tendency to use unsilicided (or silicide-blocked) resistors. In a typical silicide-blocked resistor, the center of the device is silicide-blocked and the end portions are left unblocked (see Figure 2.7). Thus, the end portions either receive the conventional silicide processing for a contact pad or receive specific optimized processing procedures for certain applications. The silicide-blocking layer is usually an oxide or nitride and is frequently chosen to leverage a preexisting layer elsewhere in the process. Existence of a silicide-blocking layer also enables devices such as silicide-blocked diffusion resistors and silicide-blocked MOS devices.

Polysilicon resistors are usually placed on a field oxide. In technologies with thin field oxides (such as local oxidation of silicon (LOCOS)), there is significant electrical interaction through the field oxide and parasitic capacitances as well as depletion of the bottom of the resistor, which produces a voltage-dependent resistance change. All these effects must be considered in the resistor design. However, such effects are significantly reduced with the thicker oxides (3000–6000 A), characteristic of STI processes, which are used in more advanced technologies. It should be noted that the sheet resistance, as well as the thermal and voltage coefficients of silicide-blocked polysilicon resistors, are very process-dependent. Implant conditions, grain boundary size, thermal activation, and end-portion silicide quality can all impact key polysilicon resistor parameters. Therefore, the reported values for the parameters of poly resistors vary widely.

Metal thin film resistors can be built at any of the traditional metal layers. In addition, TaN thin film is frequently used as a precise thin film resistor owing to its easy availability in a Cu-damascene process where it is used as a Cu-diffusion barrier or even as an

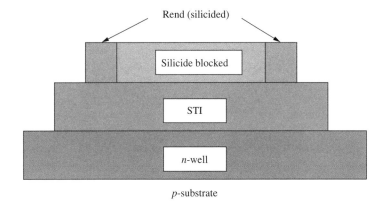

Figure 2.7 Cross section of an unsilicided poly resistor

aluminum back end (the TaN film is deposited from reactively sputtered Ta in an N_2 ambient). The parasitic capacitance of a TaN resistor fabricated in the back-end-of-the-line (BEOL) is significantly reduced relative to a polysilicon resistor. Another reason why TaN is attractive to both the process developers and circuit designers is because it exhibits a temperature coefficient of resistance (TCR)-versus-resistivity relationship that ranges from roughly 500 ppm/C at 50 Ω/sq to roughly -500 ppm/C at 400 Ω/sq and is attributed to the transition from metallic conduction (positive TCR) to hopping conduction (negative TCR).

(5) High-Q inductor fabrication

Inductors are critical components in analog/mixed-signal design. Small-valued, precise, high-Q inductors are employed in circuits such as RF transceivers. Larger, lower-Q devices have functions such as impedance matching and gain control. Significant research has been done on monolithic integration of inductors (Burghartz *et al.* 1996), and in recent years there has been increasing use of inductors in state-of-the-art CMOS processes. Spiral inductors (see Figure 2.8) can be fabricated with a conventional MOS process with negligible modifications to the design rules. A minimum of two metal layers is required, one to form the spiral and the other to form the underpass.

To minimize parasitic capacitance to the substrate, the top metal layer is the usual choice for the main spiral. The most critical factor in inductor design is the optimization of the inductor Q at the design frequency. Q, or the "quality factor," is the ratio of the imaginary to the real part of the impedance ($Q = \mathrm{Im}(Z)/\mathrm{Re}(Z)$) and represents the ratio of the useful magnetic stored energy over the average dissipation for one cycle of the signal propagation. Note that determining the geometry and area required to deliver an optimized Q at the design frequency is not a straightforward process. The most difficult factor in inductor process design is minimization of the impact of parasitic elements. Real inductors have parasitic resistance and capacitance. The parasitic resistance dissipates energy through ohmic loss, while the parasitic capacitance stores unwanted energy. At high frequencies, the skin effect causes a nonuniform current distribution in the metal

Figure 2.8   Illustration of a spiral inductor

segments, which introduces (among other factors) a frequency-dependent contribution to the parasitic resistance. Furthermore, electromagnetic effects caused by the Faraday effect introduce parasitic currents (eddy currents) in the silicon as well, adding an additional frequency dependency in the resistance. The parasitic resistance is primarily driven by ohmic resistive losses in the thin patterned metal layers. Parasitic resistance can be modulated both by design (trading off inductor area for inductor line width) and by process (increasing the thickness of the metal and/or improving a Cu-damascene polish process to minimize dishing and thus permit wider metal lines). The capacitive-induced loss is driven both by the C_{ox} between the inductor and the substrate and by the lossy properties of the substrate. (At high frequencies the current flows through C_{ox} and into the lossy substrate. The resulting dissipation adds a real component to the imaginary inductive impedance and degrades the Q.) Minimizing this capacitance typically means separating the inductor as far as possible from the lossy silicon (usually by placing the inductor in the top metal layer). Recent advancements in low-k processes for digital CMOS also carry significant benefit (up to 4X improvement in Q for SiLK∗ compared to conventional oxide ILD.) Minimizing the substrate loss is more complex. As the frequency increases to where the skin depth is on the order of the substrate thickness, eddy currents in the substrate become a major loss mechanism. (This magnetically induced loss can be thought of as transformer action between a lossy primary and a lossy secondary.) Mitigating eddy current loss can be quite difficult. There are a number of potential techniques including solid and patterned ground shields, multilevel metallizations to build vertical solenoids, as well as minimizing doping levels under the inductor. Note that since the eddy current loss is approximately proportional to the cube of the inductor diameter, strategies to minimize resistive parasitics by making large inductors (as is common in GaAs) are less effective in CMOS owing to the more conductive Si substrates. Fabrication of high-Q inductors in CMOS technology is a challenging effort.

REFERENCES

Burghartz J. N. *et al.* (1996) Monolithic spiral inductors fabricated using a VLSI Cu-damascene interconnect technology and low loss substrates, *IEDM Tech. Dig.*, **1996**, 99–102.

Kooi E. (1991) *The Invention of LOCOS*, IEEE Press, New York.

Nandakumar M. *et al.* (1998) Shallow trench isolation for advanced ULSI CMOS technologies, IEEE catalog no. 98CH34217, alternatively: *IEDM Tech. Dig.*, **1998**, 133.

Su J.-G. *et al.* (2001) Improving the RF performance of 0.18 μm CMOS with deep *n*-well implantation, *IEEE Electron Device Lett.*, **22**(10), 481–483.

Tsai H. H., Yu C. L., and Wu C. Y. (1989) A new twin-well CMOS process using nitridized-oxide-LOCOS (NOLOCOS) isolation technology, *IEEE Electron Device Lett.*, **10**(7), 307–310.

Ulrich R. K. *et al.* (2000) Getting aggressive with passive devices, *IEEE Circuits Devices*, **16**(5), 17–25.

Woerlee P. *et al.* (2001) RF-CMOS performance trends, *IEEE Trans. Electron Devices*, **48**(8), 1776–1782.

3
RF Modeling

3.1 INTRODUCTION

Advances in CMOS fabrications have resulted in deep submicron transistors with higher transit frequencies and lower noise figures. Radio-frequency (RF) designers have already started to explore the use of CMOS technology in RF circuits. This advanced performance of MOSFETs is attractive for high-frequency (HF) circuit design in view of a system-on-a-chip realization, where digital, mixed-signal base-band and HF transceiver blocks would be integrated on a single chip. Besides the ability to integrate RF circuits with other analog and logic circuits with the intention of reducing the cost by eliminating the sometimes expensive packaging, other advantages offered by silicon CMOS technologies are also interesting, such as the low cost due to the volume of wafers processed and the low power consumption feature of MOSFETs, which makes it suitable for portable applications.

To have an efficient design environment, design tools with accurate models for devices and interconnect parasitics are essential. It has been known that for analog and RF applications the accuracy of circuit simulations is strongly determined by the device models. Accurate device models become crucial to correctly predict the circuit performance.

In most of the commercially available circuit simulators, the MOS transistor models have originally been developed for digital and low-frequency analog circuit design (see, for example, Cheng *et al.* (1997a) and MOS9 Manual (2001)), which focus on the DC drain current, conductances, and intrinsic charge/capacitance behavior up to the mega-hertz range. However, as the operating frequency increases into the gigahertz range, the importance of the extrinsic components rivals that of the intrinsic counterparts. Therefore, an RF model with the consideration of the HF behavior of both intrinsic and extrinsic components in MOSFETs is extremely important for achieving accurate and predictive results in the simulation of a designed circuit.

So far, most compact MOSFET models do not include the gate resistance R_G. However, the thermal noise contributed by the gate resistance should be considered as MOS transistors approach gigahertz frequencies, and the resistive and capacitive (RC) effects at the gate should be well modeled since both of these effects are important in designing radio-frequency CMOS circuits. As shown in Figure 3.1, the gate resistance component will significantly affect the input admittance at RF, so a model without R_G cannot accurately

Device Modeling for Analog and RF CMOS Circuit Design. T. Ytterdal, Y. Cheng and T. A. Fjeldly
© 2003 John Wiley & Sons, Ltd   ISBN: 0-471-49869-6

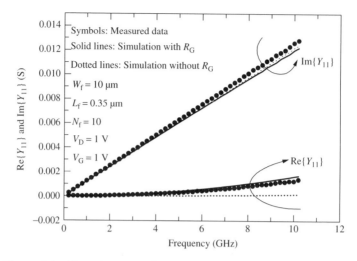

Figure 3.1 The model without the gate resistance cannot predict the measured Y_{11} characteristics. Reproduced from Cheng (2002b) MOSFET Modeling for RF IC Design, in *CMOS RF Modeling, Characterization and Applications*, Jamal D. M. and Fjeldly T. A., eds., World Scientific Publishing, Singapore

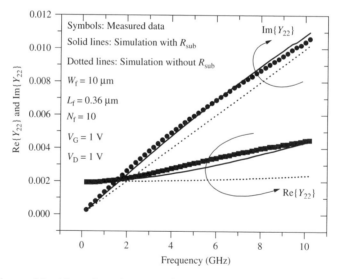

Figure 3.2 The model without the substrate resistances cannot predict measured Y_{22} characteristics. Reproduced from Cheng (2002b) MOSFET Modeling for RF IC Design, in *CMOS RF Modeling, Characterization and Applications*, Jamal D. M. and Fjeldly T. A., eds., World Scientific Publishing, Singapore

predict the HF characteristics of the device. It is very crucial because one may use this resistance for impedance matching to achieve maximum power transfer. Also, the thermal noise introduced by the gate resistance increases the noise figure of the transistor. It is an important noise source to be considered when optimizing the noise performance of an RF circuit. Furthermore, the gate resistance also reduces f_{max} (the frequency at which the maximum available power gain of the device equals to 1), which is an important device parameter in RF circuit design in addition to f_T, the frequency at which the current gain of the device equals to 1.

Another important component that almost all of the compact models implemented in commercial circuit simulators do not account for is the substrate resistance. Actually, substrate-coupling effects through the drain and source junctions and these substrate resistance components play an important role in the contribution to the output admittance, so the inclusion of these substrate components in an RF model is needed. This effective admittance of the substrate network can contribute 50% of the total output admittance (see Jen *et al.* (1998)). As shown in Figure 3.2, a MOSFET model without the substrate resistance components cannot predict the frequency dependency of the output admittance of the device, so the simulation with such a model will give misleading simulation results of the output admittance when the device operation frequency is in the gigahertz range.

3.2 EQUIVALENT CIRCUIT REPRESENTATION OF MOS TRANSISTORS

As shown in Figure 3.3, a four terminal MOSFET can be divided into two portions: intrinsic part and extrinsic part. The extrinsic part consists of all the parasitic components, such as the gate resistance R_G, gate/source overlap capacitance C_{GSO}, gate/drain overlap capacitance C_{GDO}, gate/bulk overlap capacitance C_{GBO}, source series resistance R_S, drain series resistance R_D, source/bulk junction diode D_{SB}, drain/bulk junction diode D_{DB}, and substrate resistances R_{SB}, R_{DB}, and R_{DSB}. The intrinsic part is the core of the device without including those parasitics. Even though it would be desirable to design and fabricate MOSFETs without those parasitics, they cannot be avoided in reality. Some of them may be not noticeable in DC and low-frequency operation. However, they will influence significantly the device performance at HF.

Equivalent circuits (ECs) have been an effective approach to analyze the electrical behavior of a device by representing the important components. In this section, we discuss

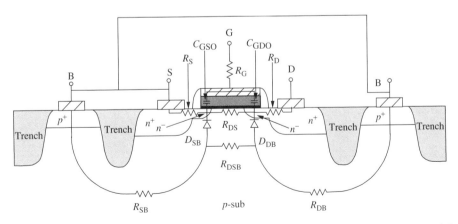

Figure 3.3   A MOSFET schematic cross section with the parasitic components. Reproduced from Cheng *et al.* (2000b) MOSFET modeling for RF circuit design, *Proceedings of the 2000 Third IEEE International Caracas Conference on Devices, Circuits and Systems*, D23/1–D23/8

the ECs for both the intrinsic device (without the parasitics) and the extrinsic device with various parasitic components.

For an intrinsic device, AC small-signal currents referring to the source of the device can be calculated by the following:

$$
\begin{bmatrix} i_{Gi} \\ i_{Di} \\ i_{Bi} \end{bmatrix} = \begin{bmatrix} j\omega C_{GGi} & -j\omega C_{GDi} & -j\omega C_{GBi} \\ G_m - j\omega C_{DGi} & G_{DS} + j\omega C_{DDi} & G_{mb} - j\omega C_{DBi} \\ -j\omega C_{BGi} & -j\omega C_{BDi} & j\omega C_{BBi} \end{bmatrix} \begin{bmatrix} v_{GSi} \\ v_{DSi} \\ v_{BSi} \end{bmatrix}
\tag{3.1}
$$

where v_{GSi}, v_{DSi}, and v_{BSi} are the AC voltages at the intrinsic gate, at the intrinsic drain, and at the intrinsic bulk (all referring to the intrinsic source); i_{Gi}, i_{Di}, and i_{Bi} are the alternating currents (AC) through the intrinsic gate, through the intrinsic drain, and through the intrinsic bulk; G_m, G_{DS}, and G_{mb} are the transconductance, channel conductance, and bulk transconductance of the device, respectively; C_{xyi} are intrinsic capacitances between the terminals with the following definitions:

$$
C_{xy} = -\frac{\partial Q_x}{\partial v_y} \quad \text{when } x \neq y,
\tag{3.2}
$$

$$
C_{xy} = \frac{\partial Q_x}{\partial v_y} \quad \text{when } x = y.
\tag{3.3}
$$

Equation (3.1) can be rewritten as the following:

$$
i_{Gi} = j\omega C_{GGi} v_{GSi} - j\omega C_{GDi} v_{DSi} - j\omega C_{GBi} v_{BSi}
\tag{3.4}
$$

$$
i_{Di} = G_m v_{GSi} + G_{DS} v_{DSi} + G_{mb} v_{BSi} - j\omega C_{DGi} v_{GSi} + j\omega C_{DDi} v_{DSi} - j\omega C_{DBi} v_{BSi}
\tag{3.5}
$$

$$
i_{Bi} = -j\omega C_{BGi} v_{GSi} - j\omega C_{BDi} v_{DSi} + j\omega C_{BBi} v_{BSi}.
\tag{3.6}
$$

In Eq. (3.1), we assume that the components between the gate and the other terminals can be considered as purely capacitive with infinite resistance, so the gate current in Eq. (3.4) does not contain any conductive current component. Similarly, the components between the bulk and the other terminals can be also considered as purely capacitive with infinite resistance, so the bulk current in Eq. (3.6) does not contain any conductive current component. Those assumptions can usually hold for an intrinsic MOSFET because of the very low leakage currents through the gate to other terminals and through the bulk to other terminals in a MOSFET fabricated with current advanced technology.

To derive an EC from the above equations, we rearrange the above equations in the following forms:

$$
i_{Gi} = j\omega C_{GSi} v_{GSi} + j\omega C_{GDi} v_{GDi} + j\omega C_{GBi} v_{GBi},
\tag{3.7}
$$

$$
i_{Di} = (G_m - j\omega C_m) v_{GSi} + j\omega C_{GDi} v_{DGi} + (G_{mb} - j\omega C_{mb}) v_{BSi}
$$

$$
+ j\omega C_{BDi} v_{DBi} + (G_{DS} + j\omega C_{SDi}) v_{DSi},
\tag{3.8}
$$

$$
i_{Bi} = j\omega C_{mgb} v_{GBi} + j\omega C_{BSi} v_{BSi} + j\omega C_{GBi} v_{BGi} + j\omega C_{BDi} v_{BDi}
\tag{3.9}
$$

where C_m, C_{mb}, and C_{mgb} are the differences of the transcapacitances between the drain and the gate, between the drain and the bulk, and between the gate and the bulk, and are given by

$$C_m = C_{DGi} - C_{GDi}, \tag{3.10}$$

$$C_{mb} = C_{DBi} - C_{BDi}, \tag{3.11}$$

$$C_{mgb} = C_{GBi} - C_{BGi}. \tag{3.12}$$

C_{SDi} and C_{BSi} are intrinsic transcapacitances between the source and the drain, and between the bulk and the source, and have the following relationships with other capacitances:

$$C_{SDi} = C_{DDi} - C_{BDi} - C_{GDi}, \tag{3.13}$$

$$C_{BSi} = C_{BBi} - C_{BGi} - C_{BDi}. \tag{3.14}$$

According to Eqs. (3.7)–(3.9), an EC referring to the source can be derived as shown in Figure 3.4, in which several current components contributed by the transcapacitances are included in the EC. As shown in Figure 3.3, parasitic capacitances such as the overlaps of gate-to-source/drain/bulk and the junction capacitances from the source/bulk and drain/bulk diodes are not negligible in a MOSFET and must be included in the EC to

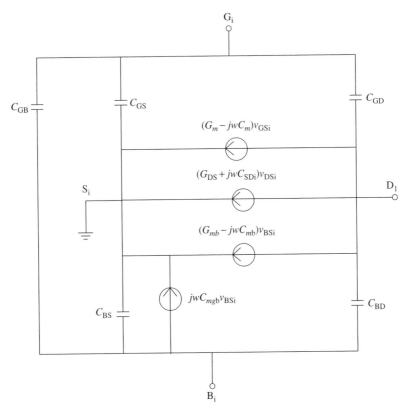

Figure 3.4 An equivalent circuit for an intrinsic MOSFET. Reproduced from Cheng (2002b) MOSFET Modeling for RF IC Design, in *CMOS RF Modeling, Characterization and Applications*, Jamal D. M. and Fjeldly T. A., eds., World Scientific Publishing, Singapore

describe the device behavior at HF. With the inclusion of those extrinsic capacitances, the EC for a MOSFET can be given in Figure 3.4, that is,

$$C_{GS} = C_{GSi} + C_{GSo}, \tag{3.15}$$

$$C_{GD} = C_{GDi} + C_{GDo}, \tag{3.16}$$

$$C_{GB} = C_{GBi} + C_{GBo}, \tag{3.17}$$

$$C_{BS} = C_{BSi} + C_{jBS}, \tag{3.18}$$

$$C_{BD} = C_{BDi} + C_{jBD}. \tag{3.19}$$

In a MOSFET model for DC and low-frequency applications, the parasitic resistances at the gate and substrate can be ignored with little influence on the simulation accuracy. Usually, the parasitic resistances at the source and drain can be treated as "virtual" components by incorporating them in the I–V equation to account for the influence of the voltage drops at those resistances (see Cheng et al. (1997b)). At HF, however, these parasitic resistances will influence the device performance significantly and they all should be modeled and included in the EC for the device.

The gate resistance is in principle a bias-independent component at DC and low frequency, but may contain the contribution of an additional component with bias dependence at HF, as discussed by Jin et al. (1998) and Cheng et al. (2001a). The parasitic resistances at the source and drain consist of several parts as we will discuss later and can be also treated as bias-independent components even though they do have some bias dependence depending on the device structure and process conditions. The resistances in the substrate can be modeled by different EC networks, such as five-resistor network, four-resistor network proposed by Liu et al. (1997), three-resistor network proposed by Cheng et al. (1998), two-resistor network by Ou et al. (1998), and one-resistor network by Tin et al. (1999), as shown in Figures 3.5 to 3.9. The four- and five-resistor networks are more accurate and can be valid up to higher frequency, but the analysis and parameter extraction of the components are very complex. The one- and two-resistor networks introduce fewer components and are easier for the analysis and parameter extraction. However,

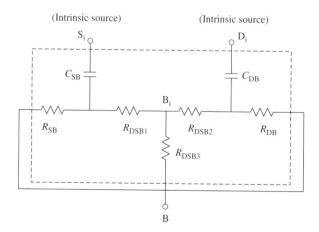

Figure 3.5   Five-resistor substrate network

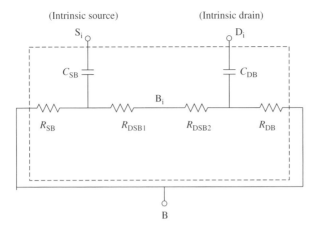

Figure 3.6 Four-resistor substrate network. Reproduced from Liu W. *et al.* (1997) R.F.MOSFET modeling accounting for distributed substrate and channel resistances with emphasis on the BSIM3v3 SPICE model, *Tech. Dig. IEDM*, 309–312

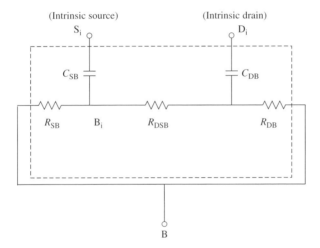

Figure 3.7 Three-resistor substrate network. Reproduced from Cheng Y. (1998) RF modeling issues of deep-submicron MOS-FETs for circuit design, *Proc. of the IEEE International Conference on Solid-State and Integrated Circuit Technology*, pp. 416–419

they may be less accurate when the operating frequency is increased. The three-resistor network is a compromise among these substrate networks. It can ensure the accuracy up to 10 GHz while maintaining a simple analysis and parameter extraction. However, it should be pointed out that the intrinsic bulk has been shifted to the end of R_{DSB}, as shown in Figure 3.7, instead of located somewhere along the resistor R_{DSB}. It has been concluded that this approximation does not influence much the simulation accuracy (see Enz and Cheng (2000)).

With further consideration of parasitic resistances at the drain, at the gate, at the source, and at the substrate, a complete lumped EC for a MOSFET at HF can be constructed and is shown in Figure 3.10.

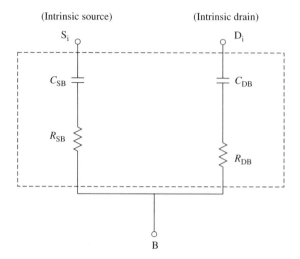

Figure 3.8 Two-resistor substrate network. Reproduced from Ou J.-J. *et al.* (1998) CMOS RF modeling for GHz communication IC's, *Proc. of the VLSI Symposium on Technology*, pp. 94, 95

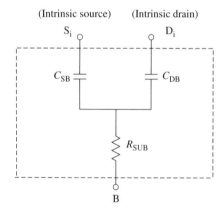

Figure 3.9 One-resistor substrate network. Reproduced from Tin S. F. *et al.* (1999) Substrate network modeling for CMOS RF circuit simulation, *Proc. IEEE Custom Integrated Circuits Conference*, pp. 583–586

The EC shown in Figure 3.10 can be used to understand and analyze the HF behavior of a MOSFET. In order to implement this EC in a SPICE simulator, a subcircuit approach has to be used.

In the subcircuit, the characteristics of the intrinsic device is described by a MOS transistor compact model implemented in the circuit simulator, and all the extrinsic components have to be located outside the intrinsic device, so that the MOS transistor symbol in the subcircuit only represents the intrinsic part of the device[1]. For example, (1) the source and drain series resistors are added outside the MOS intrinsic device to

---

[1] It may include the overlap capacitances at the source, at the drain, and at the bulk, depending on the intrinsic compact MOSFET model used in the implementation.

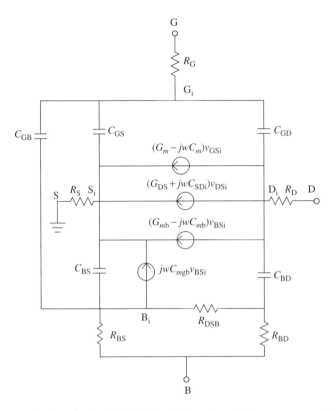

Figure 3.10   An equivalent circuit with both intrinsic and extrinsic components. Reproduced from Cheng (2002b) MOSFET Modeling for RF IC Design, in *CMOS RF Modeling, Characterization and Applications*, Jamal D. M. and Fjeldly T. A., eds., World Scientific Publishing, Singapore

make them visible in AC simulation (in most compact models, since the internal series resistances are only "virtual" resistances embedded in the $I-V$ model to account for the DC voltage drop across the source and drain resistances in calculating the drain current, they do not add any poles and are therefore invisible for AC simulation); (2) the gate resistance is added to the subcircuit model (usually R_G is not part of the MOS compact model, but plays a fundamental role in RF circuits as we discussed in Section 3.1); (3) the substrate resistors are added to account for the signal coupling through the substrate; (4) two external diodes are added in order to account for the influence of the substrate resistance at HF (the source-to-bulk and drain-to-bulk diodes are part of the compact model but their anodes are connected to the same substrate node, which will short the AC signal at HF, see Liu *et al.* (1997), so the diodes internal to the compact model should be turned off). With the above considerations, a subcircuit that represents an RF MOSFET in a circuit simulator can be defined and is shown in Figure 3.11. Note that the intrinsic substrate node should be connected at some point along the resistor R_{DSB}, but simulations have shown that connecting the intrinsic substrate to the source or the drain side has little influence on the simulated AC parameters. In some RF models (see, for example, Enz and Cheng (2000)), the intrinsic substrate has been connected to the source side in order to save one node and one component for the subcircuit model. Two external overlap capacitances,

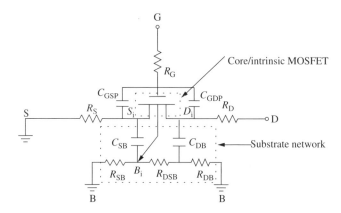

Figure 3.11   A subcircuit that can be implemented in a circuit simulator. Reproduced from Cheng (2002a) High frequency small signal AC and noise modeling of MOSFETs for RF IC design, *IEEE Trans. Electron Devices*, **49**(3), 400–408

C_{GSP} and C_{GDP} as shown in Figure 3.11, with bias dependence can be added but this is not always required, depending on the compact model used. For example, BSIM3v3 accounts for bias-dependent overlap capacitances that, if extracted correctly, have shown a sufficient accuracy. However, by adding these external capacitances, the inaccuracies of the intrinsic capacitance model appearing for short-channel devices can be corrected. In the next section, we will discuss the modeling of these intrinsic and extrinsic components shown in Figure 3.10.

3.3 HIGH-FREQUENCY BEHAVIOR OF MOS TRANSISTORS AND AC SMALL-SIGNAL MODELING

Compared with the MOSFET models for both digital and analog applications at low frequency, compact models for HF applications are more difficult to develop owing to the additional requirements of bias dependence and geometry scaling of the parasitic components as well as the requirements of accurate prediction of the distortion and noise behavior. A common modeling approach for RF applications is to build subcircuits based on the intrinsic MOSFET that has been modeled well for analog applications (see, for example, Liu *et al.* (1997), Cheng *et al.* (1998), and Enz and Cheng (2000)). The accuracy of such a model depends on how to establish subcircuits with the correct understanding of the device physics in HF operation, how to model the HF behavior of intrinsic devices and extrinsic parasitics, and how to extract parameters appropriately for the elements of the subcircuit. A reliable and physics-based parameter extraction methodology based on the appropriate characterization techniques is another important portion of the RF modeling to determine the model parameters and generate scaleable models for circuit optimization.

Currently, most RF modeling activities focus on the above subcircuit approach based on different compact MOSFET models developed for digital and low-frequency analog applications, such as EKV (Enz *et al.* 1995), MOS9 (MOS9 Manual 2001), and BSIM3v3

(Cheng *et al.* (1997b)). Several MOSFET models for RF applications have been reported (see, for example, Liu *et al.* (1997), Ou *et al.* (1998), Pehlke *et al.* (1998), Cheng *et al.* (1998), and Enz and Cheng (2000)). With added parasitic components at the gate, at the source, at the drain, and at the substrate, these models can reasonably well predict the HF AC small-signal characteristics of short-channel ($<0.5\,\mu$m) devices up to 10 GHz. However, the RF MOSFET modeling is still at a preliminary stage compared with the modeling work for digital and low-frequency analog applications. Efforts from both industry and universities are needed to bring RF MOSFET models to a mature level in further improving the RF models in describing the AC characteristics more accurately, and in improving the prediction of noise characteristics, distortion behavior, and non-quasi-static (NQS) behavior.

3.3.1 Requirements for MOSFET Modeling for RF Applications

Compared with the MOSFET modeling for digital and low-frequency analog applications, the HF modeling of MOSFETs is more challenging. All the requirements for a MOSFET model in low-frequency application, such as continuity, accuracy, and scalability of the DC and capacitance models, should be maintained in an RF model (see Cheng and Hu (1999)). In addition, there are further important requirements to the RF models:

1. The model should accurately predict bias dependence of small-signal parameters at HF operation.
2. The model should correctly describe the nonlinear behavior of the devices in order to permit accurate simulation of intermodulation distortion and high-speed large-signal operation.
3. The model should correctly and accurately predict HF noise, which is important for the design of, for example, low noise amplifiers (LNAs).
4. The model should include the NQS effect, so it can describe the device behavior at very high-frequency range in which NQS effect cannot be ignored for a model to behave correctly and will degrade the device performance significantly.
5. The components in the developed EC model should be physics-based and geometrically scaleable so that the model can be used in predictive and statistical modeling for RF applications.

   To achieve the above, the model for the intrinsic device should be derived with the inclusions of most (if not all) important physical effects in a modern MOSFET, such as normal and reverse short-channel and narrow width effects, channel-length modulation, drain-induced barrier lowering (DIBL), velocity saturation, mobility degradation due to vertical electric field, impact ionization, band-to-band tunneling, polysilicon depletion, velocity overshoot, self-heating, and channel quantization. Also, the continuities of small-signal parameters such as transconductance G_m, channel conductance G_{DS}, and the intrinsic transcapacitances must be modeled properly. Many MOSFET models, including MOS9, EKV, and BSIM3v3 have been developed for digital, analog, and mixed-signal applications. Recently, they all are extended for use in RF applications.

3.3.2 Modeling of the Intrinsic Components

Compact models including many mathematical equations for different physical mechanisms have been discussed in other chapters in this book. It has been found that the model accuracy in fittings of HF small-signal parameters and large-signal distortion of an RF MOSFET is basically determined by the DC and capacitance models. Here we only give a brief discussion on important modeling concepts without getting into the detailed equation derivation and physics analysis.

As a must for the backbone of the model, the electric field, the channel charge, and the mobility need to be modeled carefully to describe the current characteristics accurately and physically, on the basis of which, different physical effects can be added in the model.

In modeling the channel charge, physical effects such as short-channel effect, narrow width effect, nonuniform doping effect, quantization effect, and so on should be accounted for in order to describe the charge characteristics accurately in today's devices. There are two types of charge models: one can be called threshold-voltage (V_{th})-based models and the other can be called surface-potential (ψ_s)-based models (Boothroyd *et al.* 1991). ψ_s-based charge models are based on the analysis of the surface potential that will appear in the $I-V$ model to describe charge characteristics with the influence of many physical effects. V_{th}-based charge models are derived also by solving the surface potential with the consideration of those physical effects, but finally V_{th} is used instead of ψ_s in the charge (and hence $I-V$) model to account for the influence of some process parameters such as oxide thickness and doping and device parameters such as channel length and width. In both models, the continuities of the charge and its derivatives should be modeled carefully for the $I-V$ model to have good continuity and to predict correct distortion behavior of the devices (Langevelde and Klaassen 1997).

Mobility is another key parameter in MOSFET modeling. It will influence the accuracy and distortion behavior of the model significantly. The relationship between the carrier mobility and the electric field in MOSFETs has been well studied (see, for example, Liang *et al.* (1986) and Chen *et al.* (1996)). Three scattering mechanisms have been proposed to describe the dependence of mobility on the electric field. Each mechanism may be dominant under specific conditions of doping concentration, temperature, and biases as shown in Figure 3.12.

It has been realized that an accurate and physical description of a mobility model in compact MOSFET RF models for circuit simulation is essential for distortion analysis. It is also suggested that different models for electron and hole mobilities should be developed because of the difference in quantum-mechanical behavior of electrons and holes in the inversion layer in today's MOSFETs as discussed by Langevelde and Klaassen (1997).

On the basis of the charge and mobility models, complete $I-V$ equations can be developed with further inclusions of many important physical effects such as short-channel and narrow width effects, velocity saturation and overshoot, poly-depletion effect, quantization effect, and so on. In order to meet the requirements for both AC small-signal and larger-signal applications, the continuity and distortion behavior of the $I-V$ model should be ensured in deriving the equations when including these physical effects.

In real circuit operation, the device operates under time-varying terminal voltages. Depending on the magnitude of the time-varying signals, the dynamic operation can be classified as large-signal operation and small-signal operation. Both types of dynamic operation are influenced by the capacitive effects of the device.

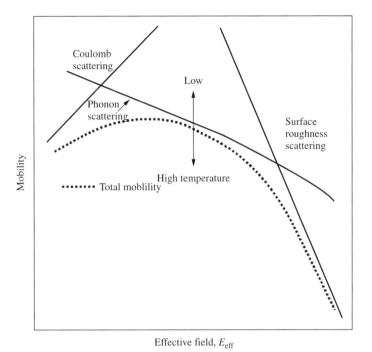

Effective field, E_{eff}

Figure 3.12 Mobility behavior influenced by different scattering mechanisms, depending on the bias and temperature conditions. Reproduced from Takagi S. *et al.* (1994) On the universality of inversion layer mobility in Si MOSFET's: part I – effects of substrate impurity concentration, *IEEE Trans. Electron Devices*, **ED-41**, 2357–2362

Many MOSFET intrinsic capacitance models have been developed. Basically, they can be categorized into two groups: (1) Meyer and Meyer-like capacitance models (see, for example, Meyer (1971)) and (2) charge-based capacitance models (see Sheu *et al.* (1984)). The advantages and shortcomings of the two groups of models have been well discussed and both of them have been implemented in circuit simulators. The Meyer and Meyer-like models are simpler than the charge-based models, so they are efficient and faster in computations. But they assume that the capacitances in the intrinsic MOSFET are reciprocal, which is not the case in real devices (see the discussion by Cheng and Hu (1999)), and earlier models based on this assumption cannot ensure charge conservation (see Yang *et al.* (1983)). Charge-based models ensure charge conservation and consider the nonreciprocal property of the capacitances in a MOSFET. These features are required to describe the capacitive effects in a MOSFET, especially for RF applications in which the influence of transcapacitances are critical and should be considered in the model. But usually the charge-based capacitance models require complex equations to describe all of the 16 capacitances in a MOSFET with four terminals, as given in the following:

$$C_{ij} = \frac{\partial Q_i}{\partial V_{ij}} \quad i \neq j \quad i, j = G, D, S, B, \tag{3.20}$$

$$C_{ij} = -\frac{\partial Q_i}{\partial V_{ij}} \quad i = j. \tag{3.21}$$

The development of an intrinsic capacitance model of modern MOSFETs is another challenging issue in RF modeling. To meet the needs in RF applications, besides ensuring charge conservation and nonreciprocity, an intrinsic MOSFET capacitance model should at least have the following features: (1) guaranteeing model continuity and smoothness in all the bias regions, (2) providing model accuracy for devices with different geometry and different bias conditions, and (3) ensuring model symmetry at $V_{DS} = 0\,\text{V}$.

Some comparisons between the MOSFET capacitance models and the measured data have been reported (see, for example, Ward (1981)). However, a complete verification of the bias and geometry dependencies of those capacitance models has not been seen. It has been found that some engineering approaches have to be used to improve the accuracy of the capacitance model if the intrinsic capacitance model cannot describe the device behavior accurately. Recently, the model continuity has been improved greatly. Many discontinuity issues in earlier capacitance models have been fixed. However, most capacitance models still cannot ensure the model symmetry when $V_{DS} = 0$. In Figures 3.13 and 3.14, the asymmetries of the capacitance model in BSIM3v3 are shown for $C_{GS} = C_{GD}$, C_{DD} and C_{SS} and for C_{BD} and C_{BS} (see, for example, Cheng and Hu (1999)). It has been known that a MOSFET should be symmetric for some capacitances at $V_{DS} = 0$, that is, $C_{DD} = C_{SS}$ and $C_{BD} = C_{BS}$. The asymmetric issue in the capacitance model is apparently nonphysical and may cause convergence and accuracy problem in the simulation. This issue may become more critical in the model for RF applications because the devices are often biased in the region of $V_{DS} \approx 0\,\text{V}$ in some applications such as switching. Efforts have been made based on the source-referenced approach, the bulk-referenced approach, and the surface potential–oriented approaches to improve the symmetry property of the models (see Tsividis (1987)). The development of advanced capacitance models with good continuity, symmetry, accuracy, and scalability is still a challenge for the model developers.

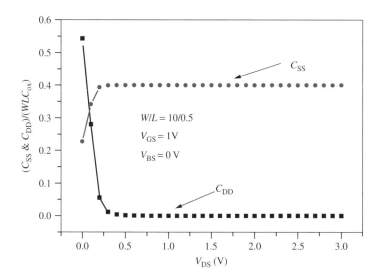

Figure 3.13 Simulated C_{SS} and C_{DD} as a function of V_{DS}. $C_{SS} \neq C_{DD}$ when $V_{DS} = 0$. Reproduced from Cheng and Hu (1999) *MOSFET Modeling & BSIM3 User's Guide*, Kluwer Academic Publishers, Norwell, MA

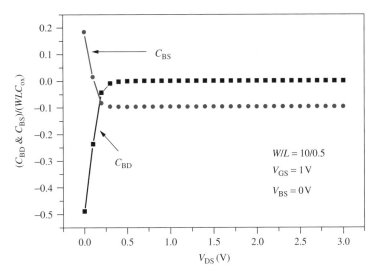

Figure 3.14   Simulated C_{SS} and C_{DD} as a function of V_{DS}. $C_{BS} \neq C_{BD}$ when $V_{DS} = 0$. Reproduced from Cheng and Hu (1999) *MOSFET Modeling & BSIM3 User's Guide*, Kluwer Academic Publishers, Norwell, MA

3.3.3 HF Behavior and Modeling of the Extrinsic Components

For an AC small-signal model at RF, the understanding and the modeling of parasitics are very important. The models for these parasitic components should be physics-based and linked to process and geometry information to ensure the scalability and prediction capabilities of the model. Also, simple subcircuits are preferred to reduce the simulation time and to make parameter extraction easier.

Besides the development of a physical and accurate intrinsic model discussed above, the following issues should be considered in developing a MOSFET model for deep submicron RF applications:

1. The gate resistance should be modeled and included in the simulation.

2. The extrinsic source and drain resistances should be modeled as real external resistors, instead of only a correction to the drain current with a virtual component.

3. Substrate coupling in a MOSFET, that is, the contribution of substrate resistance, needs to be modeled physically and accurately using appropriate substrate network for the model to be used in RF applications.

4. A bias-dependent overlap capacitance model, which accurately describes the parasitic capacitive contributions between the gate and the drain/source, needs to be included.

3.3.3.1 High-frequency behavior and modeling of gate resistance

At DC and low frequency, the gate resistance consists mainly of the polysilicon sheet resistance. The typical sheet resistance for a polysilicon gate ranges between 20 and 40 Ω/sq, and can be reduced by a factor of 10 with a silicide process, and even more with a metal stack process. At HF, however, two additional physical effects appear, which will affect the value of the effective gate resistance. One is the distributed transmission line effect on the gate and the other one is the distributed effect or NQS effect in the channel (see Jin *et al.* (1998) and Cheng *et al.* (2001a)). Both theoretical analysis with the consideration of these two HF effects and detailed experimental characterization are needed to obtain an accurate and physical gate resistance model, which is critical in predicting the HF behavior of the MOSFETs in designing an RF circuit.

In Figure 3.15, it is shown that R_G decreases first as channel length L_f increases while showing a weak bias dependence in this region, then starts to increase with L_f as L_f continues to increase above $0.4\,\mu m$ while showing a strong bias dependence. The L_f dependence of R_G varies for different V_{GS}. At lower V_{GS}, the L_f dependence of R_G is stronger. Also, R_G for the devices with longer L_f increases significantly and has stronger V_{GS} dependence. Figure 3.16 shows the per-finger-channel-width W_f dependence of R_G. It demonstrates that R_G increases as W_f decreases when $W_f < 6\,\mu m$, and the device with the same W_f has higher R_G at lower V_{GS}, which becomes more obvious when W_f narrows. Figures 3.17 and 3.18 give R_G for devices with various geometries at several V_{DS}, from which we observe similar L_f and W_f dependencies of R_G as what we found in Figures 3.15 and 3.16. However, the V_{DS} dependence of R_G becomes very weak when V_{DS} is larger than $1\,V$.

The U-shape L_f dependence of R_G in Figure 3.15 can be explained with the consideration of the distributed gate effect (DGE) and the NQS effect, in RF MOSFETs. It has been well known that the resistance of a polysilicon resistor, simulating the polysilicon gate in a MOSFET, is 3 or 12 times smaller (depending on the layout) at HF due to the DGE than that at DC but still scales with W_f/L_f. It has also been known that the NQS effect occurs in a MOSFET operated at HF in which the carriers in the channel cannot respond to the signal immediately. Thus, there is a finite channel transit time for the distributed effect of the carrier transportation in the channel due to the varying gate signal. In that case, the signal applied to the gate suffers an additional equivalent gate resistance,

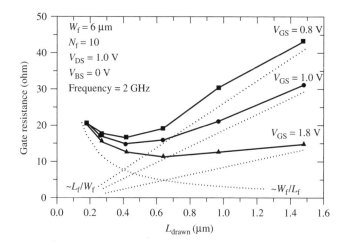

Figure 3.15   Curves of R_G versus L_f at different V_{GS}. The dotted lines illustrate approximately the dependence of $R_{G,poly}$ portion on $1/L_f$ and the dependence of $R_{G,nqs}$ portion on L_f, respectively. Reproduced from Cheng Y. *et al.* (2001a) High frequency characterization of gate resistance in RF MOSFETs, *IEEE Electron Device Lett.*, **22**(2), 98–100

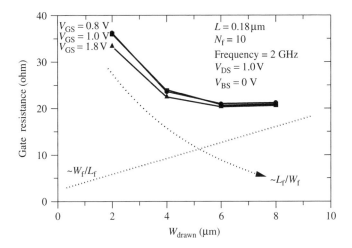

Figure 3.16 Curves of R_G versus W_f at different V_{GS}. The dotted lines illustrate approximately the dependence of $R_{G,poly}$ on W_f and the dependence of $R_{G,nqs}$ on $1/W_f$, respectively. Reproduced from Cheng Y. *et al.* (2001a) High frequency characterization of gate resistance in RF MOSFETs, *IEEE Electron Device Lett.*, **22**(2), 98–100

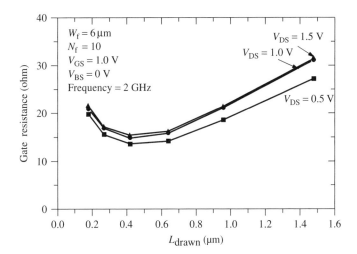

Figure 3.17 Curves of R_G versus L_f at different V_{DS}. Reproduced from Cheng Y. *et al.* (2001a) High frequency characterization of gate resistance in RF MOSFETs, *IEEE Electron Device Lett.*, **22**(2), 98–100

which is proportional to L_f/W_f, from the distributed channel resistance, which adds to the contribution from the poly-gate resistance. In other words, R_G consists of two parts: the $R_{G,poly}$ contributed by the poly-gate resistance and the $R_{G,nqs}$ due to NQS effect. As NQS effect becomes more significant, the contribution of $R_{G,nqs}$ dominates. This is the case in devices with longer L_f. Thus, we can understand the irregular geometrical dependence of R_G in Figure 3.15 when L_f is longer than 0.4 μm. It has been shown that the channel

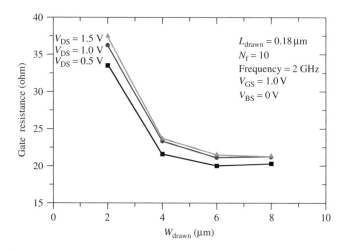

Figure 3.18 Curves of R_G versus W_f at different V_{DS}. Reproduced from Cheng Y. *et al.* (2001a) High frequency characterization of gate resistance in RF MOSFETs, *IEEE Electron Device Lett.*, **22**(2), 98–100

transit time for the NQS effect is roughly inversely proportional to $(V_{GS} - V_{th})$, where V_{th} is the threshold voltage of the device, and proportional to L_f^2 (see Tsividis (1987)). Thus, $R_{G,nqs}$ (and hence R_G) is higher in a device with longer L_f and at lower V_{GS}. As shown in Figure 3.15, the NQS effect has begun to influence R_G values in devices with relatively short L_f at RF. However, when L_f is short enough, the contribution of $R_{G,nqs}$ is smaller and $R_{G,poly}$ is dominant. In that case, R_G becomes larger as L_f tends to be shorter.

Similarly, we can understand the W_f dependence of R_G in Figure 3.16. As discussed above, the R_G portion from the distributed polysilicon gate, $R_{G,poly}$, is proportional to W_f/L_f; however, the R_G portion from the distributed channel, $R_{G,nqs}$, is proportional to L_f/W_f. As W_f becomes narrower, $R_{G,nqs}$ becomes higher, so it may dominate the total R_G when W_f reduces to some value, say $6\,\mu m$ in Figure 3.16. As W_f becomes wider, $R_{G,poly}$ becomes higher, so it may dominate the total R_G when W_f becomes larger than some specific value. Thus, as W_f changes, there exists a minimum R_G at some point of W_f, as demonstrated in Figure 3.16.

The stronger V_{GS} dependence of R_G in the narrow W_f region of Figure 3.16 can be understood because $R_{G,nqs}$ with strong bias dependence is dominant in the narrower device, while $R_{G,poly}$ without bias dependence plays a bigger role in the longer W_f region.

The V_{DS} dependence of R_G shown in Figures 3.17 and 3.18 can also be explained. It is known that the distributed effect at the gate is independent of V_{DS} and the distributed effect in the channel is stronger in the saturation region than in the linear region. Thus, according to the above analysis, we should have lower $R_{G,nqs}$ (and hence R_G) at $V_{DS} = 0.5\,V$, at which the device operates in the triode (or linear) region, than $V_{DS} = 1\,V$, at which the device operates in the saturation region, for the device with the same W_f and L_f. When V_{DS} is higher than 1 V, the device remains in saturation and the channel conductance (and hence the NQS effect) does not change much as V_{DS} increases, so R_G is insensitive to V_{DS} (>1 V) in Figures 3.17 and 3.18.

The distributed transmission line effect on the gate at HF has been studied (see, for example, Liu and Chang (1999)). It will become more severe as the gate width becomes wider at higher operation frequency. So multifinger devices (if they have wide channel widths) are used in the circuit design with narrow gate width for each finger to reduce the influence of this effect. A simple expression of gate resistance, R_G, based on that in DC or low frequency has been used to calculate the value of gate resistance with the influence of the DGE at HF. However, a factor of α is introduced, which is 1/3 or 1/12 depending on the layout structures of the gate connection to account for the distributed RC effects at RF, as given in the following:

$$R_{G,poly} = \frac{R_{Gsh}}{N_f L_f} \left(W_{ext} + \frac{W_f}{\alpha} \right). \qquad (3.22)$$

In Eq. (3.22) R_{Gsh} is the gate sheet resistance, W_f is the channel width per finger, L_f is the channel length, N_f is the number of fingers, and W_{ext} is the extension of the polysilicon gate over the active region.

Complex numerical models for the gate delay have been proposed by Abou-Allam and Manku (1997). However, the simple gate resistance model with the α factor for the distributed effect has been found accurate up to $1/2 f_T$ for a MOSFET without significant NQS effects as discussed by Enz and Cheng (2000).

For the devices with NQS effects, additional bias and geometry dependences of the gate resistance are needed to account for the NQS effect. It has been proposed that an additional resistive component in the gate should be added to represent the channel distributed RC effect, which can be "seen" by the signal applied to the gate, as shown in Figure 3.19. Thus, the effective gate resistance R_G consists of two parts:

$$R_G = R_{G,poly} + R_{G,nqs} \qquad (3.23)$$

where $R_{G,poly}$ is the distributed gate electrode resistance from the polysilicon gate material and is given by Eq. (3.22) and $R_{G,nqs}$ is the NQS distributed channel resistance seen from the gate and is a function of both biases and geometry.

Efficient and accurate modeling of the NQS effect in MOSFETs is very challenging. An R_G model with the consideration of the NQS effect has been reported (see, for example,

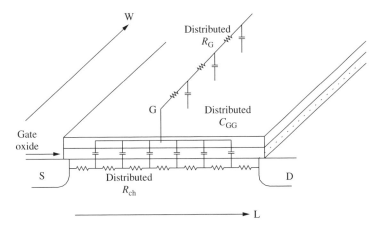

Figure 3.19 Equivalent gate resistance consists of the contributions from the distributed gate poly resistance and distributed channel resistance

Jin *et al.* (1998)). However, the following simple expression can be used to obtain the $R_{G,nqs}$ approximately in the strong inversion regime:

$$R_{G,nqs} \cong \frac{\beta}{G_m} \tag{3.24}$$

where G_m is the transconductance of the device and β is a fitting parameter with a typical value around 0.2.

3.3.3.2 Modeling of source and drain resistances

The total source and drain series resistances in a MOSFET used in Integrated Circuit (IC) designs have several components such as the via resistance, the salicide resistance, the salicide-to-salicide contact resistance, and the sheet resistance in the LDD region, as shown in Figure 3.20. However, the contact and the LDD sheet resistances usually dominate the total resistance. The typical value of the sheet resistance is around 1 kΩ/sq in the LDD region for a typical 0.25-μm CMOS technology and much smaller in more advanced technologies.

It has been known that the source/drain resistances are bias-dependent. In some compact models such as BSIM3v3 (Cheng *et al.* (1997b)), these bias dependencies are included. However, since these parasitic resistances in BSIM3v3 are treated only as virtual components in the $I-V$ expressions to account for the DC voltage drop across these resistances, they are invisible to the signal in the AC simulation. Therefore, external components for these series resistances need to be added outside an intrinsic model to accurately describe the HF noise characteristics and the AC input impedance of the device. Typically, the source/drain resistances R_D and R_S without including any bias dependence can be described by

$$R_D \cong R_{D0} + \frac{r_{dw}}{N_f W_f} \tag{3.25}$$

$$R_S \cong R_{S0} + \frac{r_{sw}}{N_f W_f} \tag{3.26}$$

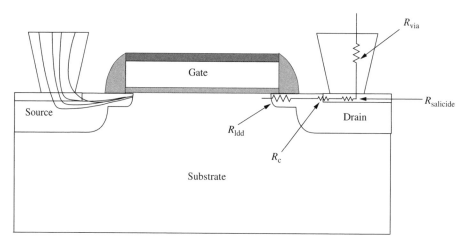

Figure 3.20   An illustration of the components of the source/drain series resistance

where r_{dw} and r_{sw} are the parasitic drain and source resistances with unit width and R_{D0} and R_{S0} account for the part of the series resistances without the width dependence. Equation (3.25) can work reasonably well in today's MOSFETs for RF applications, because the LDD region in these devices with advanced technologies (0.18 μm and less) has a very high doping concentration. Thus, the bias dependence of R_D and R_S becomes weaker compared with devices with longer channel lengths and lighter LDD doping concentrations in the older technology generation.

3.3.3.3 HF behavior and modeling of substrate resistance

Usually, the location of the substrate tie to ground the MOSFETs in low-frequency analog IC design is not regular and can be put in any suitable place in a layout of a circuit. This makes the substrate resistance a function of the distance between the active device and the substrate contacts besides the active device size and substrate contact shape. The values of the substrate resistance may be different for devices in different locations even though they have the same channel length and width. In low-frequency analog design, accurate evaluation of the substrate resistance is not required because the influence of the substrate resistance can be ignored. However, in RF ICs, the substrate resistance can contribute significantly to the device behavior, mainly output admittance, and accurate prediction of the substrate resistance becomes important.

Although it is always desirable to have a detailed distributed RC network to account for the contribution of the substrate components, it is too complex to be implemented in a compact model. Some tools using three-dimensional or quasi-three-dimensional numerical approaches to simulate the effects of the substrate resistance are available; however, a proper integration of such a tool into the design system remains an issue. Also, the accuracy of the simulation results is dependent on the accuracy of the process information provided by process simulation, which needs to be calibrated carefully (very time-consuming) to obtain the desirable accuracy. Thus, a good compromise is to use simplified lumped RC network, which is be accurate in required operation frequency range, to simulate the contribution of the substrate components.

It has been known that the contribution of the substrate resistance R_{sub}, which provides an AC path to the signal and influences the output admittance Y_{22} behavior, cannot be ignored at radio frequency (RF). An RF model without including the substrate resistance will be 20% or more off the measured data of the Y_{22} characteristic of a MOSFET. This is not desirable in RF IC design because an accurate prediction of Y_{22} is very important in designing a matching network to compensate the overall gain over a wide frequency range. Also, accurate substrate resistance is required in a power amplifier design to evaluate the overall power loss properly, and in an LNA design to predict the noise figure without underestimating the contribution of the substrate resistance. Recently, RF models with the substrate components, which include the substrate resistances and drain/source junction capacitance, have been published. However, detailed characterization of these substrate components in MOSFETs at RF has not been reported yet. It was expected that the substrate resistances may be bias-dependent due to the variations of the depletion regions below the gate and surrounding the source and drain diffusions. The HF experimental exploration of these substrate components in MOSFETs is very important to help the understanding of the device behavior and the modeling of the MOSFET at RF.

To accurately evaluate the influence of the substrate resistance in RF IC design, device structures with their own substrate contacts are preferred. This device design may also help to reduce the cross-talk caused by the substrate coupling between the devices. Its disadvantage is that it will take more space owing to the substrate contacts for each device in the circuits. However, it may be acceptable for RF IC because of the small device amounts in RF circuits compared to other circuits, say, the digital IC. It eliminates the geometry uncertainty caused by the irregular substrate contact design so that the designers can predict the contribution of the substrate resistance accurately with the developed model.

Before we discuss the model for substrate components, we first discuss the measured HF behavior of the substrate network including the substrate resistance R_{sub} and junction capacitance C_{jDB}. The details of the extraction of the substrate components will be discussed in Section 3.4.2. As we mentioned earlier, the substrate components will mainly influence the Y_{22} characteristics at HF. Figures 3.21 and 3.22 give the measured Y_{22} behavior versus frequency at various gate and drain-bias conditions. Both real and imaginary parts of Y_{22} show strong bias dependence on both gate and drain biases. Further, the characteristics of R_{sub}, C_{jDB}, and C_{GD} versus frequency at different gate and drain biases that can be extracted from measured Y_{22} according to the procedures to be discussed later are shown in Figures 3.23 to 3.26. In Figures 3.23 and 3.24, a weak gate-bias dependence of both substrate resistance and junction capacitance can be observed. It is understandable according to the device structures, and demonstrates that the modulation of the gate bias to the channel depletion layer does not influence the substrate resistance significantly. A strong gate-bias dependence of C_{GD} is understandable without any surprise. Keeping in mind the weak gate-bias dependence of the substrate resistance, the strong gate-bias dependence of $Re\{Y_{22}\}$ shown in Figure 3.22 is mainly contributed by the channel resistance R_{DS}. The bias dependence of gate-to-drain capacitance C_{GD} does not

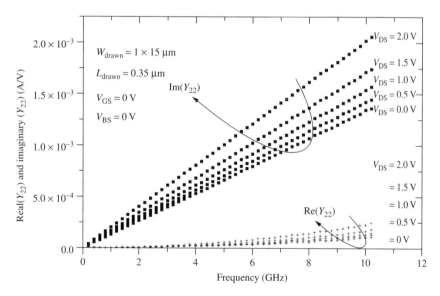

Figure 3.21 Measured Y_{22} data at different drain biases show obvious drain-bias dependence. Reproduced from Cheng *et al.* (2000a) On the high frequency characteristics of the substrate resistance in RF MOSFETs, *IEEE Electron Device Lett.*, **21**(12), 604–606

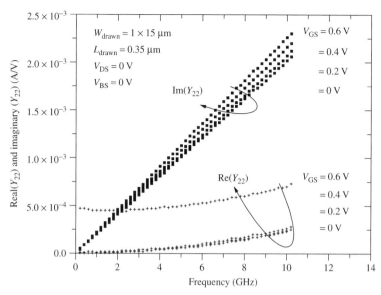

Figure 3.22 Measured Y_{22} data at different gate biases show strong gate-bias dependence. Reproduced from Cheng *et al.* (2000a) On the high frequency characteristics of the substrate resistance in RF MOSFETs, *IEEE Electron Device Lett.*, **21**(12), 604–606

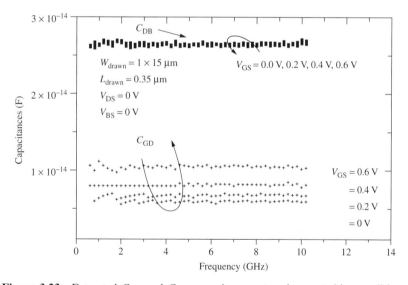

Figure 3.23 Extracted C_{GD} and C_{jDB} capacitances at various gate-bias conditions

influence the bias dependence of Re{Y_{22}} significantly; however, it is believed that C_{GD} causes the strong gate-bias dependence of Im{Y_{22}} shown in Figure 3.22. The Im{Y_{sub}} data will show a weak gate-bias dependence after de-embedding the contribution of C_{GD} from the measured Y_{22} data. The weak gate-bias dependence of junction capacitance, as shown in Figure 3.22, is consistent with the measurement results performed in low or medium frequency range.

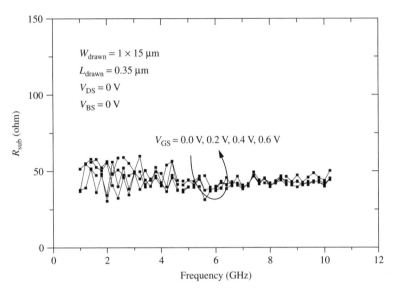

Figure 3.24   Extracted substrate resistance shows a very weak gate-bias dependence. Reproduced from Cheng Y. *et al.* (2000) On the high frequency characteristics of the substrate resistance in RF MOSFETs, *IEEE Electron Device Lett.*, **21**(12), 604–606

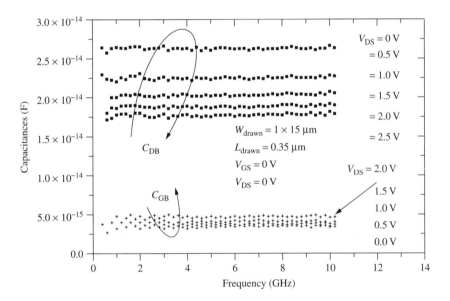

Figure 3.25   Extracted C_{GB} and C_{jDB} capacitances at various drain-bias conditions

However, the obvious drain-bias dependence of $\text{Re}\{Y_{22}\}$, shown in Figure 3.21, is believed to be caused by the contribution of both channel resistance and junction capacitance. The data still shows a dependence on drain bias after de-embedding the influence of R_{DS}, R_G, C_{GD}, and so on from $\text{Re}\{Y_{22}\}$. After further removing the influence of junction capacitance, the data (representing the substrate resistance) shows a weak dependence on

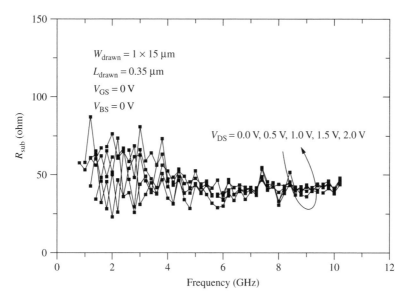

$W_{\text{drawn}} = 1 \times 15\ \mu\text{m}$

$L_{\text{drawn}} = 0.35\ \mu\text{m}$

$V_{\text{GS}} = 0\ \text{V}$

$V_{\text{BS}} = 0\ \text{V}$

$V_{\text{DS}} = 0.0\ \text{V}, 0.5\ \text{V}, 1.0\ \text{V}, 1.5\ \text{V}, 2.0\ \text{V}$

Figure 3.26 Extracted substrate resistance shows a very weak drain-bias dependence. Reproduced from Cheng *et al.* (2000a) On the high frequency characteristics of the substrate resistance in RF MOSFETs, *IEEE Electron Device Lett.*, **21**(12), 604–606

drain bias as demonstrated in Figure 3.26. However, unlike the case of varying the gate bias, the drain-bias dependence of Im{Y_{22}} is mainly due to the existence of the junction capacitance instead of C_{GD}. Much stronger drain-bias dependence of the junction capacitance C_{jDB} than of C_{GD} has been found as shown in Figure 3.25. It is also consistent with the measurement results of junction capacitances at low and medium frequency.

A simple EC for the substrate network shown in Figure 3.7 has been used to analyze the HF substrate-coupling effect and the characteristics of substrate resistance at HF (see, for example, Cheng *et al.* (2000a)). Even though a simpler substrate network has been reported by Tin *et al.* (1999), it is found that the three-resistor substrate network can ensure better model accuracy over a wider frequency range.

Figure 3.27 illustrates a lumped RC EC for the substrate components in a multifinger device with substrate ties residing at both sides of the device. Noting that all the source (or drain) terminals for different fingers are connected together, and the source terminal is grounded together with the substrate terminal, a simplified substrate network, as shown in Figure 3.7, with the following relationships can be obtained:

$$C_{\text{SB}} = \sum_{k=1}^{N_{\text{s}}} C_{\text{sb},k}, \tag{3.27}$$

$$C_{\text{DB}} = \sum_{k=1}^{N_{\text{d}}} C_{\text{db},k}, \tag{3.28}$$

$$\frac{1}{R_{\text{SB}}} = \sum_{k=1}^{N_{\text{s}}} \frac{1}{R_{\text{sb},k}}, \tag{3.29}$$

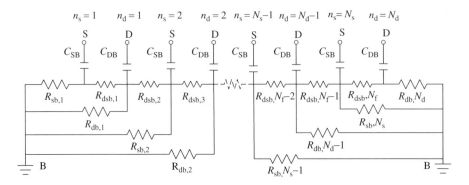

Figure 3.27 Illustration of the equivalent circuit (EC) for substrate components in a multifinger device. Reproduced from Cheng *et al.* (2002c) Parameter extraction of accurate and scaleable substrate resistance components in RF MOSFETs, *IEEE Electron Device Lett.*, **23**(4), 221–223

$$\frac{1}{R_{\mathrm{DB}}} = \sum_{k=1}^{N_{\mathrm{d}}} \frac{1}{R_{\mathrm{db},k}}, \tag{3.30}$$

$$\frac{1}{R_{\mathrm{DSB}}} = \sum_{k=1}^{N_{\mathrm{f}}} \frac{1}{R_{\mathrm{dsb},k}}, \tag{3.31}$$

where C_{SB} and C_{DB} are the total source-to-bulk and drain-to-bulk capacitances, $C_{\mathrm{SB},K}$ and $C_{\mathrm{DB},K}$ are the source-to-bulk and drain-to-bulk capacitances of each source and drain region in the multifinger device, N_{s} and N_{d} are the numbers of the source and drain regions, R_{SB}, R_{DB}, and R_{DSB} are the total equivalent resistances between the source and the substrate, between the drain and the substrate, and between the source and the drain underneath the channel in the substrate, $R_{\mathrm{SB},K}$, $R_{\mathrm{DB},K}$, and $R_{\mathrm{DSB},K}$ are the resistances, corresponding to each single source/drain.

Assuming no difference between the outer and the inner source/drain regions, we have

$$C_{\mathrm{SB}} = N_{\mathrm{s}} C_{\mathrm{sb},k}, \tag{3.32}$$

$$C_{\mathrm{DB}} = N_{\mathrm{d}} C_{\mathrm{db},k}, \tag{3.33}$$

$$R_{\mathrm{DSB}} = \frac{R_{\mathrm{dsb},k} L_{\mathrm{f}}}{N_{\mathrm{f}} W_{\mathrm{f}}} \tag{3.34}$$

where L_{f} and W_{f} are the channel length and the width per finger. $R_{\mathrm{dsb},k}$ is the sheet resistance in the substrate underneath the channel between the source and drain in a single-finger device.

Noting that the value of the substrate resistance from the outer finger subdevice is much smaller than that from the inner finger device, and also noting that the device is symmetric, we have the following:

$$\frac{1}{R_{\mathrm{SB}}} \approx \frac{1}{R_{\mathrm{sb},1}} + \frac{1}{R_{\mathrm{sb},\mathrm{Ns}}}$$

and

$$\frac{1}{R_{DB}} \approx \frac{1}{R_{db,1}} + \frac{1}{R_{db,Nd}}.$$

Also, according to the layout, the following equations have been used:

$$R_{DB} \approx \frac{r_{dbw}}{W_f}, \tag{3.35}$$

$$R_{SB} \approx \frac{r_{sbw}}{W_f} \tag{3.36}$$

where r_{dbw} and r_{sbw} are the substrate resistances with unit-channel width.

Generally, assuming that the device is symmetric with respect to the source and the drain and that it has no difference between the outer and the inner source/drain regions in a multifinger device, we have

$$R_{DSB} = \frac{r_{dsb}L_f}{N_f W_f} \tag{3.37}$$

where r_{dsb} is the sheet resistance in the substrate between the source and the drain.

Some bias dependence of the substrate resistances had been expected, on the basis of the fact that the depletion regions below the gate and surrounding the source and drain diffusions may vary at different gate and drain-bias conditions. However, it has been found that the bias dependence of the substrate resistances is actually very weak for the devices with substrate ties isolated by shallow trench from the active region, and the above simple substrate resistance network is accurate up to 10 GHz, as discussed by Cheng et al. (2000a).

3.3.3.4 High-frequency behavior and modeling of parasitic capacitances

It has been known that the gate capacitance can be directly extracted from the measured Y-parameters as discussed by Cheng et al. (2002). Figure 3.28 shows that the imaginary part of Y_{11} and Y_{12}, which can be used to extract the C_{GG} and the C_{GD} as we will discuss in Section 3.4.2. The bias dependence of the $Y-$parameters (the gate capacitance) is obvious as the gate bias varies. Strong drain-bias dependence has also been expected for the extracted capacitance data versus the gate and drain biases.

As shown in Figure 3.29, the parasitic capacitances in a MOSFET can be divided into the following components: (1) the outer fringing capacitance between the polysilicon gate and the source/drain, C_{FO}; (2) the inner fringing capacitance between the polysilicon gate and the source/drain, C_{FI}; (3) the overlap capacitances between the gate and the heavily doped S/D regions (and the bulk region), C_{GSO} and $C_{GDO}(C_{GBO})$, which are relatively insensitive to terminal voltages; (4) the overlap capacitances between the gate and the lightly doped S/D region, C_{GSOL} and C_{GDOL}, which change with biases; (5) the source/drain junction capacitances, C_{JD} and C_{JS}; and (6) the substrate capacitance, C_{SUB}. Most of them have been included in models for digital/analog applications (see the discussion by Cheng and Hu (1999)). However, additional parasitic capacitance components may have to be added to the existing models (either intrinsic or extrinsic capacitance models) if they cannot meet the accuracy requirements at RF.

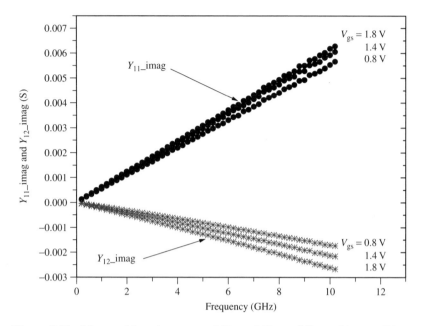

Figure 3.28   Measured imaginary part of Y_{11} and Y_{12} at different bias conditions

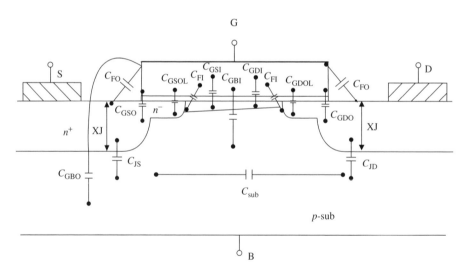

Figure 3.29   Illustration of different capacitance components in a MOSFET

In Figures 3.30 and 3.31, the capacitances C_{GDP} and C_{GSP} obtained from the total capacitances extracted from the measured S-parameters and the intrinsic capacitances simulated with the model are shown. The definitions of C_{GDP} and C_{GSP} are given in the following:

$$C_{GDP} = C_{GDtotal\_extracted} - C_{GDintrinsic\_simulated}, \tag{3.38}$$

$$C_{GSP} = C_{GStotal\_extracted} - C_{GSintrinsic\_simulated}, \tag{3.39}$$

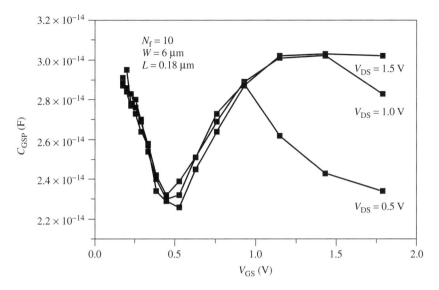

Figure 3.30   An example to show the bias dependence of extracted equivalent extrinsic capacitance between the gate and the source from measured HF data at different drain biases. Reproduced from Cheng (2002b) MOSFET Modeling for RF IC Design, in *CMOS RF Modeling, Characterization and Applications*, Jamal D. M. and Fjeldly T. A., eds., World Scientific Publishing, Singapore

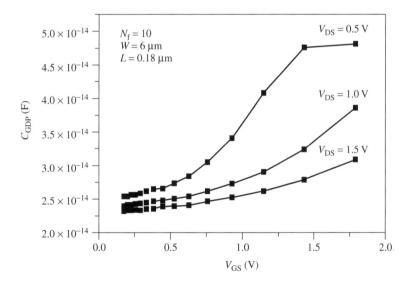

Figure 3.31   An example to show the bias dependence of extracted equivalent extrinsic capacitance between the gate and the drain from measured HF data at different drain biases. Reproduced from Cheng (2002b) MOSFET Modeling for RF IC Design, in *CMOS RF Modeling, Characterization and Applications*, Jamal D. M. and Fjeldly T. A., eds., World Scientific Publishing, Singapore

where $C_{GDtotal\_extracted}$ is the total C_{GD} capacitance extracted from the measured data, $C_{GDintrinsic\_simulated}$ is the intrinsic C_{GD} simulated by the model, $C_{GStotal\_extracted}$ is the total C_{GS} capacitance extracted from the measured data, and $C_{GSintrinsic\_simulated}$ is the intrinsic C_{GS} simulated by the model.

According to the definition of C_{GDP} and C_{GSP}, we can consider these capacitances as overlap capacitances if the intrinsic capacitance model is accurate enough. However, in some cases, C_{GDP} and C_{GSP} should not be considered as overlap capacitances since they may contain the correction to the intrinsic capacitances if the intrinsic capacitances are not properly modeled. It is clear that C_{GDP} and C_{GSP} have strong bias dependences that cannot be fitted by a constant overlap capacitance model. To improve the overall RF model accuracy, an engineering approach, adding additional capacitance components with bias dependence for C_{GDP} and C_{GSP} in the subcircuit, can be used if the capacitance model in a RF model cannot provide good accuracy over different bias regions.

The substrate capacitance is another extrinsic capacitance that should be considered in a subcircuit model for ultra-HF applications. In the above substrate RC network, we did not include the contribution of the substrate capacitance. It does not influence the model accuracy to fit the measured data up to 10 GHz. However, the substrate capacitance component may be necessary in a subcircuit model when the device operates at frequencies much higher than 10 GHz.

3.3.4 Non-quasi-static Behavior

Figure 3.32 shows the characteristics of $R_{Gsh,HF}$ $(= R_G N_f L_f / W_f)$ extracted for devices with different L_f. $R_{Gsh,cal}$ in the figure is the measured DC gate sheet resistance but divided by 3 to consider the distributed effect at HF and is a constant value independent

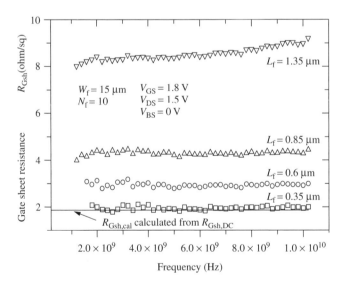

Figure 3.32 Gate sheet resistance $R_{Gsh,HF}$ versus frequency for devices with different channel lengths. Higher value of $R_{Gsh,HF}$ is obtained at HF compared with $R_{Gsh,cal}$. Reproduced from Cheng *et al.* (2001b) Frequency dependent resistive and capacitive components in RF MOSFETs, *IEEE Electron Devices Lett.*, **22**(7), 333–335

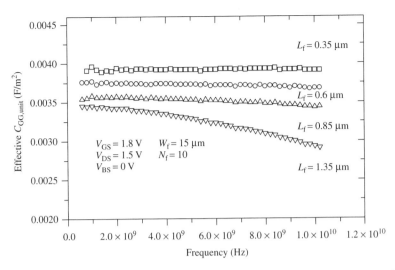

Figure 3.33 Effective unit-area gate capacitance $C_{GG,unit}$ versus frequency for devices with different channel lengths. The value of $C_{GG,unit}$ is reduced for the device with longer L_f, and also $C_{GG,unit}$ is not a constant as the frequency varies in the devices with strong NQS effect. Reproduced from Cheng *et al.* (2001b) Frequency dependent resistive and capacitive components in RF MOSFETs, *IEEE Electron Devices Lett.*, **22**(7), 333–335

of the device geometry. However, the measurements show that the $R_{Gsh,HF}$ not only is larger than the $R_{Gsh,cal}$ (even for the device with an L_f of 0.35 μm) but also increases as L_f increases. For the device with an L_f of 1.35 μm, the frequency dependency of $R_{Gsh,HF}$ is obvious, which is in contradiction to what we have seen in low and intermediate frequencies for the components in a MOSFET. Figure 3.33 shows the characteristics of $C_{GG,unit}(= C_{GG}/\{W_f L_f N_f\})$ versus frequency for devices with different channel lengths. As observed, the extracted $C_{GG,unit}$ shows some weak frequency dependency for the device with an L_f of 0.35 μm but can still be considered approximately constant over the frequency range. This is consistent with the results in low and intermediate frequencies. However, for the devices with longer L_f, the value of $C_{GG,unit}$ is smaller compared with the device with shorter L_f at the same operation frequency. Furthermore, $C_{GG,unit}$ is not a constant any more with frequency and decreases as the frequency increases, which is significant in the devices with the longest channels. Figure 3.34 shows that the normalized $G_m(= \mathrm{Re}(Y_{21})/\mathrm{Re}(Y_{21(f0)}))$ degrades seriously in devices with longer L_f as the frequency increases, where f_0 is a fixed frequency. The frequency dependency of G_m in the device with an L_f of 0.35 μm is weak; however, it becomes very strong in the device with an L_f of 1.35 μm (see Cheng *et al.* (2001b)).

As mentioned earlier, the NQS effect results in a signal delay or even a malfunction of the circuits in some cases when a MOSFET operates at HF as discussed by Oh *et al.* (1980). Typically we can see this NQS effect in a device with an L of 10 μm at about 1 MHz as discussed by Paulous *et al.* (1983). However, it is expected that the critical channel length (L_c) for NQS effect to happen will decrease as the signal frequency increases.

When the device cannot respond to the signal immediately, the distributed effect of the channel resistance should be accounted for. This distributed effect in the channel

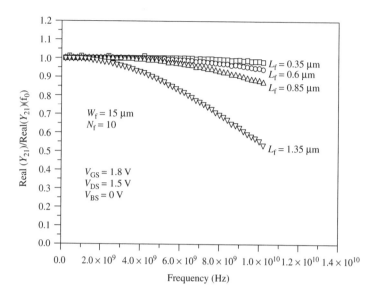

Figure 3.34 Normalized equivalent transconductance versus frequency for devices with different L_f. The degradation of G_m can be explained with the existence of NQS effect. Reproduced from Cheng *et al.* (2001b) Frequency dependent resistive and capacitive components in RF MOSFETs, *IEEE Electron Devices Lett.*, **22**(7), 333–335

or NQS effect will cause an increase in the effective gate resistance as discussed by Jin *et al.* (1998) and Cheng *et al.* (2001a). So it can be understood that in Figure 3.32 because of the existence of the NQS effect the extracted HF gate sheet resistance, $R_{Gsh,HF}$, is higher than $R_{Gsh,cal}$, a theoretically estimated value for gate sheet resistance where only the distributed effects on the gate are considered. Because the influence of the NQS effect can be ignored at low frequency but increases significantly as frequency increases, the extracted effective gate sheet resistance, $R_{Gsh,HF}$, exhibits strong frequency dependency. Similarly, the frequency dependency of $R_{Gsh,HF}$ in a 0.35-μm device is not obvious because of the weak NQS effect in this device, but becomes stronger as L_f increases. Thus, larger $R_{Gsh,HF}$ and stronger frequency dependency of $R_{Gsh,HF}$ are found in devices with larger L_f.

It is also known that the NQS effect will equivalently introduce a transcapacitance between the drain and the gate (see, for example, Cheng *et al.* (2001a)). The displacement current from this additional capacitance (referred to as C_{nqs}) can cancel partially the output current, which is equivalent to an increased delay to the signal. C_{nqs} is negative relative to the positive gate-to-source, gate-to-drain, and gate-to-bulk capacitances C_{GS}, C_{GD}, and C_{GB}, so the effective C_{GG} with NQS (the sum of C_{GS}, C_{GD}, C_{GB}, and C_{nqs}) is less than that without NQS (the sum of C_{GS}, C_{GD}, and C_{GB} only). In devices with longer L_f, the NQS effect is stronger, so $|C_{nqs}|$ is larger and hence C_{GG} is smaller. Also, as frequency increases, the NQS effect in the device is stronger, so $|C_{nqs}|$ increases and C_{GG} decreases. Thus, a frequency dependency of C_{GG} can be seen in Figure 3.33 owing to the existence of the component C_{nqs}.

The degradation of G_m at HF has been considered as an important phenomenon that should be accurately modeled to predict the circuit behavior at HF (see, for example,

Tsividis and Suyama (1993)). The reason for the degradation of G_m is considered as the contribution of the NQS effect even though it may be partially caused by the increased signal "feed-through" via C_{DG} at HF[2].

3.4 MODEL PARAMETER EXTRACTION

3.4.1 RF Measurement and De-embedding Techniques

For a model to describe the device characteristics accurately, all important model parameters should be extracted from measured data. To extract the RF model parameters, on-chip HF measurements are performed by using specifically designed test structures. Also, a de-embedding methodology has to be developed to remove the influence of the parasitics in the test structure from the measured raw data in order to obtain the data for the characteristics of the device-under-test (DUT).

Figure 3.35 illustrates the setup of an HF measurement system for on-wafer RF measurements. A controller is used to send the commands to instruments (vector network analyzer (VNA) and $I–V$ tester, etc.) and the probe station to perform the measurements for a specific DUT and to gather the measured data for postprocessing. To ensure the accuracy of the measurements, a system calibration has to be performed before conducting any measurements on the DUT. Typically, the system calibration for on-wafer measurements is done by using a so-called impedance standard substrate (ISS) that can provide high-accuracy and low-loss standards for two-port calibration procedures such as short-open-load-through (SOLT) and through-reflect-line (TRL). The SOLT calibration has been widely used because it is supported by virtually every VNA. However, TRL calibration is the most fundamental of the advanced calibrations and requires the least amount of information about the standards. Only VNAs with advanced calibration capabilities will support the TRL calibrations. ISS calibration can ensure reasonable accuracy if the substrate and the interconnect losses of the DUT are comparable to those of the ISS. Recently, however, it has been discussed that additional de-embedding of substrate parasitics in RF CMOS devices may be needed because of the high substrate losses compared with other devices such as GaAs MESFETs.

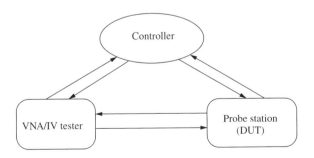

Figure 3.35   Equipment requirement of an HF measurement system

---

[2] Because the existence of C_{DG} provides a signal path, more and more signals are fed back through this capacitance as frequency increases, so the total output current (and hence the transconductance) is reduced. However, it can only explain partial G_m degradation.

Besides the system calibration discussed above, de-embedding methodology for raw data measured from the DUT has also to be developed on the basis of specific test structures designed according to de-embedding techniques. Figure 3.36(a), (b), and (c) show the test structures for the so-called two-step de-embedding procedure. Figure 3.36(a) illustrates the test structure with the DUT. The pads for port 1 and port 2 are signal pads connecting the gate and the drain terminal of the DUT and the top and the bottom ground pads connect to both source and substrate of the DUT, as illustrated further in Figure 3.37. This test structure is used for S-parameter measurements of two-port systems. Test structures for multiport systems (more than two ports) can be designed and measured also. But the measurement system with specific design consideration of the probe tips and

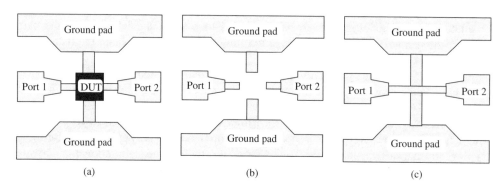

Figure 3.36 Illustrations of the test structures for a two-step calibration of S-parameter measurements: (a) test structure with the DUT; (b) open test structure; and (c) short test structure

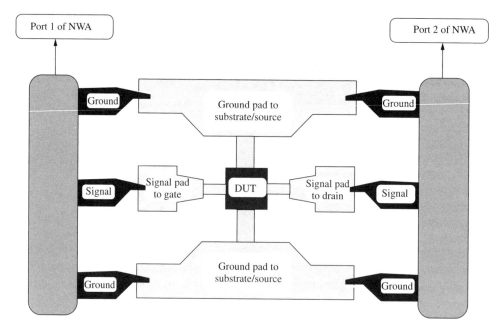

Figure 3.37 Illustration of on-wafer HF measurement for a two-port system

calibration techniques should be used. Also, the de-embedding technique of the raw data is more complex than that measured from a two-port system.

Figure 3.36(b) is the so-called "open" structure for a two-port measurement. It uses the same test structure as in Figure 3.36(a) but the DUT has been removed, so all the pads are open without any connections between them. Figure 3.36(c) shows the so-called "short" structure that is the opposite of the "open" structure in which all of the pads are shorted to each other.

Different de-embedding techniques have been developed on the basis of different calibration test structures (see discussion by Koolen *et al.* (1991) and Chen and Deen (2001)). Here, the de-embedding procedure based on the open and short calibration test structures, illustrated in Figure 3.36, is discussed as an example. This two-step de-embedding technique has been widely used in HF measurements for different technologies.

Typically, a DUT with parasitics from the test structures can be represented by the equivalent circuit in Figure 3.38, where Y_{P1}, Y_{P2}, and Y_{P3} represent the influence of the parallel parasitics and Z_{S1}, Z_{S2}, and Z_{S3} describe the influence of the series parasitics.

The parallel elements Y_{P1}, Y_{P2}, and Y_{P3} can be obtained from the measured data of the open structure, that is,

$$Y_{P3} = -Y_{12,\text{open}} = -Y_{21,\text{open}}, \tag{3.40}$$

$$Y_{P1} = Y_{11,\text{open}} + Y_{12,\text{open}}, \tag{3.41}$$

$$Y_{P2} = Y_{22,\text{open}} + Y_{21,\text{open}}. \tag{3.42}$$

The series elements Z_{s1}, Z_{s2}, and Z_{s3} can be obtained from the measured data of both open and short structures, that is,

$$\begin{bmatrix} Z_{s1} + Z_{s3}Z_{s3} \\ Z_{s3}Z_{s2} + Z_{s3} \end{bmatrix} = (Y_{\text{short}} - Y_{\text{open}})^{-1}. \tag{3.43}$$

The measured data corresponding to the transistor can be obtained according to the following equation

$$Y_{\text{transistor}} = [(Y_{\text{DUT}} - Y_{\text{open}})^{-1} - (Y_{\text{short}} - Y_{\text{open}})^{-1}]^{-1}. \tag{3.44}$$

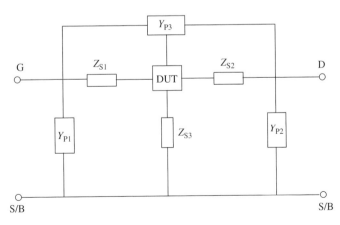

Figure 3.38   Equivalent circuit used for two-step de-embedding of measured HF data of MOSFETs

Thus, according to the above, the procedures of the two-step de-embedding technique can be given as follows:

1. Measure the s-parameters (S_{DUT}, S_{open}, and S_{short}) for DUT, open and short test structures and convert them to Y-parameters (Y_{DUT}, Y_{open}, and Y_{short}).

2. Perform the first step de-embedding by removing the parallel parasitics from both Y_{DUT} and Y_{short} according to the following equations:

$$Y_{DUT1} = Y_{DUT} - Y_{open}, \qquad (3.45)$$

$$Y_{short1} = Y_{short} - Y_{open}. \qquad (3.46)$$

3. Perform the second de-embedding by removing the series parasitics Z_{short1}, converting from Y_{short1}, from Z_{DUT1}, and from Y_{DUT1}, according to the following equation:

$$Z_{transistor} = Z_{DUT1} - Z_{short1}. \qquad (3.47)$$

Figures 3.39 to 3.42 show the data of the measured Y_{11} and Y_{22} before and after 1 step and 2 step de-embedding. Significant difference between the data before and after 1 step de-embedding has been observed. Thus, the data de-embedding with the open calibration structure is absolutely necessary to extract accurate parameters of an RF model. A minor difference between the data after the 1-step and the 2-step de-embedding indicates that the calibration with the short structure may be ignored for the MOSFETs at a frequency range up to 10 GHz. However, for the device to work at a much higher frequency range, the importance of the calibration with the short structure should be considered. Also, the short calibration may have to be used to obtain the measured data for other devices such

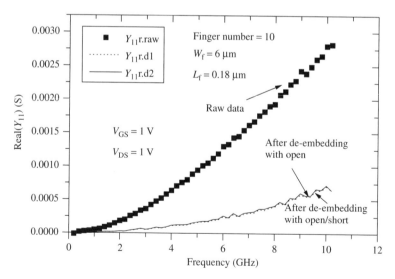

Figure 3.39   Illustration of the necessity of the de-embedding of the real part of the measured Y_{11} data. Reproduced from Cheng (2002b) MOSFET Modeling for RF IC Design, in *CMOS RF Modeling, Characterization and Applications*, Jamal D. M. and Fjeldly T. A., eds., World Scientific Publishing, Singapore

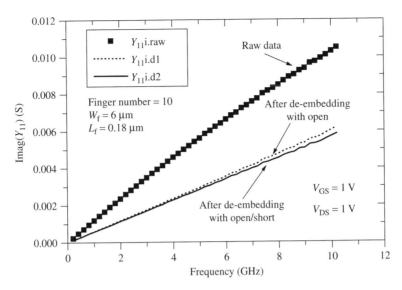

Figure 3.40 Illustration of the necessity of the de-embedding of the imaginary part of measured Y_{11}. Reproduced from Cheng (2002b) MOSFET Modeling for RF IC Design, in *CMOS RF Modeling, Characterization and Applications*, Jamal D. M. and Fjeldly T. A., eds., World Scientific Publishing, Singapore

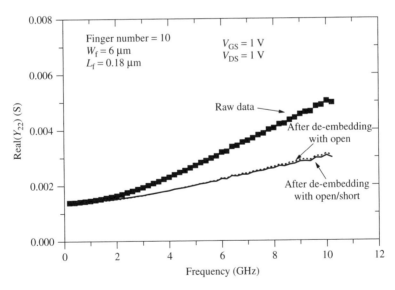

Figure 3.41 Another example to show the importance of the de-embedding of the real part of measured Y_{22}. Reproduced from Cheng (2002b) MOSFET Modeling for RF IC Design, in *CMOS RF Modeling, Characterization and Applications*, Jamal D. M. and Fjeldly T. A., eds., World Scientific Publishing, Singapore

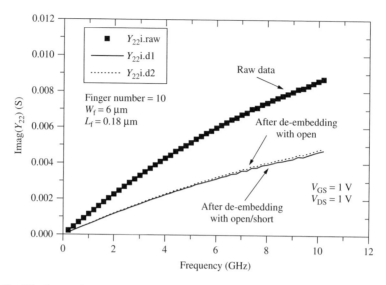

Figure 3.42   The figure shows a significant difference between the imaginary part of the measured Y_{22} before and after de-embedding. Reproduced from Cheng (2002b) MOSFET Modeling for RF IC Design, in *CMOS RF Modeling, Characterization and Applications*, Jamal D. M. and Fjeldly T. A., eds., World Scientific Publishing, Singapore

as inductors because the devices themselves are very sensitive to the influence of the series parasitics.

3.4.2 Parameter Extraction

Depending on the EC used in the model, methodologies of HF parameter extraction have been developed (see, for example, Jen *et al.* (1998) and Kolding (2000)). In the previous section, we have discussed the EC of a MOSFET for RF applications. Usually, the Y-parameter analysis of the EC is adopted to obtain the necessary equations to extract the values of some resistive and capacitive components. It has been known that the poles due to the terminal resistances (that usually are small because of the large finger numbers) are at a much higher frequency than typical transit frequencies, so that they basically can be neglected when calculating the Y-parameters and the related quantities.

The substrate resistances in the small-signal circuit of Figure 3.10 are also neglected when analyzing the Y-parameters (Y_{11}, Y_{12}, and Y_{21}, except Y_{22}) to obtain expressions that are suitable for use in parameter extraction.

The parameters related to the DC characteristics are extracted with the data from the DC measurements. The methodologies for the DC model parameter extraction have been well developed (see, for example, Cheng *et al.* (1997b)) and they are not discussed here. Next we will focus on the discussion of the extraction of the AC parameters for the components shown in Figure 3.10.

The EC given in Figure 3.10 contains too many components, especially current sources, which make the Y-parameter analysis very complex and difficult if not impossible, to obtain any useful analytical expressions for the parameter extraction. In order to extract the AC parameters, the influence from the intrinsic components has to be minimized. By considering the transistor biased in the strong inversion mode with $V_{DS} = 0\,V$, the intrinsic behavior of the transistor becomes symmetric in terms of the drain and the source terminals. Therefore, the effects of the transconductances and the transcapacitances become very small and can be neglected, that is, $G_m \approx 0$, $G_{mb} \approx 0$, $C_m \approx 0$, $C_{mb} \approx 0$, $C_{SD} \approx 0$, and the small-signal EC in Figure 3.10 can be simplified to that shown in Figure 3.43, where $R_{DS} = 1/G_{DS}$.

By applying a gate bias high enough to operate the device in strong inversion regime, the intrinsic gate-to-bulk capacitance C_{GB} is small enough and can be neglected. The EC for the Y_{11} parameter analysis is obtained, as shown in Figure 3.43, by shorting the output port and neglecting C_{GB} in Figure 3.10. Since the transistor is operating in the linear region with $V_{DS} = 0$, C_{GS} is approximately equal to C_{GD}. The structure and the equivalent effects of the circuit are fully symmetric, which makes the effect of R_{DS} very small so that it can then be neglected. Further, the following assumptions have been adopted in the Y-parameter analysis of the equivalent circuit in Figure 3.10:

1. R_G, R_S, and R_D are dominated by the contributions from the resistance of polysilicon and diffusion layers and are treated as parameters independent of bias condition and frequency.

2. The equivalent impedance from the intrinsic source/drain nodes to the external source/drain nodes are dominated by the terminal resistances R_S and R_D, that is,
$$R_S \ll \frac{1}{|j\omega C_{BS}|} \text{ and } R_D \ll \frac{1}{|j\omega C_{BD}|}.$$

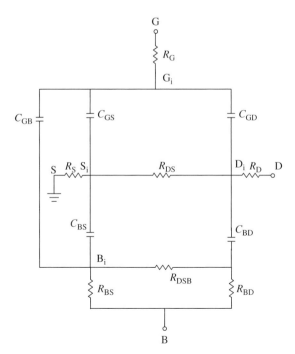

Figure 3.43 An equivalent circuit used for extracting the HF model parameters

3. The frequency range considered in this analysis is up to $10\,\mathrm{GHz}$, within which the following simplifications hold: $(\omega C_{GS} R_S)^2 \ll 1$, $(\omega C_{GD} R_D)^2 \ll 1$, $\omega^2 C_{GS} C_{GD} (R_D + R_S) R_G \ll 1$, and $(1 + j\omega C_{GG} R_G)^{-1} \ll 1 - j\omega C_{GG} R_G$, where C_{GG} is the total gate capacitance $C_{GG} = C_{GS} + C_{GD} + C_{GB}$.

On the basis of the above, the following approximate equations for the Y-parameters can be obtained:

$$Y_{11} \approx \omega^2 (C_{GG}^2 R_G + C_{GS}^2 R_S + C_{GD}^2 R_D) + j\omega C_{GG}, \tag{3.48}$$

$$Y_{12} \approx -\omega^2 C_{GG} C_{GD} R_G - j\omega C_{GD}, \tag{3.49}$$

$$Y_{21} \approx G_m - \omega^2 C_{GG} C_{GD} R_G - j\omega (C_{GD} + G_m R_G C_{GG}). \tag{3.50}$$

Direct extraction of the AC parameters can be performed from the measured data according to the above equations,

$$C_{GG} = \left| \frac{\mathrm{Im}\{Y_{11}\}}{\omega} \right|, \tag{3.51}$$

$$C_{GD} = \left| \frac{\mathrm{Im}\{Y_{12}\}}{\omega} \right|, \tag{3.52}$$

$$C_{GS} = C_{GD}, \tag{3.53}$$

$$C_{GB} = C_{GG} - C_{GS} - C_{GD}, \tag{3.54}$$

$$R_G = \left| \frac{\mathrm{Re}\{Y_{12}\}}{\mathrm{Im}\{Y_{11}\}\,\mathrm{Im}\{Y_{12}\}} \right|, \tag{3.55}$$

$$R_D = \left| \frac{\mathrm{Re}\{Y_{21}\} - \mathrm{Re}\{Y_{12}\}}{\mathrm{Im}\{Y_{12}\}^2} \right|, \tag{3.56}$$

$$R_S = \left| \frac{\mathrm{Re}\{Y_{11}\}}{\mathrm{Im}\{Y_{11}\}^2} - R_G - \frac{C_{GD}^2}{C_{GG}^2} R_D \right| \frac{C_{GG}^2}{C_{GS}^2}. \tag{3.57}$$

Depending on the measured data, which can be influenced by the design of the test structure, the calibration of the measurement system, the experience of the measurement person, and the accuracy of the de-embedding procedures, the values of R_D and R_S extracted from the S-parameter measurements may or may not equal the ones extracted from DC measurements. To ensure that the DC characteristics predicted by the model parameters extracted from DC measurements are not disturbed by the possible different R_D and R_S extracted from the measured S-parameters, it is recommended that the values of R_D and R_S extracted from DC measurements are used in extracting the AC parameters. In that case, the R_G parameter can be extracted with the following equation:

$$R_G = \left| \frac{\mathrm{Re}\{Y_{11}\} - \omega^2 (C_{GD}^2 R_D + C_{GS}^2 R_S)}{\mathrm{Im}\{Y_{11}\}^2} \right|. \tag{3.58}$$

To extract the parameters for the substrate network, additional analysis for the Y_{22} parameter ($V_{GS} = V_{DS} = 0$) is needed. Figure 3.43 gives the equivalent circuit for the device at the given bias conditions. To simplify the analysis, the influence of R_D is subtracted first from the Z_{22} corresponding to the two-port network given by Figure 3.43,

$$Z_{22'} = Z_{22} - R_D. \tag{3.59}$$

An HF small-signal EC of MOSFET is given in Figure 3.44 for the devices at the saturation-operating regime. The box surrounded by the dotted line is the RC network for the substrate components. According to the EC, it is known that the measured Y_{22} includes at least the contribution from the gate resistance R_G, drain series resistance R_D, source series resistance R_S, channel resistance R_{DS}, gate-to-source capacitance C_{GS}, and gate-to-drain capacitance C_{GD} besides the substrate components. To understand the HF behavior of the substrate components, we should either use specific test structures to measure the contributions from the substrate components only or de-embed the contributions of these components such as R_G and C_{GD}, and so on from the measured Y_{22}. Here, we adopt the latter approach. Next, we present the methodology of de-embedding the measured Y_{22} to obtain the Y_{sub} data representing the contribution of substrate components. By performing a tedious but straightforward Y-parameter analysis for the EC shown in Figure 3.43, we finally obtain the following equations:

$$\mathrm{Re}\{Y_{sub}\} = \mathrm{Re}\{y_{22'}\} - R_G(\omega C_{GD})^2 - \frac{1}{R_{DS}}, \tag{3.60}$$

$$\mathrm{Im}\{Y_{sub}\} = \mathrm{Im}\{y'_{22}\} - j\omega C_{GD} \tag{3.61}$$

where y'_{22} is the Y_{22} without the influence of R_D, Y_{sub} is the output admittance of the substrate network in Figure 3.45, $\omega = 2\pi f$ and f is the operation frequency. In the above analysis, the contributions of transconductances G_m and G_{mb} are ignored since no obvious

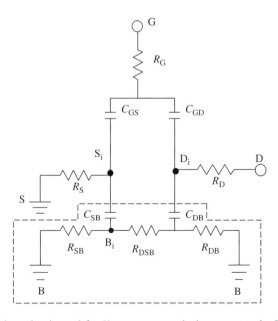

Figure 3.44 Equivalent circuit used for Y-parameter analysis to extract the HF model parameters

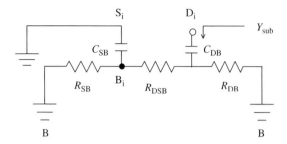

Figure 3.45   A simplified equivalent circuit of the substrate network. Reproduced from Cheng Y. *et al.* (2000) On the high frequency characteristics of the substrate resistance in RF MOSFETs, *IEEE Electron Device Lett.*, **21**(12), 604–606

current flows in the channel at the given bias conditions. Also, the influence of R_S on the total admittance is not taken into account in the analysis. This is reasonable because of the dominant contribution of C_{GS}. Furthermore, the assumptions of $\omega^2(C_{GS} + C_{GB})^2 R_G^2 \ll 1$ and $(\omega C_{GD})^2 R_G^2 \ll 1$ are used, which are generally valid in the frequency range up to 10 GHz.

The parameters of C_{GD} and R_G can be obtained as discussed earlier. Thus, the Y_{sub} data de-embedded from the measured Y_{22} data according to the above equations represents the contribution of the substrate network.

To extract the substrate components, such as the substrate resistance and junction capacitances, we further derive the following equation by doing a Y-parameter analysis of the substrate network in Figure 3.45:

$$Y_{sub} \approx \frac{R_{DB}(R_{SB} + R_{DSB})}{R_{DB} + R_{SB} + R_{DSB}}(\omega C_{DB})^2 + j\omega C_{DB} = R_{sub}(\omega C_{DB})^2 + j\omega C_{DB} \qquad (3.62)$$

where

$$\frac{(\omega C_{SB})^2 R_{SB}^3}{R_{DB} + R_{SB} + R_{DSB}} \ll 1, \qquad \frac{(\omega C_{SB})^2 R_{DB} R_{SB}^2}{R_{DB} + R_{SB} + R_{DSB}} \ll 1, \qquad \text{and} \qquad (\omega C_{SB})^2 R_{SB}^2 \ll 1.$$

These assumptions are valid in the frequency range up to 10 GHz. Therefore, we have

$$C_{DB} = \frac{\text{Im}\{Y_{sub}\}}{\omega}, \qquad (3.63)$$

$$R_{sub} = \frac{\text{Re}\{Y_{sub}\}}{\text{Im}\{Y_{sub}\}^2}. \qquad (3.64)$$

The extracted C_{DB} includes the contribution of both the intrinsic capacitance C_{BDi} and the drain junction capacitance C_{jDB}. The C_{BDi} can be separated from the extracted C_{DB} with the measured data at different V_{DS} because C_{jDB} is a function of drain bias and C_{BDi} is approximately independent of the drain bias in the saturation regime. However, typically the capacitance C_{DB} is dominated by C_{jDB}. The value of C_{jDB} at zero bias can be extracted from Eq. (3.63) with the measured data at $V_{DS} = 0$ V. The parameters to

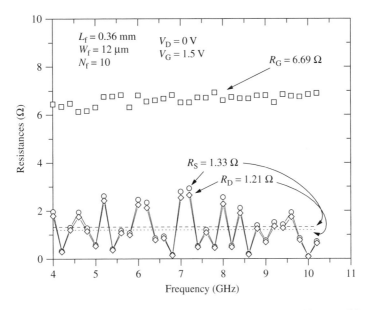

Figure 3.46   Extracted values of R_G, R_S, and R_D at a given bias condition

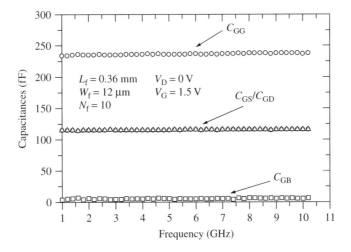

Figure 3.47   Extracted values of C_{GG}, C_{GS}, and C_{GD} at a given bias condition

describe the bias dependence of C_{jDB} can be extracted according to Eq. (3.63) with the measured data at different V_{DS}.

Figure 3.46 shows the extracted resistances as a function of frequency with the transistor at the given bias condition. It is shown that those components are frequency-independent. The extracted capacitances versus frequency are shown in Figure 3.47. For the given device in the figure, all of the capacitive components are also frequency-independent. The substrate resistance can be extracted from the slope of the Re{Y_{sub}} versus Im{Y_{sub}}^{2} as shown in Figure 3.48. The parameters for R_{sb}, R_{db}, and R_{dsb} can be

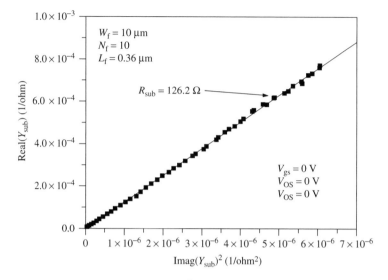

Figure 3.48   Illustration of the extraction of R_{sub} from the plot of $\text{Re}(Y_{sub})$ versus $\text{Im}(Y_{sub})^2$. Here, $\text{Re}(Y_{sub})$ is the real part of Y_{sub} and $\text{Im}(Y_{sub})$ is the imaginary part of Y_{sub}

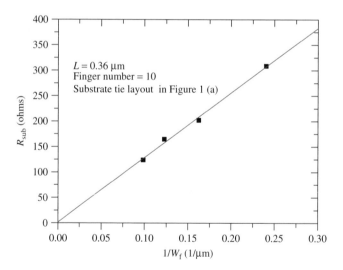

Figure 3.49   Extracted R_{sub} from devices with different per-finger-widths. r_{dbw} and r_{sbw} can be further obtained through the plots of R_{sub} versus $1/W_{eff}$ from devices with even- and odd-finger numbers. Reproduced from Cheng Y. *et al.* (2002c) Parameter extraction of accurate and scaleable substrate resistance components in RF MOSFETs, *IEEE Electron Device Lett.*, **23**(4), 221–223

extracted further once the values of R_{sub} for devices with different widths and finger numbers are obtained.

Similarly, R_{sub} in devices with different geometry can be obtained. Figure 3.49 gives the extracted R_{sub} from the devices with different widths. It can be seen that R_{sub} is approximately proportional to W_f^{-1} for devices with the same number of fingers.

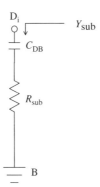

Figure 3.50   One-resistor EC for the substrate network. Reproduced from Cheng Y. *et al.* (2002c) Parameter extraction of accurate and scaleable substrate resistance components in RF MOSFETs, *IEEE Electron Device Lett.*, **23**(4), 221–223

According to the analysis above, the three-resistor substrate network can be further simplified into the one-resistor network by considering R_{sub} as an equivalent resistance of the three discrete resistors as shown in Figure 3.50. Once the R_{sub} for devices with different channel widths are determined, the values of the parameters such as r_{dbw} and r_{sbw} can be obtained. r_{sbw} is determined first from the obtained R_{sub} in odd-finger devices with different channel widths according to Eq. (3.35), and then r_{dbw} is determined from the obtained R_{sub} in even-finger devices with different channel widths after obtaining r_{sbw}. Devices with different channel lengths (besides different channel widths and fingers) should be measured at much higher frequencies than 10 GHz to extract the values of r_{dsb} accurately. Here we are interested in frequencies up to 10 GHz, and we use calculated value for r_{dsb} from the doping concentration in the substrate region (underneath the channel between the source and the drain). Depending on the processing conditions and the substrate material, r_{dsb} can be different.

3.5 NQS MODEL FOR RF APPLICATIONS

The NQS effect should be included for an RF model to accurately describe the HF characteristics of devices if the devices themselves exhibit this effect at the operating frequency. Most MOSFET models available in circuit simulators use the quasi-static (QS) approximation. In a QS model, the channel charge is assumed to be a unique function of the instantaneous biases, that is, the charge responds to a change in voltages with infinite speed. Thus, the finite charging time of the carriers in the inversion layer is ignored. In reality, the carriers in the channel do not respond to the signal immediately, and hence, the channel charge is not a unique function of the instantaneous terminal voltages (quasi-static) but a function of the history of the voltages (non-quasi-static). This problem may become pronounced in RF applications, where the input signals may have rise or fall times comparable to, or even smaller than, the channel transit time. For long channel devices, the channel transit time is roughly inversely proportional to $(V_{GS} - V_{th})$ and proportional to L^2. Because the carriers in these devices cannot follow the changes of the applied signal, the QS models may give inaccurate or anomalous simulation results that cannot be used to guide circuit design.

The modeling of the frequency-dependent components caused by the NQS effect is challenging in compact models for circuit simulation[3]. Owing to the existence of the NQS effect, a MOSFET model based on the QS approximation may not accurately describe the HF device behavior.

The NQS effect can be modeled with different approaches for RF applications: (a) R_G approach, in which a bias-dependent gate resistance is introduced to account for the distributed effects from the channel resistance as discussed earlier (see Jin *et al.* (1998)); (b) R_i approach, in which a resistance R_i (as used in modeling a MESFET (Metal Semiconductor Field Effect Transistor) or HEMT (High Electron Mobility Transistor) is introduced to account for the NQS effect (see Chen and Deen (1998)); (c) transadmittance approach, in which a voltage-control-current-source (VCCS) is connected in parallel to the intrinsic capacitances and transconductances to model the NQS effect (see Enz and Cheng (2000)); and (d) core model approach, in which the NQS effect can be modeled in the core intrinsic model (see Chan *et al.* (1998) and Cheng *et al.* (1997b)). It should be pointed out that all of these approaches would have to deal with complex implementation issues.

Both the R_G and the R_i approaches will introduce additional resistance components in the model besides the existing physical gate and channel resistances measured at DC or low frequency, so the noise characteristics of the model using either R_G or R_i approach need to be examined. Ideally, the NQS effect should be included in the core intrinsic model if the model can predict both NQS and noise characteristics without a large penalty in the model implementation and simulation efficiency.

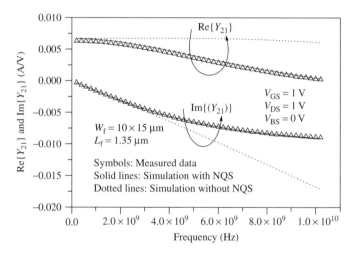

Figure 3.51   Measured and simulated results of Y_{21} for a MOSFET with 1.35-μm channel length. Model without considering the NQS effect cannot describe the HF device behavior. BSIM3v3 NQS model can predict accurately the Y_{21} characteristics even though the device has strong NQS effect. Reproduced from Cheng *et al.* (2001b) Frequency dependent resistive and capacitive components in RF MOSFETs, *IEEE Electron Devices Lett.*, **22**(7), 333–335

---

[3] Because devices with longer L_f have lower f_T and strong NQS effect, they usually are not suitable for small-signal RF applications. However, devices with longer L_f may be used in circuits such as switch or biasing circuits. It is still desirable that an RF model can simulate devices having obvious NQS effects.

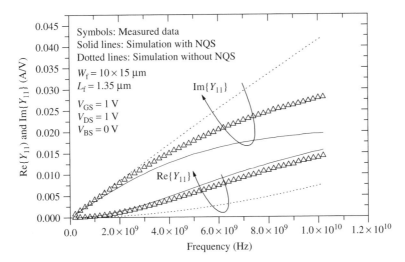

Figure 3.52 Measured and simulated results of Y_{11}. The fitting of the Y_{11} characteristics of the model needs to be improved. Reproduced from Cheng *et al.* (2001b) Frequency dependent resistive and capacitive components in RF MOSFETs, *IEEE Electron Devices Lett.*, **22**(7), 333–335

Some compact models such as BSIM3v3 with the consideration of the NQS effect have been verified with measurements for devices at the medium frequency range. Figure 3.51 shows the simulation results by using the models with and without considering the NQS effect. It is clear that the model without the NQS effect cannot predict correctly the device behavior in both Y_{11} and Y_{21}. By including the NQS effect, BSIM3v3 can predict the measured data very well in both the real and imaginary parts of Y_{21}. However, the model needs to be improved for fitting Y_{11} (see Figure 3.52) as discussed by Cheng *et al.* (2001b). The inclusion of the NQS effect would be a desirable feature for an RF model even though it remains a question whether the devices in RF circuits for small-signal applications will operate in the frequency region at which the devices show significant NQS effects.

REFERENCES

Abou-Allam E. and Manku T. (1997) A small-signal MOSFET model for radio frequency IC applications, *IEEE Trans. Computer-Aided Design Integrated Circuits Syst.*, **16**(5), 437–447.

Boothroyd A. R. *et al.* (1991) MISNAN-A physically based continuous MOSFET model for CAD applications, *IEEE Trans. CAD*, **10**, 1512–1529.

Chan M. *et al.* (1998) A robust and physical BSIM3 non-quasi-static transient and AC small signal model for circuit simulation, *IEEE Trans. Electron Devices*, **ED-45**, 834–841.

Chen C. H. and Deen M. J. (1998) High frequency noise of MOSFETs I: modeling, *Solid-State Electron.*, **42**, 2069–2081.

Chen C. H. and Deen M. J. (2001) A general noise and s-parameter de-embedding procedure for on-wafer high-frequency noise measurements of MOSFETs, *IEEE Trans. Micro-wave Theory Tech.*, **49**(5), 1004–1005.

Chen K. *et al.* (1996) MOSFET carrier mobility model based on gate oxide thickness, threshold and gate voltages, *J. Solid-State Electron. (SSE)*, **39**, 1515–1518.

Cheng Y. *et al.* (1997a) A physical and scalable BSIM3v3 I-V model for analog/digital circuit simulation, *IEEE Trans. Electron Devices*, **44**, 277–287.

Cheng Y. *et al.* (1997b) BSIM3v3.1 User's Manual, Memorandum No. UCB/ERL M97/2.

Cheng Y. (1998) RF modeling issues of deep-submicron MOS-FETs for circuit design, *Proc. of the IEEE International Conference on Solid-State and Integrated Circuit Technology*, pp. 416–419.

Cheng Y. and Hu C. (1999) *MOSFET Modeling & BSIM3 User's Guide*, Kluwer Academic publishers, Norwell, MA.

Cheng Y. *et al.* (2000a) On the high frequency characteristics of the substrate resistance in RF MOSFETs, *IEEE Electron Device Lett.*, **21**(12), 604–606.

Cheng Y. *et al.* (2000b) MOSFET modeling for RF circuit design, *Proceedings of the 2000 Third IEEE International Caracas Conference on Devices, Circuits and Systems*, D23/1–D23/8.

Cheng Y. *et al.* (2001a) High frequency characterization of gate resistance in RF MOSFETs, *IEEE Electron Device Lett.*, **22**(2), 98–100.

Cheng Y. *et al.* (2001b) Frequency-dependent resistive and capacitive components in RF MOSFETs, *IEEE Electron Device Lett.*, **22**(7), 333–335.

Cheng Y. *et al.* (2000a) High frequency small signal AC and noise modeling of MOSFETs for RF IC design, *IEEE Trans. Electron Devices*, **49**(3), 400–408.

Cheng Y. (2002b) MOSFET Modeling for RF IC Design, in *CMOS RF Modeling, Characterization and Applications*, Jamal D. M. and Fjeldly T. A., eds., World Scientific Publishing, Singapore.

Cheng Y. *et al.* (2002c) Parameter extraction of accurate and scaleable substrate resistance components in RF MOSFETs, *IEEE Electron Device Letters*, **23**(4), 221–223.

Enz C. *et al.* (1995) An analytical MOS transistor model valid in all regions of operation and dedicated to low voltage and low-current applications, *J. Analog Integrated Circuit Signal Process.*, **8**, 83–114.

Enz C. and Cheng Y. (2000) MOS transistor modeling for RF IC design, *IEEE J. Solid-State Circuits*, **35**(2), 186–201.

Jin X. *et al.* (1998), An effective gate resistance model for CMOS RF and noise modeling, *Tech. Dig. (IEDM)*, 961–964.

Jen S. H. *et al.* (1998) Accurate MOS transistor modeling and parameter extraction valid up to 10-GHz, *Proc. of the European Solid-State Device Research Conference*, Bordeaux, pp. 484–487.

Kolding T. E. (2000) A four-step method for de-embedding gigahertz on-wafer CMOS measurements, *IEEE Trans. Electron Devices*, **47**(4), 734–740.

Koolen M. C. A. M. *et al.* (1991) An improved de-embedding technique for on-wafer high-frequency characterization, *IEEE 1991 Bipolar Circuits and Technology Meeting*, pp. 191–194.

Langevelde R. van and Klaassen F. M. (1997) Effect of gate-field dependent mobility degradation on distortion analysis in MOSFETs, *IEEE Trans. Electron Devices*, **44**(11), 2044–2052.

Liang M. S. *et al.* (1986) Inversion layer capacitance and mobility of very thin gate oxide MOSFETs, *IEEE Trans. Electron Devices*, **ED-33**, 409.

Liu W. *et al.* (1997) R.F.MOSFET modeling accounting for distributed substrate and channel resistances with emphasis on the BSIM3v3 SPICE model, *Tech. Dig. International Electron Device Meeting (IEDM)*, 309–312.

Liu W. and Chang M. C. (1999) Transistor transient studies including transcapacitive current and distributive gate resistance for inverter circuits, *IEEE Trans. Circuits Syst. I: Fundam. Theory Appl.*, **45**(4), 416–422.

Mayer J. (1971) MOS models and circuit simulation, *RVA Rev.*, **32**, 42–63.

Mos9 Manual (2001) *http://www.semiconductors.philips.com/Philips_Models*.

Oh S. Y. *et al.* (1980) Transient analysis of MOS transistors, *IEEE J. Solid-State Circuits*, **DC-15**(4), 636–643.

Ou J.-J. *et al.* (1998) CMOS RF modeling for GHz communication IC's, *Proc. of the VLSI Symposium on Technology*, pp. 94,95.

Paulous J. J. *et al.* (1983) Limitations of quasi-static capacitance models for the MOS transistors, *IEEE Electron Lett.*, **EDL-4**, 221–224.

Pehlke D. R. *et al.* (1998) High-frequency application of MOS compact models and their development for scalable RF MOS libraries, *Proc. IEEE Custom Integrated Circuits Conference*, pp. 219–222.

Sheu B. J. *et al.* (1984) A compact IGFET charge model, *IEEE Trans. Circuits Syst.*, **CAS-31**, 745–748.

Takagi S. *et al.* (1994) On the universality of inversion layer mobility in Si MOSFET's: part I – effects of substrate impurity concentration, *IEEE Trans. Electron Devices*, **ED-41**, 2357.

Tin S. F. *et al.* (1999) Substrate network modeling for CMOS RF circuit simulation, *Proc. IEEE Custom Integrated Circuits Conference*, pp. 583–586.

Tsividis Y. P. (1987) *Operation and Modeling of the MOS Transistor*, McGraw-Hill, New York.

Tsividis Y. P. and Suyama K. (1993) MOSFET modeling for analog circuit CAD: problems and prospects, *Tech. Dig.*, **CICC-93**, 14.1.1–14.1.6.

Ward D. E. (1981) *Charge-Based Modeling of Capacitance in MOS Transistors*, Tech. G201-11, Stanford Electronics Laboratory, Stanford University, Stanford, CA.

Yang P. *et al.* (1983) An investigation of the charge conservation problem for MOSFET circuit simulation, *IEEE J. Solid-state Circuits*, **SC-18**, 128–138.

4
Noise Modeling

4.1 NOISE SOURCES IN A MOSFET

Both passive and active components in a circuit will generate various types of noise. To understand the noise behavior, a single MOSFET can be considered as a small circuit with different resistive, capacitive, and active components as we have seen in the previous chapter. Thus different noise sources exist in a MOS transistor as shown in Figure 4.1 with their power spectral densities (PSDs). They include (1) terminal resistance thermal noise at the gate, (2) terminal resistance thermal noise at the drain, (3) terminal resistance thermal noise at the source, (4) thermal noise and the flicker noise in the channel, (5) substrate resistance thermal noise, and (6) induced gate noise.

In principle, flicker noise is a low-frequency noise and it mainly affects the low-frequency performance of the device, so it can be ignored at very high frequency. However, the contribution of flicker noise should be considered in designing some radio-frequency (RF) circuits such as mixers, oscillators, or frequency dividers that up-convert the low-frequency noise to higher frequency and deteriorate the phase noise or the signal-to-noise ratio. Channel resistance and all terminal resistances contribute to the thermal noise at high frequency (HF), but typically channel resistance dominates in the contributions of the thermal noise from the resistances in the device. Induced gate noise is generated by the capacitive coupling of local noise sources within the channel to the gate, and usually it plays a more important role as the operation frequency goes much higher than the frequency at which channel thermal noise dominates.

4.2 FLICKER NOISE MODELING

Among all noise sources, the flicker noise is the dominant source for phase noise in silicon MOSFET circuits, especially in the low-frequency-range. It sets a lower limit on the level of signal detection and spectral purity and is one of the factors limiting the achievable dynamic range of MOS ICs, so it is important for device and circuit designers to minimize this effect in order to improve the circuit performance. As designers begin to explore circuits with low-power and low-voltage MOSFETs, the impact of low-frequency flicker noise becomes more and more crucial for providing enough dynamic range and better circuit performance.

Device Modeling for Analog and RF CMOS Circuit Design. T. Ytterdal, Y. Cheng and T. A. Fjeldly
© 2003 John Wiley & Sons, Ltd ISBN: 0-471-49869-6

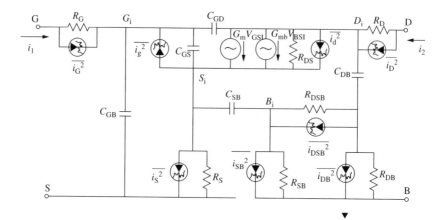

Figure 4.1   An equivalent circuit to illustrate the noise sources in a MOSFET. $i_G{}^2$, $i_S{}^2$, and $i_D{}^2$ are the noise contributions by the terminal resistances at the gate, at the source, and at the drain; $i_d{}^2$ is the noise contribution in the channel, including the flicker noise portion; $i_{DB}{}^2$, $i_{SB}{}^2$, and $i_{DSB}{}^2$ are the noise contributions by substrate resistances; and $i_g{}^2$ is the induced gate noise

4.2.1 The Physical Mechanisms of Flicker Noise

Noise at low frequencies in a MOSFET is dominated by flicker noise. Measurements generally show a spectral density of the input (gate) referred voltage noise, which is roughly inversely proportional to frequency, as shown in Figure 4.2. Therefore, flicker noise is also called $1/f$ noise. Much effort has been made in understanding the physical origin of flicker noise. However, the physical mechanism is still not very clear so far. A lot of discussions and investigations are continuing to find a universal model to explain the experimental results reported by different research groups that use devices from different manufacturers.

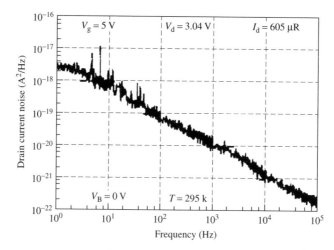

Figure 4.2   Drain current noise spectral density of an n-channel MOSFET. Reproduced from Hung *et al.* (1990b) A physical-based MOSFET noise model for circuit simulation, *IEEE Trans. Electron Devices*, **37**, 1323–1333

Although there are probably several different physical mechanisms resulting in noise in MOSFETs, there are strong indications that traps at the Si–SiO$_2$ interface play the most important role, as discussed by Jindal and Ziel (1978). Electron trapping and de-trapping can lead to conductance variations. The exact mechanism is still under discussion; however, basically, there are three different theories on the mechanism of flicker noise as follows:

1. Carrier-density fluctuation models (number fluctuations), predicting an input referred noise density independent of the gate bias voltage and proportional to the square of the oxide thickness;

2. Mobility fluctuation models, predicting an input referred noise voltage increasing with gate bias voltage and proportional to oxide thickness;

3. Correlated carrier and mobility fluctuation models, a unified model proposed by Hung *et al.* (1990a) with a functional form resembling the number fluctuation model at low bias and the mobility fluctuation model at high bias.

In the carrier density fluctuation model, the noise is explained by the fluctuation of channel-free carriers due to the random capture and emission of carriers by interface traps at the Si–SiO$_2$ interface. According to this model, the input noise is independent of the gate bias, and the magnitude of the noise spectrum is proportional to the density of the interface traps. A $1/f$ noise spectrum is predicted if the trap density is uniform in the oxide. Measurements of devices from many different CMOS processes with oxide thickness between 10 and 80 nm suggest that nMOS transistors behave as predicted by the number fluctuation model (see, for example, Vandamme (1994)). However, noise measurements of newer deep submicron transistors present a much less consistent picture. For instance, nMOS transistors also may show bias dependence, while pMOS transistors may have a noise corner frequency comparable to nMOS transistors. Also, the experimental results show a $1/f^n$ spectrum and n is not always 1 but in the range of 0.7 to 1.2. Some experimental results even show that n decreases with increasing gate bias in p-channel MOSFETs. Modified charge density fluctuation theories have been proposed to explain these experimental results. The spatial distribution of the active traps in the oxide is assumed to be nonuniform to explain the technology and the gate-bias dependence of n.

The mobility fluctuation model considers flicker noise to be the result of fluctuations in carrier mobility based on Hooge's empirical relation for the spectral density of the flicker noise in a homogeneous device. It has been proposed that the fluctuations of the bulk mobility in MOSFETs are introduced by changes in the phonon population. The mobility fluctuation models predict a gate bias–dependent noise. However, they cannot always account for the magnitude of the noise.

The unified theory for the origin of the $1/f$ noise suggests that the capture and emission of carriers by the interface traps cause fluctuation in both the carrier number and the mobility. All unified noise models assume implicitly that the mobility, limited by Coulomb scattering at trapped interface charges, does not depend on the inversion carrier density. However, recent experimental results indicate that the mobility, limited by Coulomb scattering, is proportional to the square root of the inversion carrier density (see Vandamme *et al.* (2000); Vandamme (1994)). Recently, some arguments even claim that the correlated mobility fluctuations can be neglected compared to the noise contribution from carrier number fluctuations, if the correct dependence of the Coulomb scattering–limited mobility on inversion carrier density is taken into account. As a result, the unified noise models cannot predict the experimentally observed noise as a function of gate bias in p-type MOSFETs unless nonphysical fitting parameters are used (Vandamme *et al.* (2000)). Nevertheless, even though this unified theory cannot explain all the details of the experimental data, it seems to be the most attractive model available today in circuit simulators.

4.2.2 Flicker Noise Models

It is for historic reasons that different flicker noise models have been developed on the basis of the three different approaches discussed in the preceding text. They are implemented in different simulators such as HSPICE, SPECTRE, ELDO, PSPICE, and so on. Almost all of the commercial simulators provide different options for users to select different noise models in noise simulation together with a specific compact model, such as MOS 9, EKV, and BSIM3v3, for simulations such as DC, AC, small signal, or transient analysis. For example, HSPICE includes three different models for the drain current flicker noise that are distinguished with different model levels (0–3).
   For NLEV = 0:

$$S_{id} = \frac{K_F I_{DS}^{A_F}}{C_{ox} L_{eff}^2 f} \tag{4.1}$$

where S_{id} is the drain current noise power spectral density, I_{DS} is the drain current, C_{ox} is the unit-area gate oxide capacitance, L_{eff} is the effective channel length, f is the frequency, K_F and A_F are the fitting parameters.
   For NLEV = 1:

$$S_{id} = \frac{K_F I_{DS}^{A_F}}{C_{ox} L_{eff} W_{eff} f} \tag{4.2}$$

where W_{eff} is the effective channel width.
   For NLEV = 2 and 3:

$$S_{id} = \frac{K_F G_m^2}{C_{ox} L_{eff} W_{eff} f^{A_F}} \tag{4.3}$$

where G_m is the transconductance of the device and A_F is a fitting parameter.
   In fact, some compact models have their own flicker noise models. For example, BSIM3v3 introduces two flicker noise models (Cheng *et al.* (1997)). One is the SPICE2 flicker noise model (Vladimirescu (1994)), while the other is the unified flicker noise model. The latter is a newer model developed recently and has been considered a more accurate model than the SPICE2 flicker noise model (Hung *et al.* 1990b). The reason the SPICE2 flicker noise model is included in BSIM3v3 is to provide the convenience to some BSIM3v3 users who were familiar with the SPICE2 flicker noise model before the unified BSIM3 noise model was developed and who want to continue using it in noise simulation (Cheng and Hu (1999)).
   The SPICE2 flicker noise model is

$$S_{id} = \frac{K_F I_{DS}^{A_F}}{C_{ox} L_{eff}^2 f^{E_F}} \tag{4.4}$$

where E_F is a fitting parameter.

The unified flicker noise model in BSIM3v3 is more complex. Basically, it includes a portion equivalent to the SPICE2 flicker noise model given by Eq. (4.4), but contains another portion to give a more accurate description of the flicker noise characteristics in the saturation region (Cheng and Hu (1999)).

Currently, it is a fact that many different noise models are included in circuit simulators. However, it has to be pointed out that these models in commercial simulators are not fully compatible with each other. For example, the geometry dependence between Eqs. (4.1) and (4.3) are different, and the bias dependence between them is also different. Furthermore, those flicker noise models contain different oxide thickness dependencies. Modeling engineers and circuit designers need to be aware of this when performing noise simulation. A lot of work has been done to verify the accuracy of the flicker noise models over various bias conditions (see, for example, Vandamme *et al.* (2000)), but further work is still needed to develop a better flicker noise model that can explain most (if not all) of the experiments. So a careful selection of the flicker noise model is required to make sure that the model will predict reasonable noise performance according to the circuit applications.

4.2.3 Future Work in Flicker Noise Modeling

4.2.3.1 Flicker noise modeling with the consideration of new physical mechanisms in MOSFETs with ultrathin oxides

The above physical mechanisms of flicker noise are the ones we have frequently encountered in literature. However, as the technology enters more advanced stages, new noise mechanisms may appear and play an important role. For example, it has been reported that the influence of a new mechanism on flicker noise performance should be accounted for in ultrathin oxide MOS transistors (e.g., 1.5 nm or less) owing to direct tunneling currents that will alter the characteristics of the $1/f$ noise, depending on the length of the channel and the thickness of the gate oxide, as shown in Figures 4.3 and 4.4.

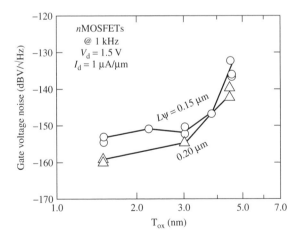

Figure 4.3 Gate oxide thickness dependence of flicker noise in *n*-channel MOSFETs with 0.15-μm and 0.2-μm gate channel lengths. Reproduced from Momose H. S. *et al.* (1998) A study of flicker noise in n- and p-MOSFETs with ultra-thin gate oxide in the direct-tunneling regime, *Tech. Dig. Int. Electron Device Meeting*, 923–926

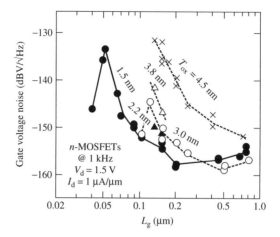

Figure 4.4   Gate length dependence of flicker noise in *n*-channel MOSFETs with various gate oxide thicknesses. Reproduced from Momose H. S. *et al.* (1998) A study of flicker noise in n- and p-MOSFETs with ultra-thin gate oxide in the direct-tunneling regime, *Tech. Dig. Int. Electron Device Meeting*, 923–926

In Figures 4.3 and 4.4, the gate length and the oxide thickness dependence of gate referred voltage noise are shown at 1-kHz operation. Figure 4.3 shows the gate oxide thickness dependence of the gate referred voltage noise in devices with 0.15-μm and 0.2-μm channel lengths. For the devices with gate lengths less than 0.2 μm, the flicker noise in a device with 1.5-nm gate oxide thickness is lower than that in devices with thicker gate oxides. It means that the noise characteristics of devices have been improved with decreasing gate oxide thickness for the devices with such short channel lengths, although the gate leakage current becomes larger in the former. A possible mechanism for the lowering of flicker noise in the devices with thinner oxides is the appearance of band-to-band tunneling. However, as also shown in Figure 4.5, for devices with channel length longer than 0.2 μm, the flicker noise in the device with 1.5-nm gate oxide is higher than that in the device with thicker oxide (2.2 nm). An understanding of this result has led to the theory that the higher flicker noise in such devices with longer (than 0.2 μm) channel length and thinner (1.5 nm) gate oxide was caused by the much larger gate leakage current as the devices with longer channel lengths have larger gate area. Further theoretical and experimental investigations on this issue are needed to fully understand the contribution of the band-to-band tunneling and gate leakage to the flicker noise characteristic in today's devices. A compact flicker noise model with the consideration of band-to-band tunneling and gate leakage has not been reported so far.

4.2.3.2 Modeling and simulation of flicker noise under switched bias conditions

It has been reported that devices under switched bias conditions show lower flicker noise than those measured at DC bias conditions (Wel *et al.* (2000) and Klumperink *et al.*

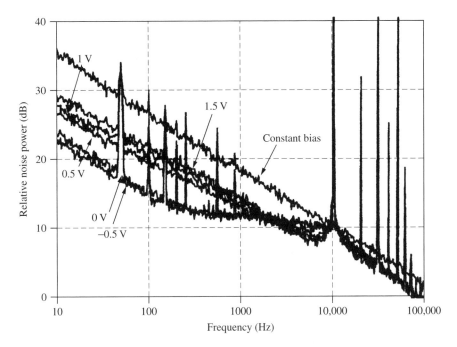

Figure 4.5 Noise reduction as a function of the "off" voltage for an nMOS, $V_{GS,on} = 2.5\,V$, $V_{th} = 1.9\,V$, $f_{switch} = 10\,kHz$, duty cycle $= 50\%$. Reproduced from Klumperink E. A. M. *et al.* (2000) Reducing MOSFET 1/f noise and power consumption by switched biasing, *IEEE J. Solid-State Circuits*, **35**(7), 994–999

(2000)). Figure 4.5 shows a typical measurement result. The noise spectrum between 10 Hz and 100 kHz is shown for constant biasing (no switching) together with noise spectra resulting from a 10-kHz switched bias signal with 50% duty cycle. For 50% duty cycle, a low-frequency noise power that is reduced by 6 dB compared to the constant-bias situation is expected. Further noise reduction is observed when the gate-source voltage in the "off" state is decreased, indicating an increasing noise reduction closer to accumulation. Figure 4.6 shows the results at various switching frequencies. All noise spectra appear to "merge" at low frequencies, with about 7 dB of intrinsic noise reduction (apart from the 6 dB related to 50% duty cycle). Even at megahertz frequencies, where the settling of the output voltages becomes incomplete, this noise reduction is found.

As switched biasing has been proposed as a technique for reducing the flicker noise in MOSFET's with reduced power consumption to benefit HF circuits (Klumperink *et al.* 2000), it becomes essential for RF MOSFET models to give a reasonable prediction of flicker noise performance of the device under such conditions. In order to do that, the flicker noise model contained in the RF model must be continuous and accurate over a wide bias range from strong inversion to accumulation and from linear to saturation regimes. Further work is needed to validate the flicker noise models with measured noise data in devices under switch-biasing conditions and to develop more advanced noise models for RF applications.

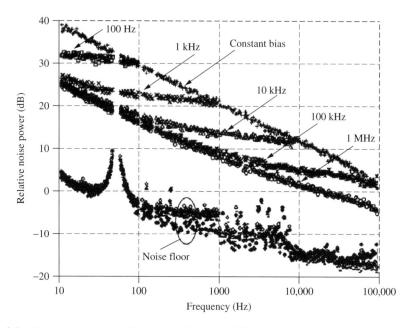

Figure 4.6 Noise reduction while switching at different frequencies for an nMOSFET, $V_{\mathrm{GS,on}} = 2.5\,\mathrm{V}$, $V_{\mathrm{GS,off}} = 0\,\mathrm{V}$, duty cycle $= 50\%$. Also shown is the noise floor under the same conditions. Reproduced from Klumperink E. A. M. *et al.* (2000) Reducing MOSFET $1/f$ noise and power consumption by switched biasing, *IEEE J. Solid-State Circuits*, **35**(7), 994–999

4.3 THERMAL NOISE MODELING

4.3.1 Existing Thermal Noise Models

At HF, although all the noise sources contribute to the total noise, the dominant contribution comes from the channel thermal noise. The channel thermal noise characteristics in MOSFETs operating in the strong inversion region have been studied for over two decades. The origin of this thermal noise has been found to be related to the random thermal motion of carriers in the channel of the device. Various models have been developed and some of them have been implemented in circuit simulators.

A simple thermal noise model has been implemented in circuit simulators since SPICE2 was developed,

$$S_{\mathrm{id}} = \frac{8k_{\mathrm{B}}TG_m}{3} \qquad (4.5)$$

where k_{B} is the Boltzman constant and T is the absolute temperature in Kelvin; G_m is the transconductance of the device.

Other similar models have also been proposed as given in the following:

$$S_{\mathrm{id}} = \frac{8k_{\mathrm{B}}TG_{\mathrm{DS}}}{3}, \qquad (4.6)$$

$$S_{\mathrm{id}} = \frac{8k_{\mathrm{B}}T(G_m + G_{\mathrm{DS}})}{3}, \qquad (4.7)$$

$$S_{id} = \frac{8k_B T (G_m + G_{DS} + G_{mb})}{3} \tag{4.8}$$

where G_{DS} and G_{mb} are the channel conductance and bulk transconductance.

Most compact models developed for circuit simulation have their own thermal noise models. For example, BSIM3v3 includes the following equation to calculate the thermal noise of the device as a user option besides the one given by Eq. (4.8):

$$S_{id} = \frac{4k_B T \mu_{eff}}{L_{eff}^2} Q_{inv} \tag{4.9}$$

where μ_{eff} is the effective carrier mobility, L_{eff} is the channel length of the device, and Q_{inv} is the total inversion charge in the channel.

It has been reported that Eq. (4.5) gives a nonphysical prediction of thermal noise at $V_{DS} = 0\,V$ (see, for example, Want et al. (1994)). Equations (4.7) and (4.8) are proposed to fix this problem even though their accuracy and physical basis need to be verified. Studies to validate the accuracy of the above noise models have been reported recently (see, for example, Chen et al. (2000)). Some discussion will be given later.

Another thermal noise model that is not implemented in all commercial circuit simulators but widely used for noise analysis by circuit designers is

$$S_{ind} = 4k_B T G_{nch},$$
$$= 4k_B T \gamma G_m \tag{4.10}$$

where G_{nch} is the channel thermal noise conductance and γ is a bias-dependent factor, which for long-channel devices is equal to unity in the linear region and to 2/3 in the saturation region.

The γ-factor has been used as a figure of merit to compare the thermal noise performance of different devices. It shows how much noise is generated by the device at the input for a given transconductance. It has been found that the γ-factor is not a constant for devices with different channel lengths and the γ-factor for short-channel device can be larger than that for long-channel device in the saturation regime owing to both velocity saturation and hot electrons. Some models have been proposed to account for the velocity saturation effect and hot carrier effects (for example, Klein (1998), Scholten et al. (1999), and Knoblinger (2000)), but they have not been implemented in any commercial circuit simulators yet. Recently, a simple thermal noise model has been proposed to account for both velocity saturation and hot carriers and can be easily implemented. This noise model was originally developed for a transistor biased in saturation and in strong inversion, but an extended expression has been proposed to cover the regions from weak to strong inversion by rewriting the noise parameter as (Enz and Cheng (2000))

$$\gamma \cong \gamma_L \left(1 + \frac{1}{G} \frac{v_{sat} \tau_r}{L_{eff}}\right) \tag{4.11}$$

where v_{sat} is the saturation velocity, τ_r is a relaxation time (of the order of ps) used as a fitting parameter, G is the normalized G_m/I_D ratio, and γ_L is the γ-factor for the long-channel device.

This simple model assumes that the carrier velocity is saturated and that the lateral field is equal to the critical field all along the channel from source to drain. Although these assumptions are questionable, the resulting model can fit the measured data over bias and geometry.

4.3.2 HF Noise Parameters

In noise model derivation and circuit simulation, the noise PSD is used as a measure for the noise output in the device. Circuit designers also prefer to use the noise PSD, the parameters related to the noisy two-port equivalent circuit. However, in measurements, the HF noise is usually characterized by several other parameters: the minimum noise factor (or minimum noise figure), the input referred noise resistance, and the optimum source admittance for which the minimum noise figure is obtained. Therefore, it is necessary to discuss these parameters to understand their physical meanings and their relationship in describing the HF noise characteristics of the device.

The noise factor is a figure of merit for the performance of a device or a circuit with respect to noise. The standard definition of the noise factor of a two-port network is the ratio of the available output noise power per unit bandwidth to the portion of that noise caused by the actual source connected to the input terminals of the device. It can be given by the following equivalent equation:

$$F = \frac{S_i/N_i}{S_o/N_o} \tag{4.12}$$

where S_i and S_o are input and output signals and N_i and N_o are input and output noise power.

The noise factor can be expressed in decibel form, which is termed as the noise figure, that is,

$$N_F = 10 \log F. \tag{4.13}$$

The noise figure of a two-port network is given by (see, for example, Pospieszalski (1986))

$$N_F = N_{F_{min}} + \frac{r_n}{g_s}|y_s - y_{opt}|^2 \tag{4.14}$$

where r_n is the equivalent normalized noise resistance of the two-port network, $y_s = g_s + jb_s$ is the normalized source admittance, and $y_{opt} = g_{opt} + jb_{opt}$ represents the normalized source admittance that results in the minimum (or optimum) noise figure $N_{F_{min}}$.

The y_s and y_{opt} can be expressed in terms of the reflection coefficients Γ_s and Γ_{opt}, the ratio of the incident to the reflected wave along a transmission line

$$y_s = \frac{1 - \Gamma_s}{1 + \Gamma_s} \tag{4.15}$$

and

$$y_{opt} = \frac{1 - \Gamma_{opt}}{1 + \Gamma_{opt}}. \tag{4.16}$$

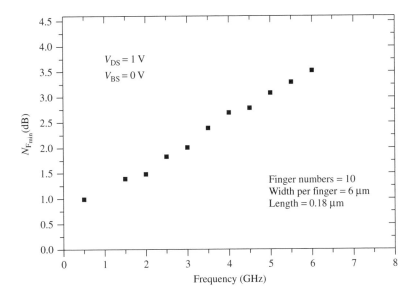

Figure 4.7   An example of measured $N_{F_{min}}$ versus frequency for a MOSFET

Thus, Eq. (4.14) becomes the following form

$$N_F = N_{F_{min}} + \frac{4\gamma_n \cdot |\Gamma_s - \Gamma_{opt}|^2}{(1 - |\Gamma_s|^2) \cdot |1 + \Gamma_{opt}|^2}. \tag{4.17}$$

In the HF noise measurements, the source reflection coefficient is varied until a minimum noise figure is reached. The value of $N_{F_{min}}$, which occurs when $\Gamma_s = \Gamma_{opt}$, is read from the noise figure meter, and the source reflection coefficient that produces $N_{F_{min}}$ is determined by a network analyzer. The noise resistance r_n is measured by reading the noise figure when $\Gamma_s = 0$.

$N_{F_{min}}$ is a function of the biases (operating current) and frequency. Each $N_{F_{min}}$ is associated with one value of Γ_{opt}. Figure 4.7 shows a typical measured characteristic of $N_{F_{min}}$ versus frequency for an RF MOSFET. Figure 4.8 gives a typical measured plot of $N_{F_{min}}$ versus bias current. According to the measured noise characteristics, MOSFET can provide a low noise figure that is attractive to the RF applications. Also, a careful selection of the bias conditions is important for the device to achieve a lowest noise performance as shown in Figure 4.8. Since an RF MOSFET with a very short channel length includes many different physical effects and contains nonnegligible parasitics, it is not very easy to optimize the noise performance of the devices in a circuit with hand calculation by using analytical equations. So it is desirable that an RF model with accurate noise prediction be developed for use in circuit simulation. Whether an RF model can accurately predict the characteristics of noise figure versus bias currents for devices with different sizes is another challenge for device model developers.

As mentioned above, circuit designers prefer to use the parameters related to a two-port network to describe the noise performance of a device and a circuit. Universal noise models have been developed for any two-port network. A noisy two-port network shown in Figure 4.9(a) can be represented by a noise-free two-port network with two noise

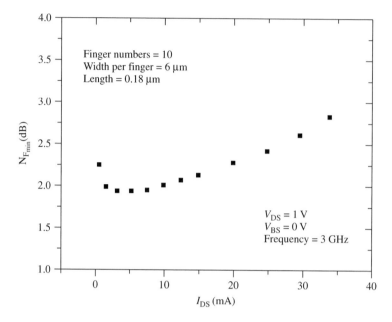

Figure 4.8   An example of measured $N_{F_{min}}$ versus I_{DS} for a MOSFET

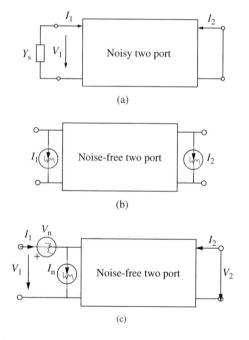

Figure 4.9   Noisy two-port and its ABCD-parameter representation

current sources, one at the input port (i_1) and the other at the output port (i_2) as shown in Figure 4.9(b). Figure 4.9(b) can also be transformed into a noise-free two-port network, shown in Figure 4.9(c), with a noise current source, $S_{in} = 4k_B TG_{in}$, and a noise voltage source, $S_{in} = 4k_B TR_{vn}$, at the input port, where the v_n and i_n are correlated to each other and the correlation relationship is described by a correlation admittance $Y_c = G_c + JB_c$. The noise source i_n can be further separated into a noise source i_{nu} that is uncorrelated to v_n and a noise source i_{nc} that is fully correlated to v_n,

$$i_n = i_{nu} + i_{nc} \tag{4.18}$$

and

$$i_{nc} = Y_c v_n. \tag{4.19}$$

The above relationship can be expressed in terms of noise PSD as follows:

$$S_{in} = S_{inu} + S_{inc}, \tag{4.20}$$

$$S_{inc} = |Y_c|^2 S_{vn}. \tag{4.21}$$

According to the two-port network given in Figure 4.9, we further have the following relationships:

$$v_n = -\frac{1}{Y_{21}} i_2, \tag{4.22}$$

$$i_n = i_1 + Y_{11} v_n, \tag{4.23}$$

$$Y_c = Y_{11} - Y_{21} \frac{S_{inc}}{S_{i_2}} = G_c + j B_c, \tag{4.24}$$

$$S_{va} = \frac{S_{i_2}}{|Y_{21}|^2} = 4k_B TR_{vn}, \tag{4.25}$$

$$S_{in} = 4k_B TG_{in}. \tag{4.26}$$

On the basis of the above relationships, the four noise parameters discussed earlier can be calculated,

$$R_n = R_{va}, \tag{4.27}$$

$$G_{opt} = \sqrt{\frac{G_{in}}{R_n} - B_c^2}, \tag{4.28}$$

$$B_{opt} = -B_c, \tag{4.29}$$

$$N_{F_{min}} = 1 + 2R_n(G_c + G_{opt}). \tag{4.30}$$

Similarly, noise parameters related to the two-port network can be calculated once we have the four noise parameters, R_n, G_{opt}, B_{opt}, and $N_{F_{min}}$,

$$R_{va} = R_n, \tag{4.31}$$

$$B_c = -B_{opt}, \tag{4.32}$$

$$G_{in} = (G_{opt}{}^2 + B_c{}^2)R_n, \tag{4.33}$$

$$G_c = \frac{N_{F_{min}} - 1}{2R_n} - G_{opt}. \tag{4.34}$$

To this point, we have established a conversion relationship between the four noise parameters obtained from the measurements and the noise parameters related to the two-port network for circuit analysis. Detailed analysis can be further performed for the noise performance of the device and a circuit based on the above derivations.

According to the noisy two-port network theory, a useful equation for the PSD of i_2 can be obtained as follows (Gonzalez (1997)):

$$S_{i_2} = S_{vn}|Y_{21}|^2 = 4k_B T R_n |Y_{21}|^2 \tag{4.35}$$

where k_B is the Boltzmann's constant, T is the absolute temperature, Y_{21} is the transadmittance from port 1 to port 2 of the noise-free two port, and R_n is the equivalent noise resistance, which is a resistance cascaded at the input port that will produce the same amount of noise PSD as i_2 does at the output port.

4.3.3 Analytical Calculation of the Noise Parameters

Figure 4.1 illustrates all noise sources in a MOSFET. However, it is too complex to be used to calculate the contribution of each noise source analytically. A simplified equivalent circuit (EC) shown in Figure 4.10 can be obtained by neglecting some components in Figure 4.1.

In Figure 4.10, the capacitances C_{BS} and C_{BD} have been neglected and the influence of the different substrate resistance components have been taken care of by R_{sub}. On the basis of the equivalent circuit, the following noise parameters are obtained:

$$R_{vn} = \frac{\lambda_{sat}}{G_m}\vartheta, \tag{4.36}$$

$$G_{in} = \lambda_{sat} G_m \theta^2 \psi, \tag{4.37}$$

$$G_c = \frac{R_G(G_m\theta)^2\psi}{\vartheta}, \tag{4.38}$$

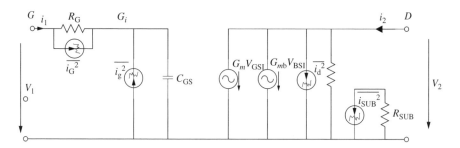

Figure 4.10   Simplified small-signal schematic for noise calculation

$$B_c = \frac{G_m \theta \chi}{\vartheta}, \tag{4.39}$$

$$\vartheta = 1 + \alpha_g + \alpha_{sub} + (G_m R_G \theta)^2 \psi, \tag{4.40}$$

$$\theta = 2\pi f \frac{C_{GS}}{G_m}, \tag{4.41}$$

where λ_{sat} is a parameter proportional to γ-factor discussed earlier, parameters ψ and χ account for the induced gate noise and its correlation to the drain noise, α_g is the ratio of the noise PSD of the gate resistance to the input referred channel noise, and α_{sub} is the ratio of the output referred substrate resistance noise PSD to the output referred channel noise,

$$\alpha_g = \frac{G_m R_G}{\lambda_{sat}}, \tag{4.42}$$

$$\alpha_{sub} = \frac{G_{mb}^2 R_{sub}}{G_m \lambda_{sat}}. \tag{4.43}$$

Parameters ψ and χ are given in the following:

$$\psi = 1 + \alpha_{sub} + \frac{\sigma_{sat}}{\lambda_{sat}} + 2c_g \sqrt{\frac{\sigma_{sat}}{\lambda_{sat}}}, \tag{4.44}$$

$$\chi = 1 + \alpha_{sub} - c_g \sqrt{\frac{\sigma_{sat}}{\lambda_{sat}}}, \tag{4.45}$$

where σ_{sat} is a bias- and geometry-dependent factor, c_g is a fitting parameter that is 0.395 for a long-channel device.

Both ψ and χ reduce to $1 + \alpha_{sub}$ when the induced gate noise is ignored ($\sigma_{sat} = 0$). The induced gate noise contributes mainly to G_{in} through the factor ψ, the gate resistance R_G contributes to R_{vn}, and the channel noise and substrate noise contribute to both G_{in} and R_{vn}. Substrate noise may typically contribute to 20% of R_{vn}, whereas R_G typically contributes to about 5% of R_{vn}. It is therefore important to account for the substrate resistance when doing noise calculation and noise optimization.

The noise parameters of an n-channel device have been measured and carefully de-embedded using the methodology presented in some literatures (see, for example, Aufinger (1996)). They are presented in Figure 4.11 and compared to the results obtained from simulation using the complete subcircuit of Figure 4.1 with the additional induced gate noise source added to the subcircuit (but not accounting for the correlation between induced gate noise and drain thermal noise). The results obtained from Eqs. (4.36) to (4.45) including the correlation between induced gate noise and channel drain noise are also shown in the figures. The meaning of the symbols (s1, s2, s3, s4) shown in Figure 4.11 are defined as follows, as discussed by Enz and Cheng (2000). s1 enables the gate resistance noise, s2 enables the substrate resistance noise, s3 enables the induced gate noise without correlation ($c_g = 0$), and s4 enables the induced gate noise correlation ($c_g = 0.395$), respectively, in the simulations. Figure 4.11 shows that the gate and substrate resistances strongly affect the minimum noise figure $N_{F_{min}}$, the optimum noise conductance G_{opt}, and the input referred noise resistance R_{vn}. The induced gate noise slightly affects $N_{F_{min}}$ and

G_{opt}, but has no effect on R_{vn}. From these results, it can be concluded that induced gate noise is not the only contributor to the minimum noise figure and that the gate resistance and, more importantly, the substrate resistance also contribute significantly.

Note that the analytical expressions for the noise parameters give reasonable results below $f_T/5$ and the discrepancies appearing at HF between the analytical and the measured results mainly come from a wrong frequency behavior due to the very simple equivalent circuit used for the derivations of Eqs. (4.36) to (4.45).

4.3.4 Simulation and Discussions

With the extracted parameters from the measured data for a 0.25-μm RF CMOS technology, the noise characteristics of the subcircuit model discussed above are verified. The four noise parameters calculated with the simulated noise characteristics are given in Figure 4.12 against the measured data for a 0.36-μm device at different bias conditions. While the RF model with extracted parameters fits accurately the measured s-parameters data, it can also predict well the HF noise characteristics of the device as shown in Figures 4.12 to 4.15 by the curves at $V_{GS} = 1$ V and $V_{DS} = 1$ V. The discrepancy in the R_n characteristics between the model and the measured data at $V_{GS} = 2$ V needs further investigation. However, obvious disagreement in the simulated and measured imaginary part of Y_{12} has been found at that bias condition, so the discrepancy in

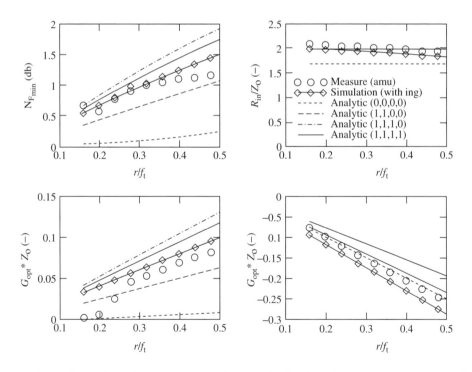

Figure 4.11 Comparison between measured and simulated noise parameters ($N_{F_{min}}$, R_n, G_{opt}, B_{opt}). Reproduced from Enz C. and Cheng Y. (2000) MOS transistor modeling for RF IC design, *IEEE Journal of Solid-State Circuits*, **35**(2), 186–201

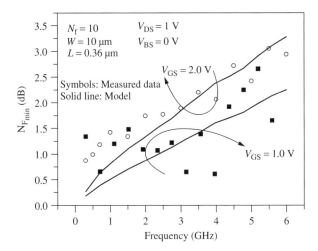

Figure 4.12 Comparisons of measured data for minimum noise figure, $N_{F_{min}}$, with simulations at different bias conditions. The N_f is the finger number of each device. The channel width per finger is $10\,\mu$m and the channel length is $0.36\,\mu$m. Reproduced from Cheng Y. *et al.* (2002) High frequency small signal AC and noise modeling of MOSFETs for RF IC design, *IEEE Trans. On Electron Devices*, **49**(3), 400–408

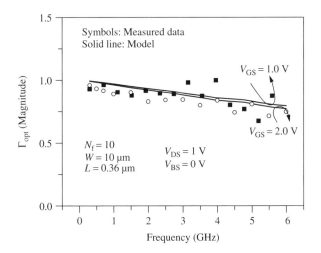

Figure 4.13 Comparisons of measured data for the magnitude of the optimized source reflection coefficient Γ_{opt}, with simulations at different bias conditions. Reproduced from Cheng Y. *et al.* (2002) High frequency small signal AC and noise modeling of MOSFETs for RF IC design, *IEEE Trans. On Electron Devices*, **49**(3), 400–408

the R_n characteristics may be caused by the inaccuracy of the capacitance model in that operation regime.

The noise characteristics of several noise models including the subcircuit RF model above are also verified with the extracted channel thermal noise with the methodology discussed by Chen (2001). Figure 4.16 shows the curves of the channel thermal noise

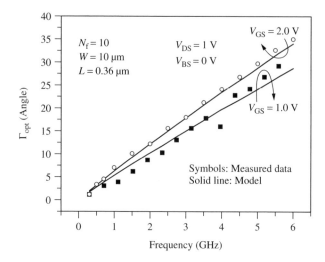

Figure 4.14 Comparisons of measured data for the phase of the optimized source reflection coefficient, Γ_{opt}, with simulations at different bias conditions. Reproduced from Cheng Y. *et al.* (2002) High frequency small signal AC and noise modeling of MOSFETs for RF IC design, *IEEE Trans. On Electron Devices*, **49**(3), 400–408

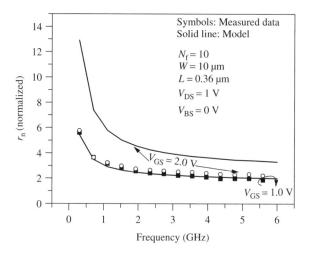

Figure 4.15 Comparisons of measured data for the noise resistance normalized to 50 Ω, r_n, with simulations at different bias conditions. Reproduced from Cheng Y. *et al.* (2002) High frequency small signal AC and noise modeling of MOSFETs for RF IC design, *IEEE Trans. On Electron Devices*, **49**(3), 400–408

versus bias current, from the measured data, and simulations of BSIM3v3 noise model (Noimod = 4) and several other noise models given by Eqs. (4.6) to (4.8). It shows that the calculated channel thermal noise based on Eqs. (4.6) to (4.8) cannot predict the channel thermal noise extracted from measured data. The subcircuit RF model with the BSIM3v3 noise model (Noimod = 4) has much better accuracy at several different bias conditions.

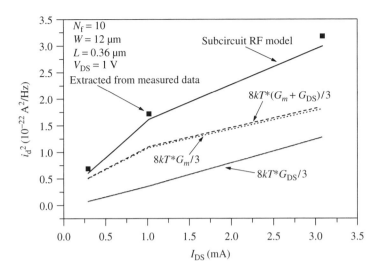

Figure 4.16   Power spectral densities of channel thermal noise versus bias current of a 0.36-μm *n*-channel MOSFET. They are extracted from the measured data and calculated from different channel thermal noise models. Reproduced from Cheng Y. *et al.* (2002) High frequency small signal AC and noise modeling of MOSFETs for RF IC design, *IEEE Trans. On Electron Devices*, **49**(3), 400–408

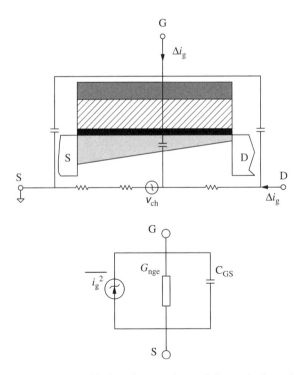

Figure 4.17   Illustration of induced gate noise and the equivalent circuit model

4.3.5 Induced Gate Noise Issue

The concept of the induced gate noise has been used for three decades (see, for example, Halladay and Ziel (1969)). But many researchers are still studying how to model it correctly (Triantis (1997)). At high frequencies, it is believed that the local channel voltage fluctuations due to thermal noise couple to the gate through the oxide capacitance and cause an induced gate noise current to flow. This noise current can be modeled by a noisy current source connected in parallel to the intrinsic gate-to-source capacitance C_{GSi} as shown in Figure 4.17. Since the physical origin of the induced gate noise is the same as that for the channel thermal noise at the drain, the two noise sources are partially correlated with a correlation factor c_g.

The power spectral density of the induced gate noise PSD is given by

$$S_{ing} = 4k_B T G_{ng} \tag{4.46}$$

and

$$G_{ng} = \frac{\sigma_{sat} (2\pi f C_{GS})^2}{G_m}. \tag{4.47}$$

Device noise simulations performed for a finger length have shown that the correlation factor remains mainly imaginary (real part about 10 times smaller than the imaginary part) and that its value is slightly smaller than the long-channel value 0.395 (it typically ranges from 0.35 to 0.3 for short-channel devices).

Currently, the induced gate noise and its correlation to the thermal noise at the drain are not yet implemented completely in compact models. One reason is the difficulty of modeling the induced gate noise and implementing it in circuit simulators. Another reason is that it is not very critical at frequencies much smaller than the f_T of the device, since at that frequency range two more important contributors to the total noise are the substrate and the gate resistances, instead of the induced gate noise, besides the channel thermal noise. A methodology to extract the induced gate noise has also been developed. However, further detailed investigations are needed to understand the induced gate noise issue and model it correctly. The subcircuit RF model discussed in this paper does not include the contribution of the induced gate noise.

REFERENCES

Aufinger K. (1996) A straightforward noise de-embedding method and its application to high-speed silicon bipolar transistors, *Proceedings of ESSDERC'96*, pp. 957–960.

Chen C. H. *et al.* (2000) Extraction of the channel thermal noise in MOSFETs, *Proceedings of the 2000 International Conference on Microelectronic Test Structures*, pp. 42–47.

Chen C. H. *et al.* (2001) Extraction of the induced gate noise, channel noise and their correlation in sub-micron MOSFETs from RF noise measurements, *IEEE Trans. Electron Devices*, **48**(12), 2884–2892.

Cheng Y. *et al.* (1997) BSIM3v3.1 User's Manual, Memorandum No. UCB/ERL M97/2.

Cheng Y. and Hu C. (1999) MOSFET Modeling & BSIM3 User's Guide, Kluwer Academic publishers, Norwell, MA.

Cheng Y. *et al.* (2002) High frequency small signal AC and noise modeling of MOSFETs for RF IC design, *IEEE Trans. On Electron Devices*, **49**(3), 400–408.

Enz C. and Cheng Y. (2000) MOS transistor modeling for RF IC design, *IEEE Journal of Solid-State Circuits*, **35**(2), 186–201.

Gonzalez G. (1997) *Microwave Transistor Amplifiers – Analysis and Design*, Second Edition, Prentice Hall, New York.

Halladay H. E. and Ziel A. Van der (1969), On the high frequency excess noise and equivalent circuit representation of the MOSFET with n-type channel, *Solid-State Electron.*, **12**, 161–176.

Hung K. K. *et al.* (1990a) A unified model for the flicker noise in metal-oxide semiconductor field-effect transistors, *IEEE Trans. Electron Devices*, **37**, 654–665.

Hung K. K. *et al.* (1990b) A physical-based MOSFET noise model for circuit simulation, *IEEE Trans. Electron Devices*, **37**, 1323–1333.

Jindal R. P. and Ziel A. Van der (1978) Phonon fluctuation model for flicker noise in elemental semiconductor, *J. Appl. Phys.* **52**, 2884.

Klein P. (1998) An analytical thermal noise model of deep-submicron MOSFETs for circuit simulation with emphasis on the BSIM3v3 SPICE model, *Proc. of the European Solid-State Dev. Res. Conf.*, pp. 460–463.

Klumperink E. A. M. *et al.* (2000) Reducing MOSFET 1/f noise and power consumption by switched biasing, *IEEE J. Solid-State Circuits*, **35**(7), 994–999.

Knoblinger G. (2000) A new model for thermal channel noise of deep submicron MOSFETs and its application in RF-CMOS design, *Digest of Technical Papers Symposium on VLSI Circuits*, pp. 150–153.

Momose H. S. *et al.* (1998) A study of flicker noise in n- and p-MOSFETs with ultra-thin gate oxide in the direct-tunneling regime, *Technical Digest of Internal Electron Device Meeting*, pp. 923–926.

Pospieszalski M. (1986) On the measurement of noise parameters of microwave two port, *IEEE Trans. Microwave Theory Tech.*, **MTT-34**(4), 456–458.

Scholten A. *et al.* (1999) Accurate thermal noise model for deep-submicron CMOS, *Tech. Dig. Int. Electron Device Meeting*, 155–158.

Triantis D. P. (1997) Induced gate noise in MOSFETs revisited: the submicron case, *Solid-State Electron.*, **41**(12), 1937–1942.

Vandamme L. K. J. (1994) 1/f noise in MOS devices, mobility or number fluctuations? *IEEE Trans. Electron Devices*, **41**, 1936–1945.

Vandamme E. P. *et al.* (2000) Critical discussion on unified 1/f noise models for MOSFETs, *IEEE Trans. Electron Devices*, **47**(11), 2146–2153.

Vladimirescu A. (1994) *The SPICE Book*, John Wiley & Sons, New York.

Wang, B. (1994) MOSFET thermal noise modeling for analog integrated circuits, *IEEE J. Solid-State Circuits*, **29**, 833–835.

Wel A. Pvan der *et al.* (2000) MOSFET 1/f noise measurement under switched bias conditions, *IEEE Electron Device Lett.*, **21**(1), 43–49.

5

Proper Modeling for Accurate Distortion Analysis

5.1 INTRODUCTION

To minimize the influence of noise in analog and RF integrated circuits, signal amplitudes are made large. This will cause high distortion effects caused by the inherent nonlinearity of the devices used to implement the circuit functions. The literature on distortion analysis is mainly focusing on operational amplifiers (OPAMPS), RF amplifiers, transconductors (voltage-to-current converters), and more recently track-and-hold amplifiers (see, for example, Shoucair and Patterson (1993); Mensink *et al.* (1996); Bruun (1998); Wambacq *et al.* (1999); Chilakapati *et al.* (2002); Hernes (2003); Limotyrakis *et al.* (2002)).

All devices available to designers of analog CMOS circuits are more or less nonlinear and hence can cause circuit functions to degrade owing to distortion effects. Especially MOS transistors, which are the most accurate and effective devices for implementing analog functions in CMOS technology, are prone to exhibit nonlinear characteristics. With the continuous downscaling of feature sizes, the transistor characteristics are more and more augmented by second-order effects such as mobility degradation due to high vertical fields, velocity saturation in the channel, and other nonideal effects that were discussed in Section 1.5.2. All these effects will introduce additional nonlinearities (see, for example, van Langevelde and Klaassen (1997a, 1997b); Pu and Tsividis (1990)).

The characterization of MOS transistors that is provided by the foundries is commonly based on fitting the modeled drain current characteristics and the small-signal capacitors to measured data. This is so-called digital characterization. In some cases, the foundries also include fitting of small-signal parameters such as transconductance g_m and channel conductance g_{ds} to satisfy the needs of analog designers. The main objective of this chapter is to illustrate that even accurately reproducing g_m and g_{ds} is not sufficient for proper estimation of distortion and intermodulation effects. For example, the third-order harmonic distortion component depends on the third-order derivative of the device characteristics.

This chapter is organized as follows. We start by defining basic terminology commonly used in analysis of distortion effects. Then in Section 5.3, we discuss the nonlinearity of the MOS devices using the terminology just described and point out important requirements of the models to be able to predict the effects of the nonlinearities. Finally,

Device Modeling for Analog and RF CMOS Circuit Design. T. Ytterdal, Y. Cheng and T. A. Fjeldly
© 2003 John Wiley & Sons, Ltd ISBN: 0-471-49869-6

in Section 5.4, we briefly discuss important methods commonly employed to calculate distortion in analog CMOS circuits.

5.2 BASIC TERMINOLOGY

For analog circuits, the linearity characteristics are often described in terms of parameters that are measured in the frequency domain for circuit excitations being one or more sinusoidal sources. In single sinusoidal excitation the input signal is of the form $A \cos(\omega t)$, where A and ω are the amplitude and the frequency of the input signal, respectively. This situation is referred to as *single-frequency excitation*.

When the amplitude of the input signal A is small enough, the output spectrum of the circuit under test contains only one frequency component above the noise floor. This component is located at the same frequency as the input signal, that is, at ω and is called the *fundamental frequency*. The amplitude of the fundamental frequency signal changes proportionally with the input amplitude. As the input signal amplitude is increased, the output spectrum also contains components at the frequencies 2ω and 3ω. These signals, called the *second and third harmonics*, originate from second- and third-order nonlinear circuit behavior, respectively, as we will see later. Harmonics higher than the third come above the noise floor at even higher values of the input amplitude. We will also learn later that the amplitude of the nth harmonic increases as the nth power of the input amplitude.

Usually, the linear response is the wanted response and it is therefore common to denote the harmonics as *distortion, nonlinear distortion*, or *harmonic distortion*.

To define and illustrate the different quantities used in distortion analysis, we consider a nonlinear system with an input signal $x(t)$ and an output signal $y(t)$. The output signal can be written as a Taylor expansion of the input signal as follows.

$$y(t) = a1 x(t) + a2 x^2(t) + a3 x^3(t) + \cdots. \tag{5.1}$$

Here, the coefficient $a1$ describes the behavior of the linearized circuit. The higher-order coefficients $a2, a3, \ldots$ are called the higher-order nonlinearity coefficients.

The *total harmonic distortion* (*THD*) of a signal is defined to be the ratio of the total power of the second- and the higher-order harmonics to the power of the fundamental component of that signal. There exist several mathematical expressions for this definition. One commonly used approach is to express *THD* in decibels:

$$THD = 10 \log \left[\sum_{n=2}^{\infty} \left(\frac{A^{n-1} an}{a1} \right)^2 \right] dB. \tag{5.2}$$

Usually, the second- and/or the third-order distortion components contribute the most to *THD* and we will denote these as HD_2 and HD_3, respectively. They are defined as the ratio of the of the second and third harmonic, respectively, to the amplitude of the fundamental response:

$$HD_2 = \frac{1}{2} A \frac{|a2|}{|a1|}, \tag{5.3}$$

$$HD_3 = \frac{1}{4}A^2 \frac{|a3|}{|a1|}. \tag{5.4}$$

These distortion components are also often referred to as harmonic amplitudes.

If the output signal $y(t)$ is a known analytic function $y(t) = f(x(t))$, then the coefficients $a1, a2, \ldots$ can be written in terms of the Taylor series coefficients. The first three coefficients are then

$$a1 = \frac{df}{dx}, \tag{5.5}$$

$$a2 = \frac{1}{2}\frac{d^2 f}{dx^2}, \tag{5.6}$$

$$a3 = \frac{1}{6}\frac{d^3 f}{dx^3}. \tag{5.7}$$

In low-pass and band-pass systems, the higher-order harmonics of single-frequency excitation often fall outside the system bandwidth and hence, are not possible to measure. To be able to characterize such systems, one usually applies an input signal containing two sinusoids at frequencies ω_1 and ω_2 that are close to each other. The nice feature of the output spectrum of such a test applied to a nonlinear system is that it contains components at the sum and differences of the two frequencies. Hence, it is easy to define the two frequencies such that at least one of the components falls inside the system bandwidth. The signals at $|\omega_1 \pm \omega_2|$ are caused by second-order nonlinear behavior and are called *second-order intermodulation products*. The output spectrum contains even higher-order intermodulation products at frequencies $|2\omega_1 \pm \omega_2|$, $|\omega_1 \pm 2\omega_2|$, and so on. When characterization of intermodulation is performed, *second-order intermodulation distortion* (IM_2) and *third-order intermodulation distortion* (IM_3) are usually measured. IM_2 and IM_3 are defined as the second- and third-order intermodulation products to the fundamental response, respectively.

Semiconductor devices are usually modeled in terms of equivalent circuits containing different basic circuit elements such as resistors, capacitors, and controlled current and voltage sources. The nonlinearity of these devices can be categorized as follows (Wambacq and Sansen 1998) and is referred to as *basic nonlinearities*:

- *Nonlinear conductance*   The current through the element is a nonlinear function of the voltage across the element.
- *Nonlinear transconductance*   The current through the element is a nonlinear function of a voltage somewhere else in the equivalent circuit.
- *Nonlinear resistance*   The voltage across the element is a nonlinear function of the current through it.
- *Nonlinear transresistance*   The voltage across the element is a nonlinear function of a current somewhere else in the equivalent circuit.
- *Multidimensional nonlinear conductance and transconductance*   The current through the element is a nonlinear function of more than one voltage.
- *Multidimensional nonlinear resistance and transresistance*   The voltage across the element is a nonlinear function of more than one current.

- *Nonlinear capacitance*   The charge stored on this element is a nonlinear function of the voltage across the element.

- *Nonlinear transcapacitance*   The charge stored on this element is a nonlinear function of a voltage somewhere else in the equivalent circuit.

As examples of the basic nonlinear elements listed above, we discuss the nonlinear conductance and the multidimensional transconductance below.

The general Taylor series description of the AC output current of a nonlinear conductance expanded around the DC operating point is given by

$$i_{\text{out}}(t) = f(V + v(t)) - f(V) = \sum_{k=1}^{\infty} \frac{1}{k!} \frac{d^k f(x(t))}{dx^k}\bigg|_{x=V} \cdot v^k(t). \tag{5.8}$$

Here, V is the DC operating point and v is the AC part of the voltage across the conductance. Usually it is sufficient to keep only the first three terms of the sum in (5.8). The term corresponding to $k = 1$ is simply the linear conductance and is given by

$$g = \frac{df(x)}{dx}\bigg|_{x=V} = a1_g. \tag{5.9}$$

The second- and third-order terms are labeled $a2_g$ and $a3_g$, respectively, and can be written as

$$a2_g = \frac{1}{2!} \frac{d^2 f(x)}{dx^2}\bigg|_{x=V}, \tag{5.10}$$

$$a3_g = \frac{1}{3!} \frac{d^3 f(x)}{dx^3}\bigg|_{x=V}. \tag{5.11}$$

Now if we are interested in the third-order harmonic amplitude HD_{3g} of the output current, it can be calculated from the definition given in (5.4) as follows:

$$HD_{3g} = \frac{1}{4} v^2 \frac{|a3_g|}{|a1_g|}. \tag{5.12}$$

Hence, to accurately model HD_3 of the nonlinear conductance, the third-order derivative of the conductance with respect to the voltage across it has to be accurately described as mentioned in the introduction section of this chapter.

Now let us turn to the multidimensional transconductance. To illustrate the added complexity of this element, we choose a transconductance with two controlling voltages. The AC output current of this element can be written in terms of a Taylor expansion at the DC operating point U, V as follows:

$$i_{\text{out}}(t) = f(U + u(t), V + v(t)) - f(U, V) = \sum_{k=1}^{\infty} \sum_{i=0}^{k} \left[\frac{\partial^k f(x, y)}{\partial x^i \partial y^{k-i}} \frac{u^i}{i!} \frac{v^{k-i}}{(k-i)!} \right]. \tag{5.13}$$

If we keep only terms up to and including third order, (5.13) can be simplified to

$$
\begin{aligned}
i_{\text{out}}(t) &\approx \frac{\partial f}{\partial x} u + \frac{1}{2} \frac{\partial^2 f}{\partial x^2} u^2 + \frac{1}{6} \frac{\partial^3 f}{\partial x^3} u^3 + \frac{\partial f}{\partial y} v + \frac{1}{2} \frac{\partial^2 f}{\partial y^2} v^2 + \frac{1}{6} \frac{\partial^3 f}{\partial y^3} v^3 \\
&\quad + \frac{\partial f}{\partial x} \frac{\partial f}{\partial y} uv + \frac{1}{2} \frac{\partial^2 f}{\partial x^2} \frac{\partial f}{\partial y} u^2 v + \frac{1}{2} \frac{\partial f}{\partial x} \frac{\partial^2 f}{\partial y^2} uv^2 \\
&= a1_u u + a2_u u^2 + a3_u u^3 + a1_v v + a2_v v^2 + a3_v v^3 \\
&\quad + a2_{uv} uv + a3_{2uv} u^2 v + a3_{u2v} uv^2.
\end{aligned}
\tag{5.14}
$$

Here we have added a subscript to the nonlinearity coefficients to indicate for which voltage the derivative has been taken.

5.3 NONLINEARITIES IN CMOS DEVICES AND THEIR MODELING

To be able to analyze the nonlinear behavior of integrated circuits, the nonlinearity of the devices involved must be known. All devices used in integrated CMOS circuits are more or less nonlinear. Here we will only discuss MOS transistors to illustrate the modeling requirements for accurate analysis of distortion effects since the same approach can be applied to other devices as well.

Equivalent circuits for MOS transistors include the following types of nonlinear elements: nonlinear conductance, multidimensional nonlinear transconductance, and nonlinear transcapacitance.

Let us use the well-known square law model[1] for the drain current of MOS transistors and derive expressions for the AC part of the drain current. If we assume that the device is biased in the saturation region above threshold, the drain current is given by

$$
I_{\text{d}} = K(V_{\text{gs}} - V_{\text{T}})^2 (1 + \lambda V_{\text{ds}}),
\tag{5.15}
$$

$$
V_{\text{T}} = V_{\text{T0}} + \gamma \left(\sqrt{\phi - V_{\text{bs}}} - \sqrt{\phi} \right),
\tag{5.16}
$$

$$
K = \frac{1}{2} \mu C_{\text{ox}} \frac{W}{L}.
\tag{5.17}
$$

Here, V_{T} is the threshold voltage, λ is the channel-length modulation parameter, V_{T0} is the threshold voltage at zero bulk-source voltage, γ is the body effect parameter, ϕ is the bulk Fermi potential, μ is the field effect mobility, C_{ox} is the oxide capacitance, W is the gate width, and L is the gate length. From the model equations above we note that we have three voltages that constitutes the independent variables of our nonlinear model, V_{gs}, V_{ds}, and V_{bs}. Following the approach described in the previous section, we

---

[1] The square law model is the level 1 MOS transistor model in SPICE, which was first published in Shichman and Hodges (1968).

can expand (5.15) into a Taylor series that gives the following expression for the AC part of the drain current:

$$
\begin{aligned}
i_d = {}& g_m v_{gs} + a2_{g_m} v_{gs}^2 + a3_{g_m} v_{gs}^3 + \cdots \\
& + g_{ds} v_{ds} + a2_{g_{ds}} v_{ds}^2 + a3_{g_{ds}} v_{ds}^3 + \cdots \\
& + g_{mb} v_{bs} + a2_{g_{mb}} v_{bs}^2 + a3_{g_{mb}} v_{bs}^3 + \cdots \\
& + a2_{g_m g_{ds}} v_{gs} v_{ds} + a3_{2g_m g_{ds}} v_{gs}^2 v_{ds} + a3_{g_m 2g_{ds}} v_{gs} v_{ds}^2 + \cdots \\
& + a2_{g_m g_{mb}} v_{gs} v_{bs} + a3_{2g_m g_{mb}} v_{gs}^2 v_{bs} + a3_{g_m 2g_{mb}} v_{gs} v_{bs}^2 + \cdots \\
& + a2_{g_{ds} g_{mb}} v_{ds} v_{bs} + a3_{2g_{ds} g_{mb}} v_{ds}^2 v_{bs} + a3_{g_{ds} 2g_{mb}} v_{ds} v_{bs}^2 + \cdots \\
& + a3_{g_m g_{ds} g_{mb}} v_{gs} v_{ds} v_{bs} + \cdots .
\end{aligned}
\tag{5.18}
$$

In (5.18) we have used conductance symbols instead of voltages as subscripts to indicate for which voltage the derivative has been taken. The expressions for the nonlinearity coefficients up to and including third order are given by

$$
g_m = \frac{\partial I_d}{\partial V_{gs}} = 2K(V_{gs} - V_T)(1 + \lambda V_{ds}),
\tag{5.19}
$$

$$
a2_{g_m} = \frac{1}{2} \frac{\partial^2 I_d}{\partial V_{gs}^2} = K(1 + \lambda V_{ds}),
\tag{5.20}
$$

$$
a3_{g_m} = \frac{1}{6} \frac{\partial^3 I_d}{\partial V_{gs}^3} = 0,
\tag{5.21}
$$

$$
g_{ds} = \frac{\partial I_d}{\partial V_{ds}} = K(V_{gs} - V_T)^2 \lambda,
\tag{5.22}
$$

$$
a2_{g_{ds}} = \frac{1}{2} \frac{\partial^2 I_d}{\partial V_{ds}^2} = 0,
\tag{5.23}
$$

$$
a3_{g_{ds}} = \frac{1}{6} \frac{\partial^3 I_d}{\partial V_{ds}^3} = 0,
\tag{5.24}
$$

$$
g_{mb} = \frac{\partial I_d}{\partial V_{bs}} = \frac{\partial I_d}{\partial V_T} \frac{\partial V_T}{\partial V_{bs}} = \frac{K(V_{gs} - V_T)(1 + \lambda V_{ds})\gamma}{\sqrt{\phi - V_{bs}}},
\tag{5.25}
$$

$$
a2_{g_{mb}} = \frac{1}{2} \frac{\partial^2 I_d}{\partial V_{bs}^2} = \frac{\partial g_{mb}}{\partial V_{bs}} = \frac{K\gamma(V_{gs} - V_{T0} + \gamma\sqrt{\phi})(1 + \lambda V_{ds})}{4(\phi - V_{bs})^{3/2}},
\tag{5.26}
$$

$$
a3_{g_{mb}} = \frac{1}{6} \frac{\partial^3 I_d}{\partial V_{bs}^3} = \frac{3K\gamma(V_{gs} - V_{T0} + \gamma\sqrt{\phi})(1 + \lambda V_{ds})}{8(\phi - V_{bs})^{5/2}},
\tag{5.27}
$$

$$
a2_{g_m g_{ds}} = \frac{\partial I_d}{\partial V_{gs}} \frac{\partial I_d}{\partial V_{ds}} = 2K(V_{gs} - V_T)\lambda,
\tag{5.28}
$$

$$
a3_{2g_m g_{ds}} = \frac{1}{2} \frac{\partial^2 I_d}{\partial V_{gs}^2} \frac{\partial I_d}{\partial V_{ds}} = K\lambda,
\tag{5.29}
$$

$$a3_{g_m 2 g_{ds}} = \frac{1}{2} \frac{\partial I_d}{\partial V_{gs}} \frac{\partial^2 I_d}{\partial V_{ds}^2} = 0, \tag{5.30}$$

$$a2_{g_m g_{mb}} = \frac{\partial I_d}{\partial V_{gs}} \frac{\partial I_d}{\partial V_{bs}} = \frac{K\gamma(1 + \lambda V_{ds})}{\sqrt{\phi - V_{bs}}}, \tag{5.31}$$

$$a3_{2 g_m g_{mb}} = \frac{1}{2} \frac{\partial^2 I_d}{\partial V_{gs}^2} \frac{\partial I_d}{\partial V_{bs}} = 0, \tag{5.32}$$

$$a3_{g_m 2 g_{mb}} = \frac{1}{2} \frac{\partial I_d}{\partial V_{gs}} \frac{\partial^2 I_d}{\partial V_{bs}^2} = \frac{K\gamma(1 + \lambda V_{ds})}{(\phi - V_{bs})^{3/2}}, \tag{5.33}$$

$$a2_{g_{ds} g_{mb}} = \frac{\partial I_d}{\partial V_{ds}} \frac{\partial I_d}{\partial V_{bs}} = \frac{K\lambda\gamma(V_{gs} - V_T)}{\sqrt{\phi - V_{bs}}}, \tag{5.34}$$

$$a3_{2 g_{ds} g_{mb}} = \frac{1}{2} \frac{\partial^2 I_d}{\partial V_{ds}^2} \frac{\partial I_d}{\partial V_{bs}} = 0, \tag{5.35}$$

$$a3_{g_{ds} 2 g_{mb}} = \frac{1}{2} \frac{\partial I_d}{\partial V_{ds}} \frac{\partial^2 I_d}{\partial V_{bs}^2} = \frac{K\gamma\lambda}{(\phi - V_{bs})^{3/2}}, \tag{5.36}$$

$$a3_{g_m g_{ds} g_{mb}} = \frac{\partial I_d}{\partial V_{gs}} \frac{\partial I_d}{\partial V_{ds}} \frac{\partial I_d}{\partial V_{bs}} = \frac{K\gamma\lambda}{\sqrt{\phi - V_{bs}}}. \tag{5.37}$$

By examining the above equations, we notice that the AC current depends on the higher-order derivatives of the drain current, which was mentioned in the introduction section of this chapter. For example, the third-order harmonic amplitude related to the transconductance contains the term $\partial^3 I_d/(\partial V_{gs})^3$. Hence, to accurately estimate $HD_{3 g_m}$, the device model should accurately reproduce the third-order derivative of the drain current with respect to the gate-source voltage. Evidently, by looking at (5.21) we conclude that the MOSFET level 1 model does not perform well in predicting $HD_{3 g_m}$ since $a3_{g_m}$ is identical to zero for this model.

It is not only the quadratic model that has problems with reproducing higher-order derivatives. Also, modern state-of-the-art models like the MM9 model from Philips (see Chapter 8) and the different BSIM3 versions also struggle to provide accurate estimates of the higher-order nonlinearity coefficients. These problems are not only caused by the expressions used in the model but also by the parameter extraction procedures employed to generate the parameter sets for the model. A thorough and excellent discussion of the performance of commonly used MOSFET models in terms of accuracy in predicting distortion effects is given in Wambacq and Sansen (1998). As an example of poor estimation of higher-order derivative, we show in Figure 5.1 the simulated transconductance and the third-order derivative of the drain current with respect to the gate-source voltage $a3_{g_m}$ versus the gate-source voltage for a 0.5-μm-long MOS transistor having a drain-source voltage of 3.0 V using the BSIM3v2.0 model. The calculated $a3_{g_m}$ exhibits several discontinuities when the gate-source voltage is around 1 V.

Several MOSFET models have been published that address the problems discussed above. These models provide enhanced accuracy in estimating the nonlinearity characteristics of modern deep submicron MOS transistors. Here, we would like to mention one such model that was published in van Langevelde (1998). In this work, the author

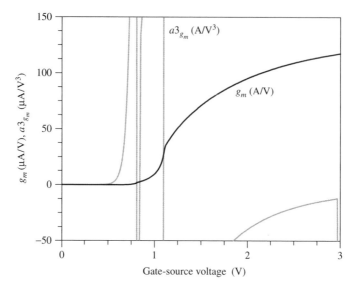

Figure 5.1   Simulated transconductance and $a3_{g_m}$ of a 0.5-μm-long MOS transistor having a drain-source voltage of 3.0 V using the BSIM3v2.0 model

presented a MOSFET model with emphasis on accurately predicting distortion effects in deep submicron MOS transistors. The model is also suitable for implementation in circuit simulators such as SPICE. According to the author, the following requirements should be satisfied for such a model:

- The model should include *physics-based* description of short-channel, high-field, and other effects that become important in deep submicron devices such as, mobility degradation, series resistance, velocity saturation, drain-induced barrier lowering, channel-length modulation, hot-carrier effects, and non-quasi-static effects.

- The model should be symmetric around $V_{ds} = 0$ in all quantities including higher-order derivatives.

In addition to the points above, we would like to add that the parameter extraction procedure should be part of the model and developed in parallel with the development of the model equations. Parameters should be extracted on the basis of measurements of not only currents, g_m and g_{ds}, but also of higher-order derivatives.

All nonlinearities discussed so far have been frequency-independent. Nonlinearities that arise from memory elements, such as capacitances, increase with frequency since the impedance of such elements decrease with frequency. In Figure 5.2 we show measured HD_2 and HD_3 of a CMOS two-stage Miller-compensated OPAMP configured as an inverting amplifier using feedback. We note that the distortion at low and moderate frequencies is low and almost constant. Then, around 10 kHz the two harmonic amplitudes suddenly start increasing at a rate of 20 dB per decade. This sudden increase is caused by capacitances inside the amplifier. In Hernes (2003) a thorough nonlinear analysis of the feedback-folded cascade Miller OPAMP is carried out. The authors presented closed-form equations for the second and third harmonics as a function of frequency and circuit

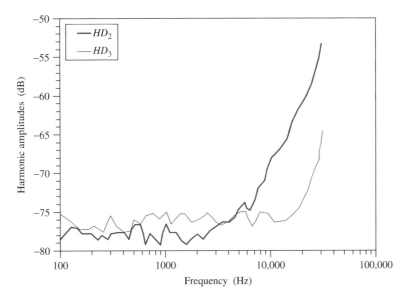

Figure 5.2   Measured HD_2 and HD_3 of a CMOS two-stage Miller-compensated operational ampli-fier configured as an inverting amplifier using feedback. Reproduced from Wambacq P., Gielen G. G. E., Kinget P. R., and Sansen W. (1999) High-frequency distortion analysis of analog inte-grated circuits, *IEEE Trans. Circuits Syst. II: Analog Digital Signal Process.*, **46**(3), 335–345 (© 1999 IEEE)

element parameters. Also, an optimization procedure for low linearity in feedback circuits was described.

5.4 CALCULATION OF DISTORTION IN ANALOG CMOS CIRCUITS

In the previous sections, we have explained how to calculate the distortion characteristics of single transistors. In this section, we will briefly discuss important methods commonly employed to calculate distortion in analog CMOS circuits, such as OPAMPS, transcon-ductors, and RF amplifiers. The methods employed are based on either numerical or symbolic techniques.

Probably the most straightforward numerical method for the calculation of harmonic components of circuit responses is to run a time-domain simulation in, for example, SPICE (Quarles *et al.* 1994) and then perform a Fast-Fourier transform (FFT) to calculate the spectrum of the output signal(s). This approach has several disadvantages. Usually the circuit exhibits transients at the beginning of the simulation before it reaches steady state. The sequence of time points that are used as input to the FFT algorithm must be free of transients, and hence, the time-domain simulation must be run at least one period of the output signal plus the time required to reach steady state. For circuits that exhibit widely separated time constants, the transients remain active for a long time and the CPU time for a single simulation may become uncomfortably long. Example of such circuits are high-Q filters and RF mixers.

Two methods that do not suffer from the above problems are the shooting method (Aprille and Trick 1972; Skelboe 1980) and the harmonic balance (HB) method (Cunningham 1958; Mees 1981). In the former method, several time-domain simulations are run over a single period. On each iteration the initial conditions are varied to match the response at the end of the simulation interval. When they match, the initial conditions found do not cause transients and the steady state response of the circuit is found.

The HB method is a frequency-domain technique in which the steady state response of the circuit is calculated directly. Today the most common frequency-domain technique for steady state analysis implemented in circuit simulators is the direct calculation of the coefficients of a trigonometric series (Kundert *et al.* 1990). Applying this method to circuits containing only linear devices is equivalent to a phasor analysis, in which superposition is utilized for each device. However, the introduction of nonlinear devices complicates matters. Superposition does not apply any more, and even worse, the calculation of the coefficients of a single device response is no longer a simple task of directly transforming the coefficients of the input stimulus. A common solution to this problem is to calculate the response coefficients approximately by transforming the stimulus, the node voltages, and the branch currents into the time domain, where all nonlinear devices can easily be evaluated. The results are then converted back into the frequency domain in order to obtain the coefficients of the response.

Frequency-domain techniques based on a direct calculation of trigonometric series coefficients are referred to as harmonic balance methods (see, for example, the review articles by Rizzoli and Neri (1988); Gilmore and Steer (1991a, 1991b)). The first term (harmonic) is obvious because trigonometric series are used. The second term (balance) is mostly for historical reasons. It stems from an approach based on balancing the currents and voltages at the boundaries between linear and nonlinear subcircuits.

All methods discussed above are numerical and usually do not present the results in a way such that designers can extract the devices that dominate the nonlinearity characteristics of the circuit. Numerical methods based on so-called Volterra series (see, for example, Schetzen (1980)) that provide such insight have been proposed by many researchers (see, for example, Weiner and Spina (1980)) and have also been implemented in SPICE (Chisholm and Nagel (1973); Roychowdhury (1989)). A disadvantage of methods based on Volterra series is that they require the circuits to be weakly nonlinear. Weakly nonlinear means that the input amplitudes should be small. As the amplitudes are increased, the analysis based on the weakly nonlinear assumption becomes less and less accurate. Since it is generally not possible to estimate for what amplitudes the assumption fails, other numerical methods are used to verify the results.

Symbolic methods are also usually based on Volterra series, and as mentioned above, are accurate only for weakly nonlinear circuits. Now, you may ask why we need symbolic methods after all. The answer is that the results obtained from symbolic methods give much better insight into the nonlinear characteristics of the circuits since such methods provide closed-form analytical expressions containing contributions of basic nonlinearities that are dominant in the harmonic or intermodulation amplitudes. As a result, it is possible to easily identify which devices dominate the nonlinear response and how to modify sizing, biasing, and topology to reduce the distortion components. In general, this is not possible in numerical methods. Work published on symbolic analysis concentrate on distortion in OPAMPS. In Wambacq *et al.* (1999), the authors present an approach based on Volterra series for the analysis at high frequencies of the nonlinear behavior of weakly nonlinear circuits with one input port, such as amplifiers, and with more than one input port, such as analog mixers and multipliers. The complexity of the representation of frequency-dependent distortion based on Volterra series grows rapidly with the size of the circuit. In Hernes (2003), the authors present a more effective method based on the phasor method (see, for example Wambacq and Sansen (1998)) instead of Volterra series. By using this approach, the authors showed that the complexity of the representation of distortion in CMOS circuits can be drastically reduced compared to previous methods.

In general, circuits exceeding a few transistors, symbolic hand calculation are not feasible. Instead, symbolic network analysis programs are utilized to automate the calculations. Both programs developed in commercially available symbolic software packages, such as Mathematica (*http://www.wolfram.com/*) and Maple (*http://www.maplesoft.com/*), and stand-alone software programs have been developed. In Hernes (2003), Maple was used for automating the symbolic calculations. The computer program ISAAC by Gielen *et al.* (1989) is an example of a stand-alone program that requires no other software packages to be installed on your computer system. A copy of the program can be obtained by visiting their Web site at *http://www.esat.kuleuven.ac.be/micas/Software/*.

REFERENCES

Aprille T. and Trick T. (1972) Steady-state analysis of nonlinear circuits with periodic inputs, *Proc. IEEE* **60**(1), 108–114.

Bruun E. (1998) Harmonic distortion in CMOS current mirrors, *Proc. of the 1998 IEEE International Symposium on Circuits and Systems, ISCAS '98*, pp. 567–570.

Chilakapati U., Fiez T. S., and Eshraghi A. (2002) A CMOS transconductor with 80-dB SFDR up to 10 MHz, *IEEE J. Solid-State Circuits*, **37**(3), 365–370.

Chisholm S., and Nagel L. (1973) Efficient computer simulation of distortion in electronic circuits, *IEEE Trans. Circuit Theory*, **20**(6), 742–745.

Cunningham W. J. (1958) *Introduction to Nonlinear Analysis*, McGraw-Hill, New York.

Gielen G., Walscharts H., and Sansen W. (1989) ISAAC: a symbolic simulator for analog integrated circuits, *IEEE J. Solid-State Circuits*, **24**(6), 1587–1597.

Gilmore R. J. and Steer M. B. (1991a) Nonlinear circuit analysis using the method of harmonic balance-a review of the art I. Introductory concepts, *Int. J. Microwave Millimeter-Wave Computer-Aided Eng.*, **1**(1), 22–37.

Gilmore R. J. and Steer M. B. (1991b) Nonlinear circuit analysis using the method of harmonic balance-a review of the art II. Advanced concepts, *Int. J. Microwave Millimeter-Wave Computer-Aided Eng.*, **1**(1), 159–180.

Hernes B. (2003) *Design Criteria for Low Distortion in Feedback Opamp Circuits*, Kluwer Academic Publishers, Boston, MA.

Kundert K. S., White J. K., and Sangiovanni-Vincentelli A. (1990) *Steady-State Methods for Simulating Analog and Microwave Circuits*, Kluwer, Boston, MA.

Limotyrakis S., Nam K., and Wooley B. A. (2002) Analysis and simulation of distortion in folding and interpolating A/D converters, *IEEE Trans. Circuits Syst. II: Analog Digital Signal Process.*, **49**(3), 161–169.

Mees A. I. (1981) *Dynamics of Feedback Systems*, John Wiley & Sons, New York.

Mensink C. H. J., Klumperink E. A. M., and Nauta B. (1996) On the reduction of the third order distortion in a CMOS triode transconductor, *Proc. of the 1996 IEEE International Symposium on Circuits and Systems, ISCAS '98*, pp. 33–36.

Pu L.-J. and Tsividis Y. P. (1990) Harmonic distortion of the four-terminal MOSFET in non-quasistatic operation, *IEE Proc. Circuits, Devices Syst.*, **137**, 325–332.

Quarles T., Newton A. R., Pederson D. O., and Sangiovanni-Vincentelli A. (1994) *SPICE3 Version 3f5 User's Manual*, Berkeley.

Rizzoli V. and Neri A. (1988) State of the art and present trends in nonlinear microwave CAD techniques, *IEEE Trans. Microwave Theory Tech.*, **36**(2), 343–365.

Roychowdhury J. (1989) SPICE3 Distortion Analysis, Memo No. UCB/ERL M89/48, Electron Research Laboratories, University of California, Berkeley.

Schetzen M. (1980) *The Volterra and Wiener Theories of Nonlinear Systems*, John Wiley & Sons, New York.

Shichman H. and Hodges D. (1968) Modeling and simulation of insulated-gate field-effect transistor switching circuits, *IEEE J. Solid-State Circuits*, **3**(3), 285–289.

Shoucair F. S. and Patterson W. R. (1993) Analysis and modeling of nonlinearities in VLSI MOSFETs including substrate effects, *IEEE Trans. Electron Devices*, **40**(10), 1760–1767.

Skelboe S. (1980) Computation of the periodic steady-state response of nonlinear networks by extrapolation methods, *IEEE Trans. Circuits Syst.*, **27**(3), 161–175.

van Langevelde R. and Klaassen F. M. (1997a) Accurate drain conductance modeling for distortion analysis in MOSFETs, *Proc. International Electron Devices Meeting, IEDM '97*, pp. 313–316.

van Langevelde R. and Klaassen F. M. (1997b) Effect of gate-field dependent mobility degradation on distortion analysis in MOSFETs, *IEEE Trans. Electron Devices*, **44**(11), 2044–2052.

van Langevelde R. (1998) *A Compact MOSFET Model for Distortion Analysis in Analog Circuit Design*, Ph.D. thesis, Eindhoven University of Technology, Holland.

Wambacq P. and Sansen W. (1998) *Distortion Analysis of Analog Integrated Circuits*, Kluwer Academic Publishers, Boston, MA.

Wambacq P., Gielen G. G. E., Kinget P. R., and Sansen W. (1999) High-frequency distortion analysis of analog integrated circuits, *IEEE Trans. Circuits Syst. II: Analog Digital Signal Process.*, **46**(3), 335–345.

Weiner D. and Spina J. (1980) *Sinusoidal Analysis and Modeling of Weakly Nonlinear Circuits*, Van Nostrand Reinhold, New York.

6

The BSIM4 MOSFET Model

6.1 AN INTRODUCTION TO BSIM4

Since MOSFET was invented, many physical effects have been studied, such as (1) vertical and lateral nonuniform doping effect (Hori *et al.* (1993)); (2) drain-induced barrier-lowering effect (Troutman (1979)); (3) normal and reverse short-channel effect (Viswanathan *et al.* (1985); Hsu *et al.* (1991); Rafferty *et al.* (1993)); (4) normal and reverse narrow width effect (Akers *et al.* (1982); Cheng *et al.* (1997b)); (5) field-dependent mobility and velocity saturation (Sodini *et al.* (1984); Talkhan *et al.* (1972)); (6) channel-length modulation (Frohman-Bentchknowsky and Grove (1969)); (7) impact ionization (Arora and Sharama (1991)); and (8) polysilicon gate depletion (Huang *et al.* (1993)). These well-known physical effects have been included in various MOSFET models such as BSIM3v3 model (Cheng *et al.* (1997a)). As CMOS technology approaches more advanced levels, many novel physical effects appear, such as (1) gate-induced drain leakage (Chen *et al.* (1987)) and gate direct tunneling leakage (Majkusiak (1990)); (2) inversion layer quantization (King *et al.* (1997)); (3) finite charge layer effect (Liu *et al.* (1999)); (4) HF influence of MOSFET parasitics (Cheng *et al.* (2002)); and (5) asymmetric source/drain resistance. In BSIM4 (Liu *et al.* (2000)), these new physical effects are modeled. Also, modeling for some physical effects included already in BSIM3v3 have been revisited in order to improve the model accuracy.

6.2 GATE DIELECTRIC MODEL

In today's MOSFETs, in which the gate oxide thickness is very thin (<3 nm), the effect of channel quantization, determining the finite charge layer thickness (FCLT) in the channel, becomes nonnegligible (King *et al.* (1998)). BSIM4 accounts for this effect in both DC and capacitance model. To activate this effect of FCLT in the simulation, two of the following three can be used as input model parameters: (a) electrical gate oxide thickness *TOXE*, (b) the physical gate oxide thickness *TOXP*, (c) and their difference $DTOX = TOXE - TOXP$. On the basis of these parameters, BSIM4 models the FCLT effect by introducing an effective gate oxide capacitance C_{oxeff} in I–V and C–V models (Liu *et al.* (1999)),

$$C_{\text{oxeff}} = \frac{C_{\text{oxe}} \cdot C_{\text{cen}}}{C_{\text{oxe}} + C_{\text{cen}}} \qquad (6.1)$$

Device Modeling for Analog and RF CMOS Circuit Design. T. Ytterdal, Y. Cheng and T. A. Fjeldly
© 2003 John Wiley & Sons, Ltd ISBN: 0-471-49869-6

where C_{oxe} and C_{cen} are called *effective gate oxide capacitance* and *centroid channel charge capacitance*, respectively, and can be given as follows:

$$C_{oxe} = \frac{EPSROX \cdot \varepsilon_0}{TOXE},$$ (6.2)

$$C_{cen} = \frac{\varepsilon_{si}}{X_{DC}},$$ (6.3)

where *EPSROX* is a model parameter, with a default value of 3.9, for the gate dielectric constant relative to vacuum; *TOXE* is the parameter for the equivalent electrical gate oxide thickness; ε_0 and ε_{si} are the permittivity of free space and silicon, respectively; X_{DC} is the equivalent DC centroid of the channel charge layer and is given by

$$X_{DC} = \frac{1.9 \times 10^{-9}\,\text{cm}}{1 + \left(\dfrac{V_{gsteff} + 4 \cdot (VTH0 - VFB - \Phi_S)}{2 \cdot TOXP} \right)}$$ (6.4)

where *VTH0* and *VFB* are model parameters for the long-channel threshold voltage at $V_{BS} = 0$ and the flat-band voltage, respectively; *TOXP* is the parameter for the equivalent physical gate oxide thickness; V_{gsteff} is similar to the one used in BSIM3v3 but introduces additional parameters to improve the model accuracy in moderate inversion region,

$$V_{gsteff} = \frac{n \cdot V_t \cdot \ln\left\{ 1 + \exp\left[\dfrac{m^*(V_{gse} - V_{th})}{n \cdot V_t} \right] \right\}}{m^* + n \cdot C_{oxe}\sqrt{\dfrac{2\Phi_s}{q \cdot NDEP \cdot \varepsilon_{si}}}\, \exp\left[-\dfrac{(1 - m^*)(V_{gse} - V_{th}) - V'_{off}}{n \cdot V_t} \right]}$$ (6.5)

where $m^* = 0.5 + arctan(MINV)/\pi$, *MINV* is a model parameter introduced to improve the model accuracy in moderate inversion region; n is the subthreshold swing parameter; V_{gse} is the effective gate voltage with the consideration of polysilicon gate depletion effect (Cheng *et al.* (1997a)); V_{th} is the threshold voltage; V'_{off} is a potential offset parameter (Cheng and Hu (1999)) and equals to $VOFF + VOFFL/L_{eff}$ to describe the channel-length dependence of V'_{off} in devices with nonuniform doping profiles; L_{eff} is the effective channel length; V_t is the thermal voltage and is equal to $k_B T/q$, Φ_s can be considered the surface potential defined as

$$\Phi_s = 0.4 + \frac{k_B T}{q} \ln\left(\frac{NDEP}{n_i} \right) + PHIN$$ (6.6)

where *NDEP* is doping concentration at the edge of the channel depletion layer at $V_{BS} = 0$, n_i is the intrinsic carrier concentration in the channel region, and *PHIN* is a model parameter to describe the nonuniform vertical doping effect on surface potential.

High-k gate dielectric can be modeled as SiO_2 (relative permittivity: 3.9) with an equivalent SiO_2 thickness. For example, 3-nm gate dielectric with a dielectric constant of 7.8 would have an equivalent oxide thickness of 1.5 nm. BSIM4 also allows the user to specify the model parameter (*EPSROX*) for gate dielectric constant different from 3.9 (SiO_2) as an alternative approach to modeling high-k dielectrics. Figure 6.1 illustrates the

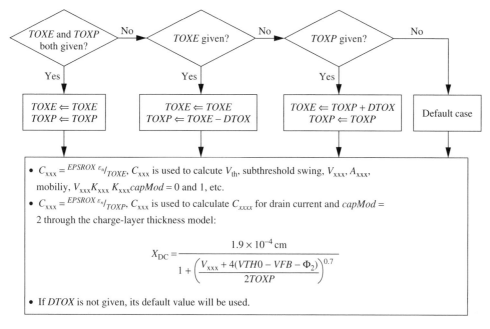

Figure 6.1   Illustration of the options for gate dielectric models in BSIM4

algorithm and options for specifying the gate dielectric thickness and calculation of the gate dielectric capacitance for BSIM4 model evaluation.

6.3 ENHANCED MODELS FOR EFFECTIVE DC AND AC CHANNEL LENGTH AND WIDTH

Two additional fitting parameters *XL* and *XW* are introduced in BSIM4 to account for the offset in channel length and width due to the processing factors such as mask and etching. In DC case, the effective channel length and width can be given

$$L_{\text{eff}} = L_{\text{drawn}} + XL - 2 \cdot dL, \tag{6.7}$$

$$dL = LINT + \frac{LL}{L^{LLN}} + \frac{LW}{W^{LWN}} + \frac{LWL}{L^{LLN} W^{LWN}}, \tag{6.8}$$

$$W_{\text{eff}} = W_{\text{drawn}} + XW - 2dW, \tag{6.9}$$

$$dW = WINT + \frac{WL}{L^{WLN}} + \frac{WW}{W^{WWN}} + \frac{WWL}{L^{WLN} W^{WWN}}$$

$$+ DWGV_{\text{gsteff}} + DWB \left(\sqrt{\Phi_{\text{s}} - V_{\text{bseff}}} - \sqrt{\Phi_{\text{s}}} \right), \tag{6.10}$$

where *LINT, LL, LW, LWL, LLN*, and *LWN* are model parameters to describe the geometry dependence of dL, in which *LINT* can be considered as the traditional "ΔL" extracted from the intercept of a straight line in an R_{DS} versus L_{drawn} plot; *WINT, WL, WW, WWL, WLN*, and *WWN* are model parameters to describe the geometry dependence of dW, in which *WINT* can be considered as the traditional "ΔW" extracted from the intercept of a straight line in a $1/R_{DS}$ versus W_{drawn} plot; W_{drawn} here and in Eqs. (6.13) and (6.15) stands for a channel width for a single-finger device. In a multifinger device, W_{drawn} is the channel width per finger, that is, $W_{drawn} = W_{total}/N_F$, where W_{total} is the total width of the device and N_F is the finger numbers of the device. The *DWG* and *DWB* parameters are used to account for both the gate and substrate bias effects; V_{gsteff} and V_{bseff} are effective gate and substrate biases. Figures 6.2 and 6.3 illustrate the definitions of *XL*, *XW*, *dW*, *dL*, W_{eff}, L_{eff} and N_F.

In AC case, the effective channel length and width are represented by L_{active} and W_{active}, defined as

$$L_{active} = L_{drawn} + XL - 2 \cdot dLC, \tag{6.11}$$

$$dLC = DLC + \frac{LLC}{L^{LLN}} + \frac{LWC}{W^{LWN}} + \frac{LWLC}{L^{LLN} W^{LWN}}, \tag{6.12}$$

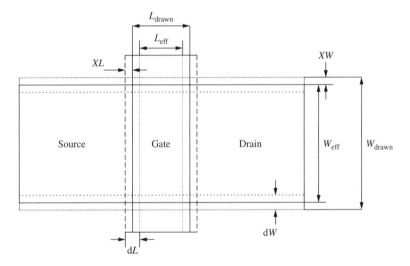

Figure 6.2   Definitions of *XL*, *XW*, *dW*, *dL*, W_{eff}, and L_{eff}

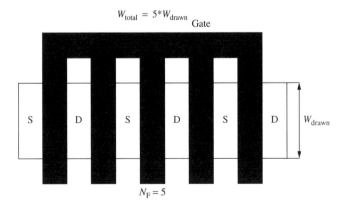

Figure 6.3   Definitions of N_F, W_{drawn}, and W_{total}

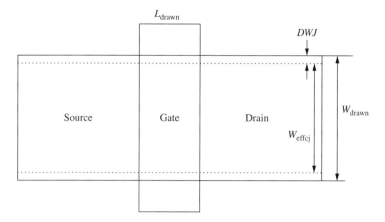

Figure 6.4 Definitions of *DWJ* and W_{effcj}

$$W_{\text{active}} = W_{\text{drawn}} + XW - 2\text{d}WC, \qquad (6.13)$$

$$\text{d}WC = DWC + \frac{WLC}{L^{WLN}} + \frac{WWC}{W^{WWN}} + \frac{WWLC}{L^{WLN}W^{WWN}}, \qquad (6.14)$$

where *DLC, LLC, LWC,* and *LWLC* are model parameters to describe the geometry dependence of d*LC*, which is similar to d*L* in DC but is a closer representation of the parasitic portion of the physical channel length at one side of the device than d*L*; *DWC, WLC, WWC,* and *WWLC* are model parameters to describe the geometry dependence of d*WC*.

Please note that the calculation of d*LC* and d*WC* are triggered by including DLC and DWC or the length/width dependence parameters (*LLC, LWC, LWLC, WLC, WWC,* and *WWLC*) in the model file. As a default, *DLC = LINT, DWC = WINT,* and *DWC, DLC, LLC, LWC, LWLC, WLC, WWC,* and *WWLC* are set to the values of their DC counterparts.

As shown in Figure 6.4, BSIM4 also introduced a new parameter *DWJ* to calculate the effective source/drain diffusion width, W_{effcj}, in modeling parasitics such as source/drain series resistance, gate resistance, gate-induced drain leakage (GIDL), and so on.

$$W_{\text{effcj}} = W_{\text{drawn}} - 2 \cdot \text{d}WJ,$$

$$= W_{\text{drawn}} - 2 \cdot \left(DWJ + \frac{WLC}{L^{WLN}} + \frac{WWC}{W^{WWN}} + \frac{WWLC}{L^{WLN}W^{WWN}} \right). \qquad (6.15)$$

6.4 THRESHOLD VOLTAGE MODEL

6.4.1 Enhanced Model for Nonuniform Lateral Doping due to Pocket (Halo) Implant

To reduce the short-channel effects (SCE), local high-dose implantation near the source/drain region edges have been employed. This is called lateral channel engineering or pocket (Halo) implantation, which causes higher doping concentration near the source/drain junctions than that in the middle of the channel as shown in Figure 6.5. A V_{th} roll-up, sometimes called reverse short-channel effect (RSCE), will usually happen as

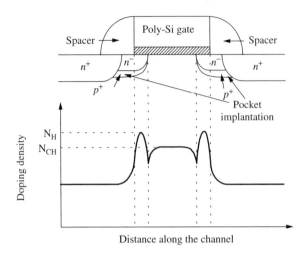

Figure 6.5   Doping profile along the channel in a MOSFET with pocket implant

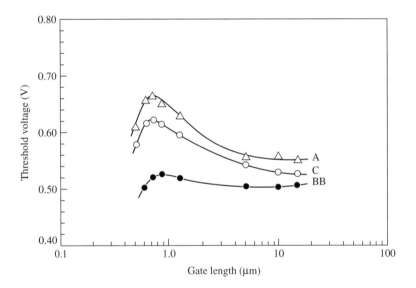

Figure 6.6   Illustration of V_{th} roll-off effect. Reproduced from Lu, C.-Y. and Sung J. M. (1989) Reverse short-channel effects on threshold voltage in submicrometer salicide devices, *IEEE Electron Device Lett.*, **10**, 446–448

the channel length becomes shorter since the effective channel doping concentration gets higher (see Figure 6.6). The pocket implantation will not only cause nonuniform lateral doping (NULD) but will also contribute to a nonuniform vertical doping, so the body effect will be influenced as well. A simple equation to consider the lateral nonuniform doping has been derived in BSIM3v3 (Cheng *et al.* (1997a)). It can model the V_{th} roll-off reasonably well as shown in Figure 6.7. An empirical term has been introduced in BSIM4 to improve the model accuracy with the consideration of the influence of NULD

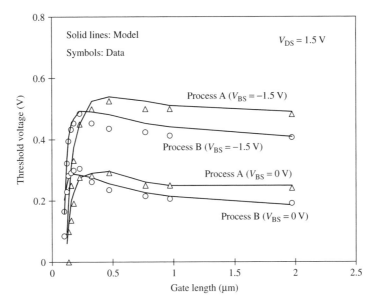

Figure 6.7 Comparison of measured data and BSIM3v3 model to simulate the V_{th} roll-off effects. Reproduced from Cheng Y., Sugii T., Chen K., and Hu C. (1997b) Modeling of small size MOSFETs with reverse short-channel and narrow width effects for circuit simulations, *Solid-State Electron.*, **41**(9), 1227–1231

to the body effect,

$$V_{th} = VTH0 + K1 \cdot \left(\sqrt{\Phi_s - V_{BS}} - \sqrt{\Phi_s} \right) \cdot \sqrt{1 + \frac{LPEB}{L_{eff}}} - K2 \cdot V_{BS}$$

$$+ K1 \cdot \left(\sqrt{1 + \frac{LPE0}{L_{eff}}} - 1 \right) \cdot \sqrt{\Phi_s} \tag{6.16}$$

where $VTH0$, $K1$, $K2$, $LPE0$, and $LPEB$ are model parameters. $LPE0$ replaces the NLX parameter in BSIM3v3 and $LPEB$ is a new parameter introduced in BSIM4.

Besides the implants discussed above, it has been known that the pocket implant also causes so-called drain-induced threshold voltage shift (DITVS) in long-channel devices, which is described by the following equation in BSIM4:

$$\Delta V_{th}(DITVS) = -n \cdot V_t \cdot \ln \left(\frac{L_{eff}}{L_{eff} + DVTP0 \cdot (1 + e^{-DVTP1 \cdot V_{DS}})} \right) \tag{6.17}$$

where $DVTO0$ and $DVTP1$ are model parameters, L_{eff} is the effective channel length, and V_{DS} is the drain-to-source bias.

6.4.2 Improved Models for Short-channel Effects

Compared with the long-channel devices, short-channel devices show a significant dependence on channel length and drain voltage. Various models have been developed to describe the so-called short-channel effects. Typically, it can be summarized by the following equation:

$$\Delta V_{th}, SCE = \Delta V_{th}(L_{eff}) + \Delta V_{th}(V_{DS}) \tag{6.18}$$

where $\Delta V_{th}(L_{eff})$ is the change in threshold voltage caused by the SCE without the influence of drain/source voltage V_{DS} and $\Delta V_{th}(V_{DS})$ is the change in threshold voltage caused by the influence of V_{DS}.

In a short-channel device with nonzero V_{DS}, the depletion layer in the channel region will be modulated as drain bias varies, as shown in Figure 6.8. On the basis of a quasi-two-dimensional analysis of the Poisson equation, the influence of SCE on the threshold voltage can be described by the following equation (see Liu *et al.* (1993)):

$$\Delta V_{th}, SCE = -\theta_{th}(L_{eff})[2(V_{bi} - \Phi_s) + V_{DS}] \tag{6.19}$$

where V_{bi} is the built-in voltage of the source/drain junctions and $\theta_{th}(L_{eff})$ is given by

$$\theta_{th}(L_{eff}) = \frac{1}{2\cosh\left(\dfrac{L_{eff}}{l_t}\right) - 1} \tag{6.20}$$

where l_t is called the characteristic length (see Cheng *et al.* (1997a)).

Unlike BSIM3v3, in which an approximated form of Eq. (6.20) has been used (Cheng *et al.* (1997a)), BSIM4 implemented Eq. (6.20) without approximation. However, a set of fitting parameters similar to those in BSIM3v3 has been introduced to increase the model flexibility for different technologies, such as *DVT0*, *DVT1*, *DVT2*, *DSUB*, *ETA0*, and *ETAB*, as shown in the following model equations:

$$\Delta V_{th}, SCE = -\frac{DVT0}{2 \cdot \cosh\left(DVT1 \cdot \dfrac{L_{eff}}{l_t}\right) - 1}(V_{bi} - \Phi_s) - \frac{(ETA0 + ETAB \cdot V_{BS}) \cdot V_{DS}}{2 \cdot \cosh\left(DSUB \cdot \dfrac{L_{eff}}{l_{t0}}\right)} \tag{6.21}$$

where

$$V_{bi} = \frac{k_B T}{q} \cdot \ln\left(\frac{NDEP \cdot NSD}{n_i{}^2}\right), \tag{6.22}$$

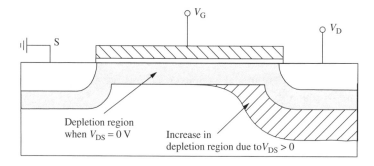

Figure 6.8   An increase in depletion layer caused by the applied drain bias

$$l_t = \sqrt{\frac{\varepsilon_{si} \cdot TOXE \cdot X_{dep}}{EPSROX}} (1 + DVT2 \cdot V_{BS}), \tag{6.23}$$

$$l_{t0} = \sqrt{\frac{\varepsilon_{si} \cdot TOXE \cdot X_{dep0}}{EPSROX}}, \tag{6.24}$$

where X_{dep} is depletion layer width in the channel with the influence of V_{bs}; X_{dep} is depletion layer width in the channel when $V_{BS} = 0$,

$$X_{dep} = \sqrt{\frac{2 \cdot \varepsilon_{si} \cdot (\Phi_s - V_{BS})}{q \cdot NDEP}}, \tag{6.25}$$

$$X_{dep0} = \sqrt{\frac{2 \cdot \varepsilon_{si} \cdot \Phi_s}{q \cdot NDEP}}. \tag{6.26}$$

Figure 6.9 shows the model results with SCE only, with NULP effect, and with both SCE and NULP effects. It clearly demonstrates that the V_{th} roll-off is a combined result, contributed by both SCE and NULP.

6.4.3 Model for Narrow Width Effects

Narrow width effect was found initially in devices with LOCOS isolation, in which additional contribution of charges in the depletion region in the edge of the field implant region will impact the threshold voltage of the device as shown in Figure 6.10. As the channel width decreases, the influence becomes significant, resulting in a higher threshold voltage. This behavior is called "normal" narrow width effect. In contrast with the "normal" narrow width effect, the so-called "reverse" narrow width effect has been reported in devices with shallow trench isolation (STI) as illustrated in Figure 6.11. A decrease in threshold voltage has been observed as the channel width decreases.

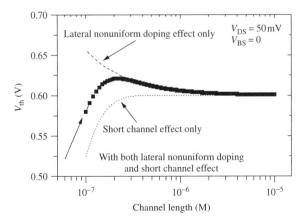

Figure 6.9 Illustration of BSIM4 threshold voltage model in predicting different characteristics of V_{th} versus L in MOSFETs with different channel/pocket doping conditions

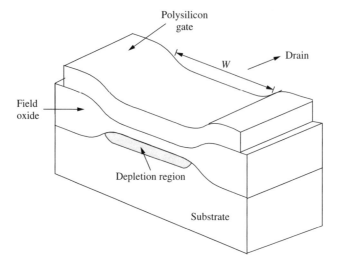

Figure 6.10   Illustration of MOSFET with LOCOS isolation

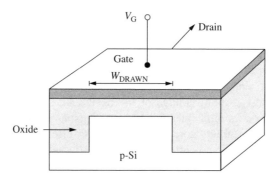

Figure 6.11   Illustration of MOSFET with STI (shallow trench isolation)

To model narrow width effects, similar to BSIM3v3, BSIM4 adopts an empirical approach based on the following general form for narrow width effects:

$$\Delta V_{\mathrm{th,W}} \propto \frac{T_{\mathrm{ox}}}{W_{\mathrm{eff}}} \Phi_{\mathrm{s}}. \tag{6.27}$$

By introducing several fitting parameters, $K3$, $K3B$, $W0$, $DVT0W$, $DVT1W$ and $DVT2W$, the narrow width effect is modeled in BSIM4 by

$$\Delta V_{\mathrm{th,W}} = (K3 + K3B \cdot V_{\mathrm{BS}}) \frac{TOXE}{W'_{\mathrm{eff}} + W0} \cdot \Phi_{\mathrm{s}}$$

$$+ \frac{DVT0W}{2 \cdot \cosh\left(DVT1W \dfrac{L_{\mathrm{eff}} \cdot W'_{\mathrm{eff}}}{l_{t0}}\right) - 1} (V_{\mathrm{bi}} - \Phi_{\mathrm{s}}) \tag{6.28}$$

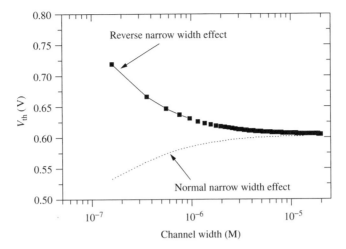

Figure 6.12 Calculated result of BSIM3v3/BSIM4 V_{th} model in predicting characteristics of V_{th} versus channel width with extracted different model parameters

where the second term accounts for the narrow width effect in small devices with both narrow width and short-channel length.

Figure 6.12 shows the calculated result of threshold voltage versus channel width with different model parameters for both "normal" and "reverse" narrow width effects.

6.4.4 Complete Threshold Voltage Model in BSIM4

The complete BSIM4 threshold voltage model is

$$
\begin{aligned}
V_{th} = {} & VTH0 + \left(K_{1ox} \cdot \sqrt{\Phi_s - V_{bseff}} - K1 \cdot \sqrt{\Phi_s}\right) \cdot \sqrt{1 + \frac{LPEB}{L_{eff}}} - K_{2ox} \cdot V_{bseff} \\
& + K_{1ox} \cdot \left(\sqrt{1 + \frac{LPE0}{L_{eff}}} - 1\right) \cdot \sqrt{\Phi_s} + (K3 + K3B \cdot V_{bseff})\frac{TOXE}{W'_{eff} + W0} \cdot \Phi_s \\
& + \left[\frac{DVT0W}{2 \cdot \cosh\left(DVT1W\dfrac{L_{eff} \cdot W'_{eff}}{l_{t0}}\right) - 1} + \frac{DVT0}{2 \cdot \cosh\left(DVT1\dfrac{L_{eff}}{l_t}\right) - 1}\right] \\
& \cdot (V_{bi} - \Phi_s) - \frac{1}{2 \cdot \cosh\left(DSUB\dfrac{L_{eff}}{l_{t0}}\right) - 1}(ETA0 + ETAB \cdot V_{bseff}) \cdot V_{DS} \\
& - n \cdot V_t \cdot \ln\left[\frac{L_{eff}}{L_{eff} + DVTP0 \cdot (1 + e^{-DVTP1 \cdot V_{DS}})}\right]
\end{aligned}
\tag{6.29}
$$

where

$$K_{1ox} = K1 \cdot \frac{TOXE}{TOXM},$$ (6.30)

$$K_{2ox} = K2 \cdot \frac{TOXE}{TOXM},$$ (6.31)

$$V_{bseff} = V_{bc} + 0.5 \cdot [(V_{BS} - V_{bc} - \delta_1) + \sqrt{(V_{BS} - V_{bc} - \delta_1)^2 - 4 \cdot \delta_1 \cdot V_{bc}},$$ (6.32)

where $\delta_1 = 0.001$, V_{bc} is the upper bound of V_{bseff} and is defined by the following equation:

$$V_{bc} = 0.9 \cdot \left(\Phi_s - \frac{K1^2}{4 \cdot K2^2} \right).$$ (6.33)

The temperature dependence of V_{th} is modeled by

$$V_{th}(T) = V_{th}(TNOM) + \left(KT1 + \frac{KT1L}{L_{eff}} + KT2 \cdot V_{bseff} \right) \cdot \left(\frac{T}{TNOM} - 1 \right)$$ (6.34)

where $KT1$ and $KT1L$ and $KT2$ are model parameters.
Figure 6.13 illustrates the dependence of V_{th} on temperature.

6.5 CHANNEL CHARGE MODEL

BSIM4 uses a unified expression for the channel charge Q_{ch} from the strong inversion and subthreshold regions. The equation of the unified charge model in BSIM4 is similar to that in BSIM3v3, but additional improvements have been introduced to enhance the model accuracy in the transition region from the subthreshold to strong inversion. With the

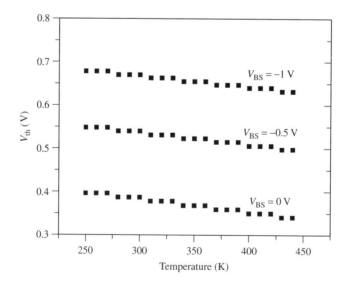

Figure 6.13   Calculated characteristics of V_{th} versus temperature

consideration of the influence of both drain and body biases (the details of the derivation can be found in Cheng and Hu (1999); Cheng et al. (1998)), the channel change density along the channel, at a location y, can be described by

$$Q_{ch}(y) = C_{oxeff} \cdot V_{gsteff} \cdot \left[1 - \frac{V_F(y)}{V_b}\right] \tag{6.35}$$

where C_{oxeff} and V_{gsteff} have been discussed earlier. $V_F(y)$ is the quasi-Fermi potential at given point y along the channel with respect to the source; the expression of V_b is given by

$$V_b = \frac{V_{gsteff} + 2 \cdot V_t}{A_{bulk}} \tag{6.36}$$

where A_{bulk} is a factor to describe the bulk charge effects and is expressed as

$$A_{bulk} = \left\{1 + F_{NUD} \cdot \left[\frac{A_0 \cdot L_{eff}}{L_{eff} + 2 \cdot \sqrt{XJ \cdot X_{dep}}} \cdot (1 - AGS)\right.\right.$$
$$\left.\left. \cdot V_{gsteff} \cdot \left(\frac{L_{eff}}{L_{eff} + 2 \cdot \sqrt{XJ \cdot X_{dep}}}\right)^2 + \frac{B0}{W'_{eff} + B1}\right]\right\} \cdot \frac{1}{1 + KETA \cdot V_{bseff}} \tag{6.37}$$

where $A0$, AGS, $B0$, $B1$, and $KETA$ are model parameters; F_{NUD} is used in BSIM4 to model the nonuniform doping effects,

$$F_{NUD} = \frac{\sqrt{1 + LPEB/L_{eff}} K_{1ox}}{2\sqrt{\Phi_s - V_{bseff}}} + K_{2ox} - K3B \frac{TOXE}{W_{eff}' + W0} \Phi_s. \tag{6.38}$$

Equation (6.35) can become a well-used piecewise equation in strong inversion and subthreshold regions, respectively, as shown in Figure 6.14 (Cheng et al. (1998)). In strong inversion region, threshold voltage, V_{th}, and bulk charge factor, A_{bulk}, are the main factors determining the channel charge characteristics. In subthreshold region, besides these two factors, the subthreshold swing factor, n, and the potential offset parameter, V'_{off}, are also important for influencing channel charge and hence the subthreshold conduction.

Both V_{off} and n have been mentioned in previous section. The expression of V'_{off} has been introduced earlier also when discussing $V_{gsteff} \cdot n$ is subthreshold conduction or subthreshold swing factor. Basically, it is a function of the channel length and the S_i–S_iO_2 interface property. In BSIM4, the n factor is given by

$$n = 1 + NFACTOR \cdot \frac{C_{dep}}{C_{oxe}} + \frac{CIT}{C_{oxe}} + \frac{(CDSC + CDSCD \cdot V_{DS} + CDSCB \cdot V_{bseff})}{2 \cdot C_{oxe} \cdot \cosh\left(DVT1 \cdot \frac{L_{eff}}{l_t}\right) - 1} \tag{6.39}$$

where $NFACTOR$, CIT, $CDSC$, $CDSCD$, $CDSCB$ are model parameters. The second term is the portion of the subthreshold swing factor derived for long-channel devices, with $NFACTOR$ introduced. The third term is to model the contribution of the interface state. The last term is to describe the contribution of the coupling between drain/source and channel as shown in Figure 6.15. The capacitance coupling effect is described by an exponential function of the channel length. $CDSCD$ and $CDSCB$ are model parameters

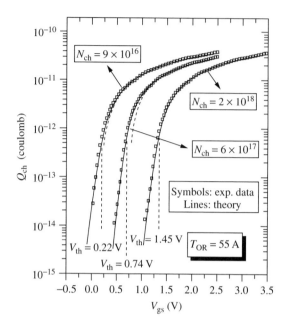

Figure 6.14 Channel charge model fits the measurement data taken from devices with different N_{ch} well. The model covers weak, moderate, and strong inversion regions of nMOSFETs. Reproduced from Cheng Y. *et al.* (1998) A unified MOSFET channel charge model for device modeling in circuit simulation, *IEEE Trans. Computer-Aided Design Integrated Circuits Syst.*, **17**, 641–644

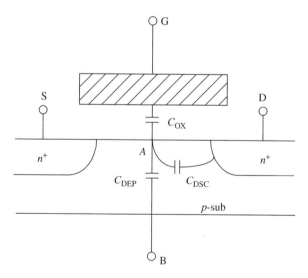

Figure 6.15 Illustration of the gate oxide capacitance, channel depletion capacitance, and C_{dsc} coupling capacitances

introduced to describe the drain and body-bias dependence of the capacitance coupling effect. Note that it shares the *DVT1* parameter for DIBL effect in threshold voltage model to describe the length dependence.

6.6 MOBILITY MODEL

Three scattering mechanisms, phonon scattering, coulomb scattering, and surface rough-ness scattering, have been found to explain the mobility behavior and account for the bias and geometry dependence of mobility. Each mechanism may be dominant under some specific conditions of the doping concentration, temperature, and biases. For example, mobility depends on the gate oxide thickness, channel doping concentration, thresh-old voltage, gate and bulk biases. A unified mobility model is given by Sabnis and Clemens (1979):

$$\mu_{\text{eff}} = \frac{\mu_0}{1 + (E_{\text{eff}}/E_0)^{\nu}} \tag{6.40}$$

where μ_0 is the low-field mobility, E_0 is called critical electric field, ν is a constant, the value of which depends on the device type and technology. E_{eff} is an effective field defined empirically by

$$E_{\text{eff}} = \frac{Q_{\text{B}} + Q_{\text{INV}}/2}{\varepsilon_{\text{si}}} \tag{6.41}$$

where Q_{B} and Q_{INV} are the charge density in the bulk and in the channel, respectively.

BSIM4 provides two mobility models that have been used in BSIM3v3. In addition, it also introduces a new mobility model based on the universal model given in Eq. (6.40). Users can select a different mobility model by defining a model parameter, **mobMod**, in the model parameter file (also called model card). When **mobMod** $= 0$ and 1, mobility models from BSIM3v3 are used. When **mobMod** $= 2$, the universal mobility model is used. The detailed equations are given as follows:

When **mobMod** $= 0$,

$$\mu_{\text{eff}} = \frac{U_0(T)}{1 + (U_A(T) + U_C(T) \cdot V_{\text{bseff}}) \cdot \left(\dfrac{V_{\text{gsteff}} + 2 \cdot V_{\text{th}}}{TOXE}\right) + U_B(T) \cdot \left(\dfrac{V_{\text{gsteff}} + 2 \cdot V_{\text{th}}}{TOXE}\right)^2}. \tag{6.42}$$

When **mobMod** $= 1$,

$$\mu_{\text{eff}} = \frac{U_0(T)}{1 + \left[\begin{array}{c} U_A(T) \cdot \left(\dfrac{V_{\text{gsteff}} + 2 \cdot V_{\text{th}}}{TOXE}\right) \\ + U_B(T) \cdot \left(\dfrac{V_{\text{gsteff}} + 2 \cdot V_{\text{th}}}{TOXE}\right)^2 \end{array}\right] \cdot (1 + U_C(T) \cdot V_{\text{bseff}})}. \tag{6.43}$$

When **mobMod** $= 2$,

$$\mu_{\text{eff}} = \frac{U_0(T)}{1 + (U_A(T) + U_C(T) \cdot V_{\text{bseff}}) \cdot \left[\dfrac{V_{\text{gsteff}} + C_0 \cdot (VTH0 - VFB - \Phi_{\text{s}})}{TOXE}\right]^{EU}} \tag{6.44}$$

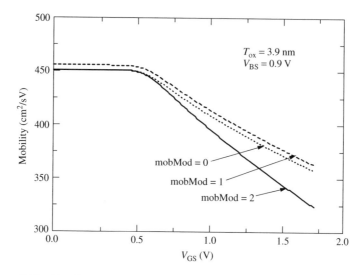

Figure 6.16 V_{GS} dependence of mobility behavior of different ***mobMod*** options

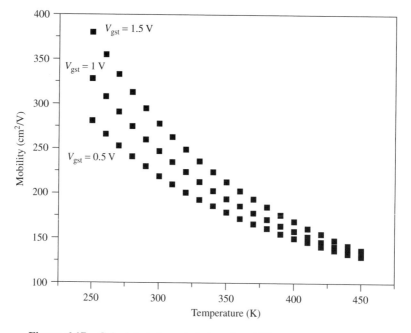

Figure 6.17 Calculated characteristics of mobility versus temperature

where C_0 is a constant and it is 2 for nMOS devices and 2.5 for pMOS devices; *U0, UA, UB, UC,* and *EU* are model parameters.

Figure 6.16 shows the calculated mobility behavior for different ***mobMod*** options to demonstrate the bias dependence of the mobility model when selecting different mobility model options. They approach a constant as V_{GS} tends to V_{th} even though this number is not exactly equal to U_0.

The temperature dependence of mobility is described by the following equations:

$$U_0(T) = U0 \cdot \left(\frac{T}{TNOM}\right)^{UTE},$$

$$(6.45)$$

$$U_A(T) = UA + UA1 \cdot \left(\frac{T}{TNOM} - 1\right),$$

$$(6.46)$$

$$U_B(T) = UB + UB1 \cdot \left(\frac{T}{TNOM} - 1\right),$$

$$(6.47)$$

$$U_C(T) = UC + UC1 \cdot \left(\frac{T}{TNOM} - 1\right),$$

$$(6.48)$$

where *UA, UB*, and *UC* are model parameters to describe the mobility behavior at temperature *TNOM*.

Figure 6.17 shows the calculated characteristics of mobility versus temperature at several different V_{GS}.

6.7 SOURCE/DRAIN RESISTANCE MODEL

The total source and drain series resistances in a MOSFET used in integrated circuit (IC) designs have several components such as the via resistance, the salicide resistance, the salicide-to-salicide contact resistance, and the sheet resistance in LDD region, as shown in Figure 6.18. However, the contact and LDD sheet resistances usually dominate the total resistance. The typical value of the sheet resistance is around $1\,k\Omega/sq$ in LDD region for a typical 0.25-μm CMOS technology and much smaller in more advanced technologies.

It has been known that the source/drain resistances are bias-dependent. In some compact models such as BSIM3v3 (Cheng *et al.* (1997a)), these bias dependencies are included.

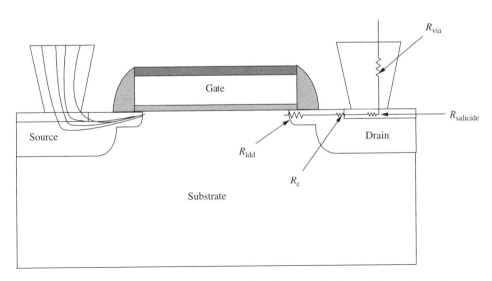

Figure 6.18   Different resistance components for the source/drain parasitic series resistances

However, since these parasitic resistances in BSIM3v3 are treated only as virtual components in the $I-V$ expressions to account for the DC voltage drop across these resistances, they are invisible to the signal in the AC simulation. Therefore, external components for these series resistances need to be added outside an intrinsic model to accurately describe the HF noise characteristics and the input AC impedance of the device (Enz and Cheng (2000)). In BSIM4, similar to that in BSIM3v3, the source/drain series resistances are modeled by a bias-independent component and a bias-dependent component. However, the users have the option to select the source/drain resistance model according to their model applications to decide where to put the source/drain series resistances. In other words, in addition to the resistance model in BSIM3v3, which embeds the source/drain resistance in the $I-V$ equation and assumes that series resistance at the source side equals the one at the drain side, BSIM4 introduces an asymmetric source/drain resistance model, which allows that the bias-dependent resistances at the source and the drain do not have to be equal and that they are physically connected between the external and the internal source/drain nodes, as shown in Figure 6.19. This asymmetric external source/drain resistance model is needed in simulating the high-frequency small-signal AC and noise behavior. A parameter called **rdsMod** is introduced in BSIM4 to select different source/drain resistance models. When **rdsMod** $= 0$, symmetric source/drain BSIM3v3 resistance model is used and when **rdsMod** $= 1$, the external asymmetric source/drain resistance model is selected. The detailed equations are given below.

When **rdsMod** $= 0$,

$$R_{\text{DS}} = \frac{R_{DSWMIN}(T) + R_{DSW}(T) \cdot \left[PRWB \cdot \left(\sqrt{\Phi_s - V_{\text{bseff}}} - \sqrt{\Phi_s} \right) + \dfrac{1}{1 + PRWG \cdot V_{\text{gsteff}}} \right]}{(10^6 \cdot W_{\text{effcj}})^{WR}}$$

(6.49)

where $PRWB$, $PRWG$, and WR are model parameters; $R_{DSWMIN}(T)$ and $R_{DSW}(T)$ are the parameters with temperature dependence, given in the following:

$$R_{DSW}(T) = RDSW + PRT \cdot \left(\frac{T}{TNOM} - 1 \right),$$

(6.50)

$$R_{DSWMIN}(T) = RDSWMIN + PRT \cdot \left(\frac{T}{TNOM} - 1 \right),$$

(6.51)

where $RDSWMIN$ and $RDSW$ are model parameters to describe the source/drain resistance at $TNOM$; PRT is a model parameter for the temperature dependence.

When **rdsMod** $= 1$,

$$R_{\text{D}} = \frac{R_{DWMIN}(T) + R_{DW}(T) \cdot \left[-PRWB \cdot V_{\text{BD}} + \dfrac{1}{1 + PRWG \cdot (V_{\text{GD}} - V_{\text{fbSD}})} \right]}{N_{\text{F}} \cdot (10^6 \cdot W_{\text{effcj}})^{WR}},$$

(6.52)

$$R_{\text{S}} = \frac{R_{SWMIN}(T) + R_{SW}(T) \cdot \left[-PRWB \cdot V_{\text{BS}} + \dfrac{1}{1 + PRWG \cdot (V_{\text{GS}} - V_{\text{fbSD}})} \right]}{N_{\text{F}} \cdot (10^6 \cdot W_{\text{effcj}})^{WR}},$$

(6.53)

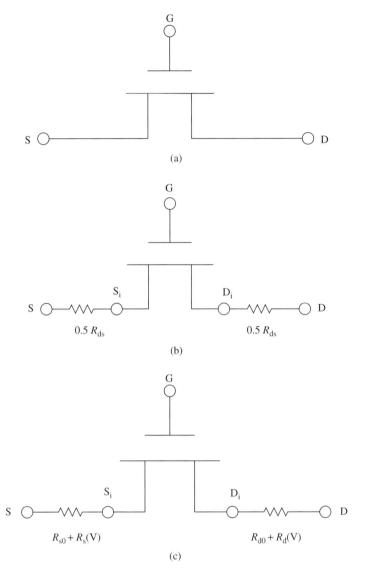

Figure 6.19 (a) Intrinsic model without any series source/drain resistance; (b) symmetric source/drain series resistance model **rdsMod** = 0; and (c) asymmetric external source/drain resistance model in BSIM4 (**rdsMod** = 1)

where V_{fbSD} is the calculated flat-band voltage between the gate and the source/drain diffusion regions,

$$V_{\text{fbSD}} = \frac{K_B T}{q} \ln \left(\frac{NGATE}{NSD} \right)$$

(6.54)

where $NGATE$ and NSD are model parameters for the doping concentration in the gate and source/drain regions.

In addition, diffusion source drain resistance R_{sdiff} and R_{ddiff} are also modeled to account for the layout difference in source/drain regions in different MOSFET applications. A parameter **rgeoMod** is introduced to turn on/off the components R_{sdiff} and R_{ddiff} in simulation. When **rgeoMod** = 0, R_{sdiff} and R_{ddiff} are set to zero; if **rgeoMod** = 1, R_{sdiff} and R_{ddiff} will be calculated differently, depending on which geometry-dependent parameters are given. We will discuss some details later when introducing the layout-dependent parasitics.

The temperature dependence of the *RDWMIN, RSWMIN, RDW,* and *RSW* are given by the following equations:

$$R_{DW}(T) = RDW + PRT \cdot \left(\frac{T}{TNOM} - 1 \right), \tag{6.55}$$

$$R_{SW}(T) = RSW + PRT \cdot \left(\frac{T}{TNOM} - 1 \right), \tag{6.56}$$

$$R_{DWMIN}(T) = RDWMIN + PRT \cdot \left(\frac{T}{TNOM} - 1 \right), \tag{6.57}$$

$$R_{SWMIN}(T) = RSWMIN + PRT \cdot \left(\frac{T}{TNOM} - 1 \right). \tag{6.58}$$

6.8 *I–V* MODEL

The details of the derivation of the *I–V* model can be found in Cheng and Hu (1999), Huang *et al.* (1994), and Cheng *et al.* (1997a). Here, we focus our discussion on understanding of the *I–V* model with different source/drain resistance options, without giving the derivation details of the *I–V* model equations. Depending on which resistance model is used in the model (**rdsMod** = 0 or 1, and $R_{DS}(V) = 0$ or $R_{DS}(V) \neq 0$), the *I–V* model and implementation are different. When $R_{DS}(V) = 0$, the *I–V* model is used only for an intrinsic device without series source/drain resistance. Since people usually do not use this option when extracting model parameters, we do not discuss this option in detail and only give a brief discussion when we discuss the *I–V* model, when **rdsMod** = 1, where the device is modeled by an intrinsic device and the external source/drain series resistance components are modeled by introducing two more nodes as shown in Figure 6.19(c).

6.8.1 *I–V* Model When rdsMod = 0 ($R_{DS}(V) \neq 0$)

When **rdsMod** = 0 and $R_{DS}(V) \neq 0$, the *I–V* model with so-called "virtual" series source/drain resistance components is used, in which the series source/drain resistance components are embedded in the *I–V* equation instead of "real" physical resistance components in the model implementation, so the impact of the source/drain resistance components is modeled in DC but not in AC and noise simulation (Enz and Cheng

2000). The complete single equation channel current model is given by

$$
I_{DS} = \frac{I_{ds0}}{1 + \dfrac{R_{DS} \cdot I_{ds0}}{V_{dseff}}} \cdot \left[1 + \frac{1}{C_{clm}} \cdot \ln\left(\frac{V_A}{V_{ASAT}} \right) \right] \cdot \left(1 + \frac{V_{DS} - V_{dseff}}{V_{ADIBL}} \right)
$$

$$
\cdot \left(1 + \frac{V_{DS} - V_{dseff}}{V_{ADITS}} \right) \cdot \left(1 + \frac{V_{DS} - V_{dseff}}{V_{ASCBE}} \right) \tag{6.59}
$$

where the contributions of velocity saturation, channel-length modulation (CLM), drain-induced barrier lowering (DIBL), and substrate current–induced body effect (SCBE) to the channel current and conductance have been included. We will give some detailed discussion below.

In Eq. (6.59), I_{ds0} is the channel current for an intrinsic device (without including the source/drain resistance) in the regions from strong inversion to subthreshold,

$$
I_{ds0} = \frac{W_{eff} \cdot \mu_{eff} \cdot C_{oxeff} \cdot V_{gsteff} \cdot V_{dseff} \cdot \left(1 - \dfrac{V_{dseff}}{2 \cdot V_b} \right)}{L_{eff} \cdot \left(1 + \dfrac{\mu_{eff} \cdot V_{dseff}}{2 \cdot V_{SAT}(T) \cdot L_{eff}} \right)} \tag{6.60}
$$

where L_{eff} and W_{eff} are given by Eqs. (6.7) and (6.9); V_{dseff} is introduced to ensure a smooth transition from triode to saturation region and is expressed as

$$
V_{dseff} = V_{dsat} - \frac{1}{2} \cdot \left(V_{dsat} - V_{DS} - DELTA + \sqrt{(V_{dsat} - V_{DS} - DELTA)^2 + 4 \cdot DELTA \cdot V_{dsat}} \right) \tag{6.61}
$$

where *DELTA* is a model parameter; V_{dsat} is the saturation voltage and is formulated as

$$
V_{dsat} = \frac{-b - \sqrt{b^2 - 4 \cdot a \cdot c}}{2 \cdot a}, \tag{6.62}
$$

$$
a = A_{bulk}^2 \cdot W_{eff} \cdot VSAT \cdot C_{oxe} \cdot R_{DS} + A_{bulk} \cdot \left(\frac{1}{\lambda} - 1 \right), \tag{6.63}
$$

$$
b = -\left[(V_{gsteff} + 2 \cdot V_t) \cdot \left(\frac{2}{\lambda} - 1 \right) + \frac{2 \cdot A_{bulk} \cdot VSAT \cdot L_{eff}}{\mu_{eff}} \right.
$$

$$
\left. + 3 \cdot A_{bulk} \cdot (V_{gsteff} + 2 \cdot V_t) \cdot W_{eff} \cdot VSAT \cdot C_{oxe} \cdot R_{DS} \right], \tag{6.64}
$$

$$
c = (V_{gsteff} + 2 \cdot V_t) \frac{2 \cdot VSAT \cdot L_{eff}}{\mu_{eff}} + 2 \cdot (V_{gsteff} + 2 \cdot V_t) \cdot W_{eff} \cdot VSAT \cdot C_{oxe} \cdot R_{DS}, \tag{6.65}
$$

$$
\lambda = A1 \cdot V_{gsteff} + A2. \tag{6.66}
$$

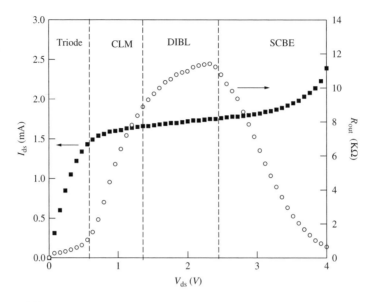

Figure 6.20   Output conductance behavior of a MOSFET at different bias regions

As given in Eq. (6.59), BSIM4 models the channel current and conductance by using the Early voltage concept well known in bipolar device modeling. One assumption has been made to derive the model, that is, the contribution of each physical mechanism such as CLM, DIBL, SCBE, and so on to channel current and conductance can be calculated independently. Figure 6.20 illustrates the contribution of each physical effect to the output conductance (resistance). The details of the derivation and analysis can be found in Huang (1992), Cheng *et al.* (1997a), and Cheng *et al.* (1999); here we introduce briefly the model equations and discuss the enhancements of the output conductance model in BSIM4.

The temperature dependence of saturation velocity is described by

$$V_{\text{SAT}}(T) = VSAT - AT \cdot \left(\frac{T}{TNOM} - 1 \right) \tag{6.67}$$

where $VSAT$ is the model parameter for saturation velocity at $TNOM$ and AT is a model parameter defining the temperature dependence of the saturation velocity.

V_A in Eq. (6.59) is the sum of the contribution of the Early voltage at the saturation voltage point and the contribution of the Early voltage from the CLM effect,

$$V_A = V_{\text{ASAT}} + V_{\text{ACLM}} \tag{6.68}$$

and $VASAT$ and $VACLM$ are given by

$$V_{\text{ASAT}} = \cfrac{\cfrac{2 \cdot VSAT \cdot L_{\text{eff}}}{\mu_{\text{eff}}} + V_{\text{dsat}} + 2 \cdot R_{\text{DS}} \cdot VSAT \cdot C_{\text{oxe}} \cdot W_{\text{eff}}}{R_{\text{DS}} \cdot VSAT \cdot C_{\text{oxe}} \cdot W_{\text{eff}} \cdot A_{\text{bulk}} - 1 + \cfrac{2}{\lambda}} \cdot V_{\text{gsteff}} \cdot \left[1 - \cfrac{A_{\text{bulk}} \cdot V_{\text{dsat}}}{2 \cdot (V_{\text{gsteff}} + 2 \cdot V_t)} \right], \tag{6.69}$$

$$V_{ACLM} = C_{\text{clm}} \cdot (V_{\text{DS}} - V_{\text{dsat}}), \tag{6.70}$$

$$C_{\text{clm}} = \frac{F_{\text{pocket}}}{PCLM \cdot litl} \cdot \left(L_{\text{eff}} + \frac{\mu_{\text{eff}} \cdot V_{\text{dsat}}}{2 \cdot VSAT}\right)$$

$$\cdot \left(1 + PVAG \cdot \frac{\mu_{\text{eff}} \cdot V_{\text{gsteff}}}{2 \cdot VSAT \cdot L_{\text{eff}}}\right) \cdot \left(1 + \frac{R_{\text{DS}} \cdot I_{\text{ds}_0}}{V_{\text{dseff}}}\right), \tag{6.71}$$

$$Litl = \sqrt{\frac{\varepsilon_{\text{si}} \cdot TOXE \cdot XJ}{EPSROX}}, \tag{6.72}$$

$$F_{\text{pocket}} = \frac{1}{1 + FPROUT \dfrac{\sqrt{L_{\text{eff}}}}{V_{\text{gsteff}} + 2 \cdot V_{\text{t}}}}, \tag{6.73}$$

where *PCLM, PVAG, FPROUT* are model parameters introduced to improve the accuracy. The Early voltage contributed by DIBL effect is formulated as

$$V_{ADIBL} = \frac{V_{\text{gsteff}} + 2 \cdot V_{\text{t}}}{\theta_{\text{rout}} \cdot (1 + PDIBLCB \cdot V_{\text{bseff}})} \cdot \left(1 - \frac{A_{\text{bulk}} \cdot V_{\text{dsat}}}{A_{\text{bulk}} \cdot V_{\text{dsat}} + V_{\text{gsteff}} + 2 \cdot V_{\text{t}}}\right)$$

$$\cdot \left(1 + PVAG \frac{\mu_{\text{eff}} \cdot V_{\text{gsteff}}}{2 \cdot VSAT \cdot L_{\text{eff}}}\right), \tag{6.74}$$

$$\theta_{\text{rout}} = PDIBLC2 + \frac{PDIBLC1}{2 \cdot \cosh\left(\dfrac{DROUT \cdot L_{\text{eff}}}{l_{\text{t0}}}\right) - 2}, \tag{6.75}$$

where *PDIBLC1, PDIBLC2, DROUT* are model parameters that have been introduced in BSIM3v3. However, the cosh function has been used to replace the exponential function in BSIM3v3 for the length dependence of V_{ADIBL}.

The Early voltage due to the contribution of the substrate current is calculated by

$$V_{ASCBE} = \frac{L_{\text{eff}}}{PSCBE2} e^{PSCBE1 \cdot Litl / V_{\text{DS}} - V_{\text{dsat}}} \tag{6.76}$$

where *PSCBE*1 and *PSCBE*2 are model parameters. The dependence of Early voltage on SCBE is determined by V_{dsat} and *Litl* built in the equation.

An enhancement in the output conductance model in BSIM4 is the inclusion of the modeling of the drain-induced threshold shift (DITS) caused by pocket implantation. The influence of DITS on the Early voltage can be expressed as

$$V_{ADITS} = \frac{F_{\text{pocket}}}{PDITS} \cdot [1 + (1 + PDITSL \cdot L_{\text{eff}}) \cdot e^{PDITSD \cdot V_{\text{DS}}}] \tag{6.77}$$

where *PDITS, PDITSL,* and *PDITLD* are model parameters.

6.8.2 *I–V* Model When *rdsMod* = 1($R_{\text{DS}}(V) = 0$)

When **rdsMod** = 1, the *I–V* characteristics of the device are described by a subcircuit that consists of an intrinsic portion and external series source and drain resistance components.

For the intrinsic device $(R_{DS}(V) = 0)$, in the linear region $(V_{DS} < V_{dsat})$, the I–V equation has been given in Eq. (6.60) and rewritten as follows:

$$I_{ds0} = \frac{W_{eff} \cdot \mu_{eff} \cdot C_{oxeff} \cdot V_{gsteff} \cdot V_{dseff} \cdot \left(1 - \dfrac{V_{dseff}}{2 \cdot V_b}\right)}{L_{eff} \cdot \left(1 + \dfrac{\mu_{eff} \cdot V_{dseff}}{2 \cdot V_{SAT}(T) \cdot L_{eff}}\right)} \tag{6.78}$$

where V_{dsat} is the intrinsic saturation voltage and is given by

$$V_{dsat} = \frac{E_{sat} \cdot L \cdot (V_{gsteff} + 2 \cdot V_t)}{A_{bulk} \cdot E_{sat} \cdot L + V_{gsteff} + 2 \cdot V_t}, \tag{6.79}$$

$$E_{sat} = \frac{2 \cdot VSAT}{\mu_{eff}}. \tag{6.80}$$

In saturation region, the Eqs. (6.70) to (6.77) are used, except that the parameter R_{DS} is set to zero. The model covers from strong inversion, through the weak inversion transition, to subthreshold region.

6.9 GATE TUNNELING CURRENT MODEL

As the gate oxide thickness is scaled down to 3 nm and below, gate leakage current due to direct carrier tunneling becomes significant. The carriers to conduct the gate leakage can be either electron or holes or both, depending on the type of the gate and the bias conditions.

In BSIM4, the gate tunneling current is modeled by several different components, the tunneling current between the gate and the substrate (I_{GB}), the current between the gate and the channel (I_{GC}), and the currents between the gate and the source/drain diffusion regions (I_{GS} and I_{GD}) (Liu et al. (2000)). I_{GC} can be further partitioned between the source and the drain, that is, $I_{GC} = I_{GCS} + I_{GCD}$. Figure 6.21 illustrates these gate tunneling current components in an nFET.

Depending on the model applications, the gate leakage model can be turned on or off by model selectors **igcMod** and **igbMod**. When **igcMod** $= 1$, BSIM4 calculates the contribution of I_{GC}, I_{GS}, and I_{GD} in the simulation. When **igbMod** $= 1$, I_{GB} will be calculated. When the selectors are set to zero in the model file, no gate tunneling currents are calculated in the simulation.

6.9.1 Gate-to-substrate Tunneling Current I_{GB}

Depending on the bias conditions, devices can operate in accumulation or inversion regimes, in which the gate tunneling current I_{GB} can be significant.

In accumulation region, $I_{GB,acc}$ is determined by electron tunneling from conduction band and can be expressed as (Liu et al. (2000))

$$I_{GB,acc} = W_{eff} \cdot L_{eff} \cdot A \cdot T_{oxratio} \cdot V_{GB} \cdot V_{aux} \cdot \exp[-B \cdot TOXE$$
$$\cdot (AIGBACC - BIGBACC \cdot V_{oxacc}) \cdot (1 + CIGBACC \cdot V_{oxacc})] \tag{6.81}$$

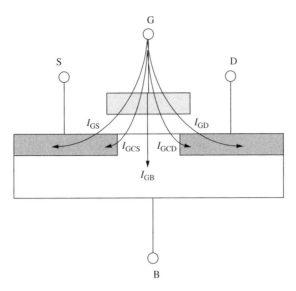

Figure 6.21   Gate-to-source/drain/bulk tunneling components in a MOSFET

where the physical constants $A = 4.97232 \times 10^{-7}$ A/V^2, $B = 7.45669 \times 10^{11}$, *AIGBACC*, *BIGBACC*, and *CIGBACC* are model parameters, and

$$T_{\text{oxratio}} = \left(\frac{TOXREF}{TOXE}\right)^{NTOX} \cdot \frac{1}{TOXE^2}, \tag{6.82}$$

$$V_{\text{aux}} = NIGBACC \cdot V_t \cdot \ln\left[1 + \exp\left(-\frac{V_{\text{GB}} - V_{\text{fbzb}}}{NIGBACC \cdot V_t}\right)\right], \tag{6.83}$$

$$V_{\text{oxacc}} = V_{\text{fbzb}} - V_{\text{FBeff}}, \tag{6.84}$$

where *NIGBACC, TOXREF*, and *NTOX* are model parameters; V_{oxacc} is the voltage across the oxide of the device in accumulation and has a form of Eq. (6.84) to avoid the discontinuity of V_{oxacc} from accumulation through depletion to inversion. V_{fbzb} is the flat-band voltage calculated from zero-bias V_{th}

$$V_{\text{fbzb}} = V_{\text{th}}|_{V_{\text{bs}}=V_{\text{ds}}=0} - \Phi_s - K1 \cdot \sqrt{\Phi_s} \tag{6.85}$$

and

$$V_{\text{FBeff}} = V_{\text{fbzb}} - \frac{1}{2} \cdot \left(V_{\text{fbzb}} - V_{\text{GB}} - 0.02 + \sqrt{(V_{\text{fbzb}} - V_{\text{GB}} - 0.02)^2 + 0.08 \cdot V_{\text{fbzb}}}\right). \tag{6.86}$$

Depending on the bias conditions, devices can operate in accumulation or inversion regimes, in which the gate tunneling current I_{GB} can be significant.

In inversion region, $I_{\text{GB,inv}}$ is determined by electron tunneling from Valence band and can be expressed as

$$I_{\text{gb,inv}} = W_{\text{eff}} \cdot L_{\text{eff}} \cdot A_{\text{inv}} \cdot T_{\text{oxratio}} \cdot V_{\text{GB}} \cdot V_{\text{aux,inv}} \cdot \exp[-B_{\text{inv}} \cdot TOXE$$
$$\cdot (AIGBINV - BIGBINV \cdot V_{\text{oxinv}}) \cdot (1 + CIGBINV \cdot V_{\text{oxinv}})] \tag{6.87}$$

where the physical constants $A_{inv} = 3.75956 \times 10^{-7}$ A/V^2, $B_{inv} = 9.82222 \times 10^{11}$, *AIG-BINV*, *BIGBINV*, and *CIGBINV* are model parameters, and

$$V_{aux,inv} = NIGBINV \cdot V_t \cdot \ln\left[1 + \exp\left(\frac{V_{oxdepinv} - EIGBINV}{NIGBINV \cdot V_t}\right)\right], \qquad (6.88)$$

$$V_{ox,depinv} = K_{1ox}\sqrt{\Phi_s} + V_{gsteff}, \qquad (6.89)$$

where *NIGBINV* and *EIGBINV* are model parameters; $V_{ox,depinv}$ is the voltage across the oxide of the device in depletion and inversion regions and has a form of Eq. (6.89) to avoid the discontinuity of $V_{ox,depinv}$ from accumulation through depletion to inversion.

6.9.2 Gate-to-channel and Gate-to-S/D Currents

The gate-to-channel current in an *n*FET is determined by the electron tunneling from conduction band and the gate-to-channel current in a *p*FET is determined by the hole tunneling from valence band. The following equation can be used to describe the gate-to-channel current (I_{GC}) in both *n*FETs and *p*FETs,

$$I_{GC} = W_{eff} \cdot L_{eff} \cdot A \cdot T_{oxratio} \cdot V_{gse} \cdot V_{aux,gc} \qquad (6.90)$$
$$\cdot \exp[-B \cdot TOXE \cdot (AIGC - BIGC \cdot V_{oxdepinv}) \cdot (1 + CIGC \cdot V_{oxdepinc})]$$

where *AIGC*, *BIGC*, and *CIGC* are model parameters; $A = 4.97232$ A/V^2 for *n*FETs and 3.42537 A/V^2 for *p*FETs, $B = 7.45669 \times 10^{11}$ g$^{1/2}$/(F$^{1/2}$ s) for *n*FETs and 1.6645×10^{12} g$^{1/2}$/(F$^{1/2}$ s) for *p*FETs; and

$$V_{aux,gc} = NIGC \cdot V_t \cdot \ln\left[1 + \exp\left(\frac{V_{gse} - VTH0}{NIGC \cdot V_t}\right)\right] \qquad (6.91)$$

where *NIGC* is a model parameter.

In the above discussion, we did not account for the influence of the drain bias. With the consideration of the V_{ds} bias, I_{GC} is partitioned into two components between the drain and the source. The gate-to-channel tunneling current partitioned to the source, I_{GCS}, is given by

$$I_{GCS} = I_{GC}\frac{PIGCD \cdot V_{DS} + \exp(-PIGCD \cdot V_{DS}) - 1 + 10^{-4}}{(PIGCD \cdot V_{DS})^2 + 2 \cdot 10^{-4}} \qquad (6.92)$$

and the gate-to-channel tunneling current partitioned to the drain, I_{GCD}, is given by

$$I_{GCD} = I_{GC}\frac{1 - (PIGCD \cdot V_{DS} + 1) \cdot \exp(-PIGCD \cdot V_{DS}) + 10^{-4}}{(PIGCD \cdot V_{DS})^2 + 2 \cdot 10^{-4}} \qquad (6.93)$$

where *PIGCD* is a model parameter whose default value is given by

$$PIGCD = \frac{B \cdot TOXE}{V_{gsteff}^2}\left(1 - \frac{V_{dseff}}{2 \cdot V_{gsteff}}\right). \qquad (6.94)$$

As mentioned earlier, in addition to the gate-to-channel tunneling current, gate tunneling current between the gate and the source/drain region (I_{GS} and I_{GD}) should be accounted for in modeling of the gate tunneling,

$$I_{GS} = W_{\text{eff}} \cdot DLCIG \cdot A \cdot T_{\text{oxratioEdge}} \cdot V_{GS} \cdot V_{GS}' \cdot \exp\left[-B \cdot TOXE \right.$$
$$\left. \cdot POXEDGE \cdot (AIGSD - BIGSD \cdot V_{GS}') \cdot (1 + CIGSD \cdot V_{GS}')\right] \quad (6.95)$$

and

$$I_{GD} = W_{\text{eff}} \cdot DLCIG \cdot A \cdot T_{\text{oxratioEdge}} \cdot V_{GD} \cdot V_{GD}' \cdot \exp[-B \cdot TOXE$$
$$\cdot POXEDGE \cdot (AIGSD - BIGSD \cdot V_{GD}') \cdot (1 + CIGSD \cdot V_{GD}')] \quad (6.96)$$

where $DLCIG$, $AIGSD$, $BIGSD$, $CIGSD$, and $POXEDGE$ are model parameters; and

$$T_{\text{oxratioEdge}} = \left(\frac{TOXREF}{TOXE \cdot POXEDGE}\right)^{NTOX} \cdot \frac{1}{(TOXE \cdot POXEDGE)^2}, \quad (6.97)$$

$$V_{GS}' = \sqrt{(V_{GS} - V_{\text{fbSD}})^2 + 10^{-4}}, \quad (6.98)$$

$$V_{GD}' = \sqrt{(V_{GD} - V_{\text{fbSD}})^2 + 10^{-4}}, \quad (6.99)$$

where V_{fbSD} is the flat-band voltage between the gate and the S/D diffusions and is given by the following equation:

$$V_{\text{fbSD}} = \frac{k_B T}{q} \cdot \ln\left(\frac{NGATE}{NSD}\right). \quad (6.100)$$

6.10 SUBSTRATE CURRENT MODELS

The modeling of substrate current is important in today's MOSFETs, especially for analog circuit design. In addition to the junctions diode current and gate-to-body tunneling current, two other current components dominate the contribution of the total substrate current in a MOSFET: one is the substrate current due to impact ionization of the channel current and the other is the gate-induced drain leakage (GIDL) current. Next, we will discuss the models for these substrate current components.

6.10.1 Model for Substrate Current due to Impact Ionization of Channel Current

The substrate current due to impact ionization of the channel current in BSIM4 uses the same model equation as that in BSIM3v3.2,

$$I_{ii} = \frac{ALPHA0 + ALPHA1 \cdot L_{\text{eff}}}{L_{\text{eff}}}(V_{DS} - V_{\text{dseff}}) \cdot \exp\left(-\frac{BETA0}{V_{DS} - V_{\text{dseff}}}\right) \cdot I_{\text{dsa}} \quad (6.101)$$

where *ALPHA0* and *BETA0* are model parameters for impact ionization coefficients; *ALPHA1* is also a model parameter that is introduced to improve the model scalability, and

$$I_{dsa} = \frac{I_{ds0}}{1 + \dfrac{R_{DS} \cdot I_{ds0}}{V_{dseff}}} \cdot \left[1 + \frac{1}{C_{clm}} \cdot \ln\left(\frac{V_A}{V_{ASAT}}\right) \right]$$

$$\cdot \left(1 + \frac{V_{DS} - V_{dseff}}{V_{ADIBL}}\right) \cdot \left(1 + \frac{V_{DS} - V_{dseff}}{V_{ADITS}}\right) \tag{6.102}$$

where the detailed equations of I_{ds0}, R_{ds}, V_{dseff}, V_{ASAT}, C_{clm}, V_{ADIBL}, V_{ADITS}, and V_A have been discussed previously when discussing the I–V model.

6.10.2 Models for Gate-induced Drain Leakage (GIDL) and Gate-induced Source Leakage (GISL) Currents

The models for GIDL and GISL currents can be formulated as (Liu *et al.* (2000))

$$I_{GIDL} = AGIDL \cdot W_{effCj} \cdot \frac{V_{DS} - V_{gse} - EGIDL}{3 \cdot T_{oxe}}$$

$$\cdot \exp\left(-\frac{2 \cdot T_{oxe} \cdot BGIDL}{V_{DS} - V_{gse} - EGIDL}\right) \cdot \frac{V_{DB}^3}{CGIDL + V_{DB}^3}, \tag{6.103}$$

$$I_{GISL} = AGIDL \cdot W_{effCj} \cdot \frac{-V_{DS} - V_{gse} - EGIDL}{3 \cdot T_{oxe}}$$

$$\cdot \exp\left(-\frac{3 \cdot T_{oxe} \cdot BGIDL}{-V_{DS} - V_{gse} - EGIDL}\right) \cdot \frac{V_{SB}^3}{CGIDL + V_{SB}^3}, \tag{6.104}$$

where *AGIDL*, *BGIDL*, *CGIDL*, and *EGIDL* are model parameters. *CGIDL* is introduced to account for the body-bias dependence of I_{GIDL} and I_{GISL}.

6.11 CAPACITANCE MODELS

BSIM4 provides three options for selecting different capacitance models. With only a minor change in separating CKAPPA parameter into different parameters for gate-source and gate-drain overlap capacitances, BSIM4 maintains the model equations and model parameters in BSIM3v3. However, the option **capMod** = 1 in BSIM3v3 is no longer supported in BSIM4. So users should be aware of this fact when selecting the capacitance models in BSIM4 because the meaning of **capMod** parameter has been changed in BSIM4 from that in BSIM3v3. In other words, when one selects **capMod** = 1 in BSIM4, the capacitance model is actually the one for **capMod** = 2 in BSIM3v3 and when one selects **capMod** = 2 in BSIM4, the capacitance model is actually the one for **capMod** = 3 in

BSIM3v3. The following table gives the correspondence between the ***capMod*** parameter and the capacitance models/parameters in BSIM3v3 and BSIM4.

In this section, we will discuss the charge and capacitance models. The charge model is the basis of the capacitance model. The space charge of a MOS structure consists of three fundamental components: the charge on the gate electrode, Q_G, the charge in the bulk depletion layer, Q_B, and the mobile charge in the channel region, Q_{INV}. Generally, the following relationship holds in a MOSFET:

$$Q_G + Q_{INV} + Q_B = 0 \qquad (6.105)$$

and

$$Q_{INV} = Q_D + Q_S \qquad (6.106)$$

where Q_D and Q_S are the channel associated with the drain node and the source node, respectively.

Capacitance between any two of the four terminals (gate, source, drain, and bulk) is defined as

$$C_{ij} = \frac{\partial Q_i}{\partial V_{ij}} \quad i \neq j; \quad i, j = G, D, S, B, \qquad (6.107)$$

$$C_{ij} = -\frac{\partial Q_i}{\partial V_{ij}} \quad i = j; \quad i, j = G, D, S, B. \qquad (6.108)$$

Depending on the bias conditions, the device can operate in accumulation, depletion, and strong inversion regions, which are further divided into linear and saturation regimes. By selecting different ***capMod*** options, different model equations can be used to describe the characteristics of the charge and the capacitances.

6.11.1 Intrinsic Capacitance Models

*(1) Capacitance model option 1 (**capMod** = 0)*

When ***capMod*** $= 0$ is input in the model file, the piecewise long-channel charge and capacitance model is used in the simulation. The capacitance model for this selection is simpler than other capacitance models (***capMod*** $= 1$ or 2). However, discontinuity in both charge and capacitance characteristics exists at the boundary of different operation regimes such as from accumulation to depletion and from depletion to strong inversion.

If $V_{GS} < VFBCV + V_{BS}$, the device operates in the accumulation region,

$$QG = W_{active} \cdot L_{active} \cdot C_{oxe} \cdot (V_{GS} - V_{bseff} - VFBCV), \qquad (6.109)$$

$$Q_B = -Q_G, \qquad (6.110)$$

$$Q_{INV} = 0, \qquad (6.111)$$

where W_{active} and L_{active} are the effective channel width and channel length of the device; *VFBCV* is a user-defined model parameter for the flat-band voltage.

If $VFBCV + V_{BS} < V_{GS} < V_{th}$, the device is in the subthreshold region and the charge expression becomes

$$Q_B = -W_{active} \cdot L_{active} \cdot C_{oxe} \cdot \frac{K_{1ox}{}^2}{2} \left(-1 + \sqrt{1 + \frac{4 \cdot (V_{GS} - VFBCV - V_{bseff})}{K_{1ox}{}^2}}\right),$$

(6.112)

$$Q_G = -Q_B,$$

(6.113)

$$Q_{INV} = 0.$$

(6.114)

If $V_{GS} > V_{th}$, the device is in strong inversion. Similar to the DC case, the device operates either in the linear or in the saturation regime, depending on whether the drain-to-source bias is lower or higher than the saturation voltage given below:

$$V_{dsat,cv} = \frac{V_{GS} - V_{th}}{A_{bulk,cv}},$$

(6.115)

$$A_{bulk,cv} = A_{bulk} \cdot \left[1 + \left(\frac{CLC}{L_{active}}\right)^{CLE}\right],$$

(6.116)

$$V_{th} = VFBCV + \Phi_s + K 1_{ox} \cdot \sqrt{\Phi_s - V_{bseff}},$$

(6.117)

where the detailed form of A_{bulk} has been given in the discussion of the $I-V$ model; CLC and CLE are model parameters.

When $V_{GS} > V_{th}$, the device operates in strong inversion region. When $V_{DS} < V_{dsat,cv}$, one can derive the following equations for the charges at the gate and bulk of the device in linear region.

$$Q_G = W_{active} L_{active} C_{oxe}$$

$$\cdot \left(V_{GS} - VFBCV - \Phi_s - \frac{V_{DS}}{2} + \frac{A_{bulk,cv} \cdot V_{DS}{}^2}{12 \cdot \left(V_{GS} - V_{th} - \frac{A_{bulk,cv} \cdot V_{DS}}{2}\right)}\right),$$

(6.118)

$$Q_B = W_{active} L_{active} C_{oxe}$$

$$\cdot \left(VFBCV - V_{th} - \Phi_s + \frac{(1 - A_{bulk,cv}) \cdot V_{DS}}{2} - \frac{(1 - A_{bulk,cv}) \cdot A_{bulk,cv} \cdot V_{DS}{}^2}{12 \cdot \left(V_{GS} - V_{th} - \frac{A_{bulk,cv} \cdot V_{DS}}{2}\right)}\right).$$

(6.119)

The inversion charges can be further partitioned into $Q_{INV} = Q_S + Q_D$. Different charge partition schemes, controlled by a model parameter $XPART$, are adopted to partition the Q_S and Q_D into different ratio, 0/100, 50/50, and 40/60 with $XPART = 1$ (or any value larger 0.5), 0.5, and 0 (or any value smaller than 0.5).

When $XPART = 0$ (or <0.5), the 40/60 charge partition scheme is used in modeling the channel charge partitioning into drain and source.

$$Q_D = -W_{active} \cdot L_{active} \cdot C_{oxe} \cdot \left[\frac{V_{GS} - V_{th}}{2} - \frac{A_{bulk,cv}}{2} \cdot V_{DS} \right.$$

$$\left. + \frac{A_{bulk,cv} \cdot V_{DS} \cdot \left[\frac{(V_{GS} - V_{th})^2}{6} - \frac{A_{bulk,cv} \cdot (V_{GS} - V_{th})}{8} + \frac{(A_{bulk,cv} \cdot V_{DS})^2}{40} \right]}{\left(V_{GS} - V_{th} - \frac{A_{bulk,cv}}{2} \cdot V_{DS} \right)^2} \right],$$

$$\tag{6.120}$$

$$Q_S = -(Q_G + Q_B + Q_D). \tag{6.121}$$

When $XPART = 0.5$, the 50/50 charge partition scheme is adopted in partitioning the channel charge into drain and source.

$$Q_D = Q_S = -W_{active} \cdot L_{active} \cdot C_{oxe} \cdot \left[\frac{V_{GS} - V_{th}}{2} - \frac{A_{bulk,cv}}{4} \cdot V_{DS} \right.$$

$$\tag{6.122}$$

$$\left. + \frac{\left(A_{bulk,cv} \cdot V_{DS} \right)^2}{24 \cdot \left(V_{GS} - V_{th} - \frac{A_{bulk,cv}}{2} \cdot V_{DS} \right)} \right].$$

When $XPART = 1$ (or >0.5), the 0/100 charge partition scheme is used, and the charge at the drain and source can be formulated as

$$Q_D = -W_{active} \cdot L_{active} \cdot C_{oxe} \cdot \left[\frac{V_{GS} - V_{th}}{2} + \frac{A_{bulk,cv}}{4} \cdot V_{DS} \right.$$

$$\left. - \frac{\left(A_{bulk,cv} \cdot V_{DS} \right)^2}{24 \cdot \left(V_{GS} - V_{th} - \frac{A_{bulk,cv}}{2} \cdot V_{DS} \right)} \right], \tag{6.123}$$

$$Q_S = -(Q_G + Q_B + Q_D). \tag{6.124}$$

When $V_{DS} > V_{dsat,cv}$, the device operates in saturation region. One can derive the following equations for the charges at the gate and bulk:

$$Q_G = W_{active} L_{active} C_{oxe} \cdot \left(V_{GS} - VFBCV - \Phi_s - \frac{V_{dsat,cv}}{3} \right), \tag{6.125}$$

$$Q_B = W_{\text{active}} L_{\text{active}} C_{\text{oxe}} \cdot \left(VFBCV - V_{\text{th}} + \Phi_s + \frac{(1 - A_{\text{bulk,cv}}) \cdot V_{\text{dsat,cv}}}{3} \right). \quad (6.126)$$

Similar to the case in the linear region, the inversion charges can be further partitioned into $Q_{\text{INV}} = Q_S + Q_D$. When $XPART = 0$ (or <0.5), the 40/60 charge partition scheme is used in partitioning the channel charge into drain and source.

$$Q_D = -\frac{4 \cdot W_{\text{active}} \cdot L_{\text{active}} \cdot C_{\text{oxe}}}{15} \cdot (V_{\text{GS}} - V_{\text{th}}), \quad (6.127)$$

$$Q_S = -(Q_G + Q_B + Q_D). \quad (6.128)$$

When $XPART = 0.5$, the 50/50 charge partition scheme is adopted,

$$Q_D = Q_S = -W_{\text{active}} \cdot L_{\text{active}} \cdot C_{\text{oxe}} \cdot \left(\frac{V_{\text{GS}} - V_{\text{th}}}{3} \right). \quad (6.129)$$

When $XPART = 1$ (or >0.5), the charge at the drain and source can be formulated as

$$Q_D = 0, \quad (6.130)$$

$$Q_S = -(Q_G + Q_B). \quad (6.131)$$

(2) Capacitance model option 2 (*capMod* = 1)

In the capacitance model with **capMod** $= 0$, piecewise equations are used in different operation regimes, so the model has discontinuities and nonsmooth transitions at the boundaries between accumulation and depletion and between subthreshold region and strong inversion region. In the capacitance model with **capMod** $= 1$, which was the **capMod** $= 2$ model in BSIM3v3, the discontinuities are removed by introducing smooth functions at the boundaries of the operation regimes in deriving the charge/capacitance equations.

While charge and capacitance characteristics of a MOSFET transit between the accumulation and depletion regimes, a smooth function for the effective flat-band voltage is introduced as given in the following:

$$V_{\text{FBeff}} = V_{\text{fbzb}} - \frac{1}{2} \cdot \left(V_{\text{fbzb}} - V_{\text{GB}} - 0.02 \right.$$

$$\left. + \sqrt{(V_{\text{fbzb}} - V_{\text{GB}} - 0.02)^2 + 0.08 \cdot V_{\text{fbzb}}} \right), \quad (6.132)$$

$$V_{\text{fbzb}} = V_{\text{th}}|_{V_{\text{BS}}=V_{\text{DS}}=0} - \Phi_s - K1 \cdot \sqrt{\Phi_s}. \quad (6.133)$$

Another smooth function $V_{\text{gsteff,cv}}$, similar to the V_{gsteff} in DC model but with a simpler form, is used to ensure the continuous transition from depletion to strong inversion,

$$V_{\text{gsteff,cv}} = NOFF \cdot n \cdot V_t \cdot \ln \left[1 + \exp \left(\frac{V_{\text{gse}} - V_{\text{th}} - VOFFCV}{NOFF \cdot n \cdot V_t} \right) \right] \quad (6.134)$$

where *NOFF* and *VOFFCV* are model parameters.

To smooth out the transition of charge/capacitance between the linear and saturation regions, one smooth function, similar to Eq. (6.61) in *I–V* model, is used with a constant value (0.02) for δ_4, a *DELTA*-like parameter,

$$V_{\text{cveff}} = V_{\text{dsat,cv}} - \frac{1}{2} \cdot \left(V_{\text{dsat,cv}} - V_{\text{DS}} - \delta_4 + \sqrt{(V_{\text{dsat,cv}} - V_{\text{DS}} - \delta_4)^2 + 4 \cdot \delta_4 \cdot V_{\text{dsat,cv}}} \right)$$

(6.135)

where $V_{\text{dsat,cv}}$ is the saturation voltage given by Eq. (6.115).

With the considerations of the charge/capacitance model continuities, the **capMod** $= 1$ model can be derived and the following relationships hold:

$$Q_{\text{G}} = -(Q_{\text{INV}} + Q_{\text{B}}),$$

(6.136)

$$Q_{\text{B}} = -(Q_{\text{ACC}} + Q_{\text{SUB0}} + \delta Q_{\text{SUB}}),$$

(6.137)

$$Q_{\text{ACC}} = -W_{\text{active}} \cdot L_{\text{active}} \cdot C_{\text{oxe}} \cdot (V_{\text{FBeff}} - V_{\text{fbzb}}),$$

(6.138)

$$Q_{\text{SUB0}} = -W_{\text{active}} \cdot L_{\text{active}} \cdot C_{\text{oxe}} \cdot \frac{K_{\text{1ox}}^2}{2}$$
$$\cdot \left(-1 + \sqrt{1 + \frac{4 \cdot (V_{\text{gse}} - V_{\text{FBeff}} - V_{\text{gsteff,cv}} - V_{\text{bseff}})}{K_{\text{1ox}}^2}} \right),$$

(6.139)

$$\delta Q_{\text{SUB}} = W_{\text{active}} \cdot L_{\text{active}} \cdot C_{\text{oxe}}$$
$$\cdot \left[\frac{1 - A_{\text{bulk,cv}}}{2} \cdot V_{\text{cveff}} - \frac{(1 - A_{\text{bulk,cv}}) \cdot A_{\text{bulk,cv}} \cdot V_{\text{cveff}}^2}{12 \cdot \left(V_{\text{gsteff,cv}} - \frac{A_{\text{bulk,cv}}}{2} \cdot V_{\text{cveff}} \right)} \right],$$

(6.140)

$$Q_{\text{INV}} = -W_{\text{active}} \cdot L_{\text{active}} \cdot C_{\text{ox}}$$
$$\cdot \left[V_{\text{gsteff,cv}} - \frac{A_{\text{bulk,cv}}}{2} \cdot V_{\text{cveff}} + \frac{A_{\text{bulk,cv}}^2 \cdot V_{\text{cveff}}^2}{12 \cdot \left(V_{\text{gsteff,cv}} - \frac{A_{\text{bulk,cv}}}{2} \cdot V_{\text{cveff}} \right)} \right].$$

(6.141)

When *XPART* $= 0.5$, the 50/50 charge partition scheme is used. The charges at the source and drain terminal can be described by

$$Q_{\text{S}} = Q_{\text{D}} = \frac{Q_{\text{INV}}}{2} = -W_{\text{active}} \cdot L_{\text{active}} \cdot C_{\text{ox}} \cdot \left[V_{\text{gsteff,cv}} - \frac{A_{\text{bulk,cv}}}{2} \cdot V_{\text{cveff}} \right.$$

(6.142)

$$\left. + \frac{A_{\text{bulk,cv}}^2 \cdot V_{\text{cveff}}^2}{12 \cdot \left(V_{\text{gsteff,cv}} - \frac{A_{\text{bulk,cv}}}{2} \cdot V_{\text{cveff}} \right)} \right].$$

When *XPART* $= 0$ (or <0.5), the 40/60 charge partition scheme is used,

$$Q_S = -\frac{W_{active} \cdot L_{active} \cdot C_{oxe}}{2 \cdot \left(V_{gsteff,cv} - \frac{A_{bulk,cv}}{2} \cdot V_{cveff}\right)^2} \cdot \left[V_{gsteff,cv}{}^3 - \frac{4}{3} \cdot V_{gsteff,cv}{}^2 \cdot A_{bulk,cv} \cdot V_{cveff}\right.$$

$$+ \frac{2}{3} \cdot V_{gsteff,cv} \cdot (A_{bulk,cv} \cdot V_{cveff})^2 - \frac{2}{15} \cdot (A_{bulk,cv} \cdot V_{cveff})^3 \left.\right], \tag{6.143}$$

$$Q_D = -(Q_G + Q_B + Q_S). \tag{6.144}$$

When *XPART* $= 1$ (or >0.5), the 0/100 charge partition scheme is adopted,

$$Q_S = -\frac{W_{active} \cdot L_{active} \cdot C_{oxe}}{2} \cdot \left[V_{gsteff,cv} + \frac{1}{2} \cdot A_{bulk,cv} \cdot V_{cveff}\right.$$

$$- \frac{(A_{bulk,cv} \cdot V_{cveff})^2}{12 \cdot (V_{gsteff,cv} - \frac{A_{bulk,cv}}{2} \cdot V_{cveff})} \left.\right], \tag{6.145}$$

$$Q_D = -(Q_G + Q_B + Q_S). \tag{6.146}$$

(3) Capacitance model option 3 (*capMod* $= 2$)

The capacitance model with **capMod** $= 2$ in BSIM4 is compatible with the capacitance model with **capMod** $= 3$ in BSIM3v3 as shown in Table 6.1. This model accounts for the quantization effect by considering finite charge thickness in deriving the charge and capacitance equations. Similarly, the following equations hold:

$$Q_G = -(Q_{INV} + Q_B), \tag{6.147}$$

$$Q_B = -(Q_{ACC} + Q_{SUB0} + \delta Q_{SUB}). \tag{6.148}$$

Table 6.1 BSIM4 capacitance model options

Capacitance models in BSIM4	Model details referencing to BSIM3v3
CapMod $= 0$	Copied from **CapMod** $= 0$ model in BSIM3v3
CapMod $= 1$	Copied from **capMod** $= 2$ model in BSIM3v3
CapMod $= 2$ (default model)	Intrinsic capacitance model copied from **capMod** $= 3$ and overlap/fringing capacitance model copied from **capMod** $= 2$ in BSIM3v3

With the consideration of the finite charge thickness in the channel, the charge in the accumulation region can be derived as

$$Q_{ACC} = W_{active} \cdot L_{active} \cdot C_{oxeff} \cdot V_{gbacc} \qquad (6.149)$$

where C_{oxeff} is the effective gate oxide capacitance given in Eq. (6.1); V_{gbacc} is a smooth function for the effective gate-to-body voltage introduced to ensure the model continuity from accumulation to depletion and is given by

$$V_{gbacc} = \frac{1}{2} \cdot \left[V_{fbzb} + V_{bseff} - V_{gse} - 0.02 + \sqrt{(V_{fbzb} + V_{bseff} - V_{gse} - 0.02)^2 + 0.08 \cdot V_{fbzb}} \right].$$
$$(6.150)$$

The other charge components for Q_B defined in Eq. (6.148) are with the following forms:

$$Q_{SUB0} = -W_{active} \cdot L_{active} \cdot C_{oxeff} \cdot \frac{K_{1ox}^2}{2}$$

$$\cdot \left(-1 + \sqrt{1 + \frac{4 \cdot (V_{gse} - V_{FBeff} - V_{gsteff,cv} - V_{bseff})}{K_{1ox}^2}} \right), \qquad (6.151)$$

$$\delta Q_{SUB} = W_{active} \cdot L_{active} \cdot C_{oxeff} \cdot$$

$$\left[\frac{1 - A_{bulk,cv}}{2} \cdot V_{cveff} - \frac{\left(1 - A_{bulk,cv}\right) \cdot A_{bulk,cv} \cdot V_{cveff}^2}{12 \cdot \left(V_{gsteff,cv} - \varphi_\delta - \frac{A_{bulk,cv}}{2} \cdot V_{cveff} \right)} \right]. \qquad (6.152)$$

where φ_δ is the potential difference between the real surface potential and the $2\phi_B$ constant and is expressed as follows:

$$\varphi_\delta = V_t \ln \left(\frac{V_{gsteff,cv} \cdot \left(V_{gsteff,cv} + 2 \cdot K_{1ox} \cdot \sqrt{2 \cdot \phi_B}\right)}{MOIN \cdot K_{1ox} \cdot V_t^2} + 1 \right). \qquad (6.153)$$

The inversion charge can be expressed as

$$Q_{INV} = -W_{active} \cdot L_{active} \cdot C_{oxeff} \cdot \left[V_{gsteff,cv} - \varphi_\delta - \frac{1}{2} \cdot A_{bulk,cv} \cdot V_{cveff} \right.$$

$$\left. + \frac{(A_{bulk,cv} \cdot V_{cveff})^2}{12 \cdot \left(V_{gsteff,cv} - \varphi_\delta - \frac{A_{bulk,cv} \cdot V_{cveff}}{2} \right)} \right]. \qquad (6.154)$$

When *XPART* = 0.5, the channel charge partitioning to the source is formulated as

$$
Q_S = Q_D = \frac{1}{2} Q_{INV} = -\frac{W_{\text{active}} \cdot L_{\text{active}} \cdot C_{\text{oxeff}}}{2} \cdot \left[V_{\text{gsteff,cv}} - \varphi_\delta - \frac{1}{2} A_{\text{bulk,cv}} \cdot V_{\text{cveff}} \right.
$$

$$
\left. + \frac{(A_{\text{bulk,cv}} \cdot V_{\text{cveff}})^2}{12 \cdot \left(V_{\text{gsteff,cv}} - \varphi_\delta - \dfrac{A_{\text{bulk,cv}} \cdot V_{\text{cveff}}}{2} \right)} \right] . \tag{6.155}
$$

When *XPART* = 0 (or <0.5), the channel charge at the source according to the 40/60 partition scheme is

$$
Q_S = -\frac{W_{\text{active}} \cdot L_{\text{active}} \cdot C_{\text{oxeff}}}{2 \cdot \left(V_{\text{gsteff,cv}} - \varphi_\delta - \dfrac{A_{\text{bulk,cv}} \cdot V_{\text{cveff}}}{2} \right)^2} \cdot \left[(V_{\text{gsteff,cv}} - \varphi_\delta)^3 - \frac{4}{3} \cdot (V_{\text{gsteff,cv}} - \varphi_\delta)^2 \right.
$$

$$
\left. \cdot A_{\text{bulk,cv}} \cdot V_{\text{cveff}} + \frac{2}{3} \cdot (V_{\text{gsteff,cv}} - \varphi_\delta) \cdot (A_{\text{bulk,cv}} \cdot V_{\text{cveff}})^2 - \frac{2 \cdot (A_{\text{bulk,cv}} \cdot V_{\text{cveff}})^3}{15} \right]
$$

$$
\tag{6.156}
$$

$$
Q_D = -(Q_G + Q_B + Q_S). \tag{6.157}
$$

When *XPART* = 1 (or >0.5), the channel charge at the source terminal can be given by

$$
Q_S = -\frac{W_{\text{active}} \cdot L_{\text{active}} \cdot C_{\text{oxeff}}}{2} \cdot \left[V_{\text{gsteff,cv}} - \varphi_\delta + \frac{1}{2} \cdot A_{\text{bulk,cv}} \cdot V_{\text{cveff}} \right.
$$

$$
\left. - \frac{(A_{\text{bulk,cv}} \cdot V_{\text{cveff}})^2}{12 \cdot \left(V_{\text{gsteff,cv}} - \varphi_\delta - \dfrac{A_{\text{bulk,cv}} \cdot V_{\text{cveff}}}{2} \right)} \right] \tag{6.158}
$$

$$
Q_D = -(Q_G + Q_B + Q_S). \tag{6.159}
$$

6.11.2 Fringing/Overlap Capacitance Models

Fringing capacitance consists of a bias-independent outer fringing capacitance and a bias-dependent inner fringing capacitance. Inner fringing capacitance is more complex to model. Only the bias-independent outer fringing capacitance is included in BSIM4 by introducing a model parameter *CF*. If *CF* is not given in the model file, it is calculated by

$$
CF = \frac{2 \cdot EPSROX \cdot \varepsilon_0}{\pi} \cdot \ln \left(1 + \frac{4 \cdot 10^{-7}}{TOXE} \right) . \tag{6.160}
$$

The overlap capacitance was modeled with a bias-independent component in previous older MOSFETs. It is in the case of the overlap capacitance model that **capMod** $= 0$ in BSIM4. However, it has been found that the overlap capacitance shows a strong bias dependence. Depending on the bias condition between the gate and the source/drain, the overlap region between the gate and the source/drain can be in accumulation, or depletion. The bias-dependent overlap capacitance was modeled in the capacitance models when **capMod** $= 1$ and 2.

When **capMod** $= 0$ is given in the model file, the gate-to-source, gate-to-drain, and gate-to-bulk overlap charges are given by

$$Q_{\text{overlap,s}} = W_{\text{active}} \cdot CGSO \cdot V_{\text{GS}}, \tag{6.161}$$

$$Q_{\text{overlap,d}} = W_{\text{active}} \cdot CGDO \cdot V_{\text{GD}}, \tag{6.162}$$

$$Q_{\text{overlap,}b} = L_{\text{active}} \cdot CGBO \cdot V_{\text{GB}}. \tag{6.163}$$

When **capMod** $= 1$ or 2, in the model file, the gate-to-source charge is given by

$$\frac{Q_{\text{overlap,s}}}{W_{\text{active}}} = CGSO \cdot V_{\text{GS}} + CGSL$$

$$\cdot \left[V_{\text{GS}} - V_{\text{gs,overlap}} - \frac{CKAPPAS}{2} \cdot \left(-1 + \sqrt{1 - \frac{4 \cdot V_{\text{gs,overlap}}}{CKAPPAS}} \right) \right], \tag{6.164}$$

$$V_{\text{gs,overlap}} = \frac{1}{2} \cdot \left(V_{\text{GS}} + 0.02 - \sqrt{(V_{\text{GS}} + 0.02)^2 + 0.08} \right), \tag{6.165}$$

and the gate-to-drain overlap charge is given by

$$\frac{Q_{\text{overlap,d}}}{W_{\text{active}}} = CGDO \cdot V_{\text{GD}} + CGDL$$

$$\cdot \left[V_{\text{GD}} - V_{\text{gd,overlap}} - \frac{CKAPPAD}{2} \cdot \left(-1 + \sqrt{1 - \frac{4 \cdot V_{\text{gd,overlap}}}{CKAPPAD}} \right) \right], \tag{6.166}$$

$$V_{\text{gd,overlap}} = \frac{1}{2} \cdot \left(V_{\text{GD}} + 0.02 - \sqrt{(V_{\text{GD}} + 0.02)^2 + 0.08} \right), \tag{6.167}$$

and the gate-to-bulk overlap charge is given by

$$\frac{Q_{\text{overlap,}b}}{L_{\text{active}}} = CGBO \cdot V_{\text{GB}} \tag{6.168}$$

Table 6.2 Calculation of the default values of *CGSO, CGDO,* and *CGBO* parameters

Conditions	Calculation
If *CGSO* is not given and *DLC* > 0	$CGSO = DLC\, C_{\text{oxe}} - CGSL\ (= 0 \text{ if } <0)$
If CGDO is not given and *DLC* > 0	$CGDO = DLC\, C_{\text{oxe}} - CGDL\ (= 0 \text{ if } <0)$
If *CGSO* is not given and *DLC* ≤ 0	$CGSO = 0.6XJ C_{\text{oxe}}$
If *CGDO* is not given and *DLC* ≤ 0	$CGDO = 0.6XJ C_{\text{oxe}}$
If *CGBO* is not given	$CGBO = 2DWC\, C_{\text{oxe}}$

where *CGSO, CGDO, CGBO, CGSL, CGDL, CKAPPAS*, and *CKAPPAD* are model parameters. The default values for *CGSL* and *CGDL* are zero. If *CGDO, CGSO*, and *CGBO* are not given in the model file, they are calculated by the following expressions in Table 6.2.

6.12 HIGH-SPEED (NON-QUASI-STATIC) MODEL

As the circuit clock rate gets faster and faster, the need for an accurate prediction of device/circuit behavior near the cutoff frequency or under very rapid transient operations increases. In BSIM4, a charge-deficit non-quasi-static (NQS) model, based on the one in BSIM3v3 (Chan *et al.* (1994)) but with many improvements, is included. Model selectors such as ***trnqsMod*** and ***acnqsMod*** are introduced to turn on or off the NQS model, depending on the applications. The default values for ***trnqsMod*** and ***acnqsMod*** are zero. When ***trnqsMod*** = 1, the charge-deficit NQS model is used in transient simulation. When ***acnqsMod*** = 1, the charge-deficit NQS model is used in AC simulation. The transient and AC NQS models are developed from the same fundamental physics, that is, the channel/gate charge response to the external signal is relaxation time (τ)–dependent and the transcapacitances and transconductances for the AC analysis can be expressed as a function of $j\omega\tau$. However, in the model implementation, the AC NQS model does not require the internal NQS charge node that is needed in the transient NQS model.

6.12.1 The Transient NQS Model

Figure 6.22(a) illustrates how the NQS effects in a MOSFET are modeled in BSIM3v3 (Chan *et al.* (1994)) and Figure 6.22(b) gives the RC equivalent circuit of the charge-deficit NQS model for transient simulation. An internal node $Q_{\text{def}}(t)$ in addition to the ones for a typical four-terminal MOSFET is added to keep track of the amount of deficit/surplus channel charge necessary to reach equilibrium. The resistance R is determined from the RC time constant (τ). The current source $i_{\text{cheq}}(t)$ represents the equilibrium channel charging effect. The capacitor C with a typical value of 10^{-9} F is implemented to improve the simulation accuracy and the Q_{def} can be expressed as

$$Q_{\text{def}} = V_{\text{def}} \cdot C_{\text{fact}}. \tag{6.169}$$

With the consideration of both the transport and the charging component, the total current related to the terminals D, G, S can be written as

$$I_{\text{D,G,S}} = I_{\text{D,G,S}}(DC) + \frac{\partial Q_{\text{d,g,s}}(t)}{\partial t}. \tag{6.170}$$

On the basis of the relaxation time approach, the terminal charge and the corresponding charging current are modeled by

$$Q_{\text{def}}(t) = Q_{\text{cheq}}(t) - Q_{\text{ch}}(t) \tag{6.171}$$

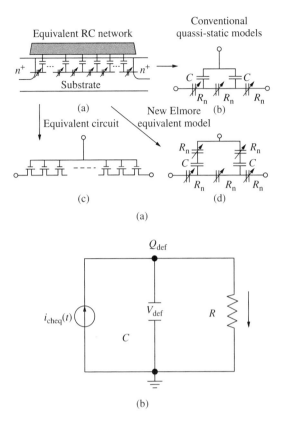

Figure 6.22   (a) Quasi-static and non-quasi-static models in circuit simulations and (b) equivalent circuit for charge-deficit NQS model for transient analysis

and

$$\frac{\partial Q_{\text{def}}(t)}{\partial t} = \frac{\partial Q_{\text{cheq}}(t)}{\partial t} - \frac{Q_{\text{def}}(t)}{\tau}, \tag{6.172}$$

$$\frac{\partial Q_{\text{d,g,s}}(t)}{\partial t} = D, G, S_{\text{xpart}} \frac{Q_{\text{def}}(t)}{\tau}, \tag{6.173}$$

where D, G, S_{xpart} are charge-deficit NQS channel charge partitioning numbers for terminals D, G, and S, respectively; $D_{\text{xpart}} + S_{\text{xpart}} = 1$ and $G_{\text{xpart}} = -1$.

The transit time τ is equal to the product of R_{ii} and $W_{\text{eff}}L_{\text{eff}}C_{\text{oxe}}$, where R_{ii} is the intrinsic-input resistance (IIR), including both the drift and the diffusion components of the channel conduction, and is given by

$$R_{\text{ii}} = \frac{1}{XRCRG1 \cdot \left(\dfrac{I_{\text{DS}}}{V_{\text{dseff}}} + XRCRG2 \cdot \dfrac{W_{\text{eff}} \cdot \mu_{\text{eff}} \cdot C_{\text{oxeff}} \cdot K_{\text{B}} \cdot T}{q \cdot L_{\text{eff}}} \right)}. \tag{6.174}$$

6.12.2 The AC NQS Model

For small-signal simulation, by substituting Eq. (6.171) into Eq. (6.173), $Q_{ch}(t)$ can be transformed into the following in the frequency domain:

$$\Delta Q_{ch}(t) = \frac{\Delta Q_{cheq}(t)}{1 + j \cdot \omega \cdot \tau}. \tag{6.175}$$

On the basis of the above equation, the transcapacitances C_{Gi}, C_{Si}, and C_{Di} (i stands for any of the G, D, S, and B terminals of the device) and the channel transconductances G_m, G_{DS}, and G_{mbs} all become complex quantities. For example, the G_m has the form

$$G_m = \frac{G_{m0}}{1 + \omega^2 \cdot \tau^2} - j \cdot \left(\frac{\omega \cdot \tau \cdot G_{m0}}{1 + \omega^2 \cdot \tau^2} \right) \tag{6.176}$$

and

$$C_{DG} = \frac{G_{DG0}}{1 + \omega^2 \cdot \tau^2} - j \cdot \left(\frac{\omega \cdot \tau \cdot C_{DG0}}{1 + \omega^2 \cdot \tau^2} \right) \tag{6.177}$$

where G_{m0} and C_{DG0} are the DC transconductance and transcapacitance at operation bias condition.

6.13 RF MODEL

6.13.1 Gate Electrode and Intrinsic-input Resistance (IIR) Model

It has been known that the gate resistance should be included in a MOSFET model for RF applications. BSIM4 provides four options for modeling gate electrode resistance (bias-independent) and intrinsic-input resistance (IIR, bias-dependent). The IIR model can be considered as a first-order NQS model. Thus, this model should not be used together with the charge-deficit NQS model. A model parameter *rgateMod* is introduced to select different gate resistance models in BSIM4.

When *rgateMod* $= 0$ (zero resistance), no gate resistance is included in the simulation. This is the default selection in the model and is the "classic" MOSFET model for digital and analog applications for three decades, as shown in Figure 6.23.

When *rgateMod* $= 1$, a resistor with constant resistance is added in the model by introducing an internal gate node as shown in Figure 6.24. The component R_{geltd} is given by

$$R_{geltd} = \frac{RSHG \cdot \left(XGW + \dfrac{W_{effcj}}{3 \cdot NGCON} \right)}{NGCON \cdot (L_{drawn} - XGL) \cdot N_F} \tag{6.178}$$

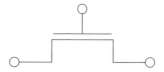

Figure 6.23   The model without gate resistance (*rgateMod* $= 0$)

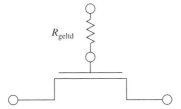

Figure 6.24   Illustration of the model with a constant resistance at the gate (***rgateMod*** = 1)

where *RSHG, NGCON, XGW*, and *XGL* are model parameters for gate sheet resistance, numbers of gate contacts, distance from the gate contact to the channel edge, and the offset of the gate length due to variation in patterning, respectively. N_F is an instance parameter for the number of fingers of the devices.

When ***rgateMod*** = 2, an intrinsic-input resistance model with variable resistance is used in the simulation. The gate resistance is the sum of the electrode gate resistance given by Eq. (6.178) and the intrinsic-input resistance given by Eq. (6.174). Only one internal gate node is created as in the case for ***rgateMod*** = 1 and no additional node is added for the additional bias-dependent gate resistance, as given in Figure 6.25. Since this model includes the first-order consideration of the NQS effect, the model selectors for NQS models should be turned off by setting ***trnqsMod*** = 0 and ***acnqsMod*** = 0.

When ***rgateMod*** = 3, the model equations remain the same for gate resistance, but in the model implementation, an additional node is added for the intrinsic-input gate resistance as shown in Figure 6.26, so the AC current through the overlap capacitance is

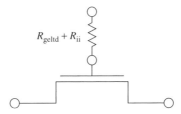

Figure 6.25   The model with both a constant and a bias-dependent gate resistance (***rgateMod*** = 2)

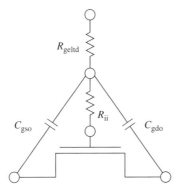

Figure 6.26   Model with two additional nodes for the gate resistance components

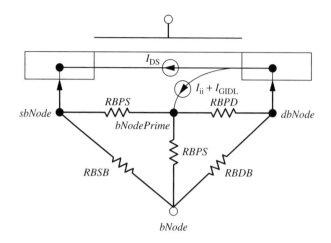

Figure 6.27   The five-resistance substrate network

parallel to the current component through the R_{ii} instead of passing through it, which is the case when **rgateMod** = 2. Similar to the case when **rgateMod** = 2, **trnqsMod** = 0 and **acnqsMod** = 0 should be selected in the model file.

6.13.2 Substrate Resistance Network

At high frequency, signal coupling through the substrate should be accounted for in the circuit simulation. To do that, the substrate components in a MOSFET should be modeled. BSIM4 offers a flexible built-in substrate resistance network. A model selector **rbodyMod** is used to turn on this option in the simulation. When **rbodyMod** = 0, the "classic" MOSFET model without substrate resistance network will be used in the simulation. When **rbodyMod** = 1, a five-resistance substrate network, as shown in Figure 6.27, is introduced. In the model implementation, a minimum conductance, *GBMIN*, is created in parallel to each resistance component to prevent any infinite resistance values during the simulation. Please note that additional nodes, such as *sbNode, dbNode*, and *bNodePrime* in Figure 6.27, have to be introduced for the model with the five-resistor substrate network. In the model implementation, the internal body node **bNodePrime** has been assumed as the reference point for the substrate defined in the intrinsic model. Thus, the impact ionization current I_{ii} and the GIDL current I_{GIDL} will flow into **bNodePrime**.

6.14 NOISE MODEL

The noise model has been improved in BSIM4 compared with BSIM3v3. The following noise sources are modeled in BSIM4, such as flicker noise, channel thermal noise, induced gate noise and the correlation with channel thermal noise, the thermal noise due to the resistances at the terminals, and the shot noise due to the gate tunneling current. Different model selectors have been introduced to use the noise models needed in the simulation, as discussed below.

6.14.1 Flicker Noise Models

Two flicker noise models are included in BSIM4. A flicker noise model selector *fnoiMod* is introduced to use a specific model in the simulation. When *fnoiMod* = 0, a simple flicker noise model that is convenient for hand calculation and model parameter extraction is used. When *fnoiMod* = 1, a unified physical flicker noise model is used. Both the models come from BSIM3v3, but the unified model has been improved in BSIM4 in several aspects such as the noise characteristics predicted by the model transit smoothly over different bias regions and also the bulk charge effect has been accounted for. The default model, if the user does not define any specific noise model in the simulation, is the unified flicker noise model.

When *fnoiMod* = 0, the spectral drain current noise power density is given by

$$S_{id}(f) = \frac{KF \cdot I_{DS}^{AF}}{C_{oxe} \cdot L_{eff} \cdot f^{EF}} \tag{6.179}$$

where *KF*, *AF*, and *EF* are model parameters and f is the operation frequency of the device.

When *fnoiMod* = 1, the spectral drain current noise power density is formulated as

$$S_{id}(f) = \frac{S_{id,inv}(f) \cdot S_{id,sub}(f)}{S_{id,inv}(f) + S_{id,sub}(f)} \tag{6.180}$$

where $S_{id,inv}(f)$ and $S_{id,sub}(f)$ are the spectral drain current noise power density of the device in the inversion and the subthreshold regions, respectively.

The spectral drain current noise power density in the inversion region is expressed as

$$S_{id,inv}(f) = \frac{k_B \cdot T \cdot q^2 \cdot I_{DS} \cdot \mu_{eff}}{f^{Ef} \cdot L_{eff}^2 \cdot C_{oxe} \cdot 10^{10}} \cdot \left[NOIA \cdot \ln\left(\frac{N_0 + N^*}{N_l + N^*}\right) + NOIB \cdot (N_0 - N_l) \right.$$

$$+ \frac{NOIC}{2} \cdot (N_0^2 - N_l^2) \right] + \frac{K_B \cdot T \cdot I_{DS}^2 \cdot \Delta L_{clm}}{f^{Ef} \cdot L_{eff}^2 \cdot W_{eff} \cdot 10^{10}}$$

$$\cdot \frac{NOIA + NOIB \cdot N_l + NOIC \cdot N_l^2}{(N_l + N^*)^2} \tag{6.181}$$

where *NOIA*, *NOIB*, and *NOIC* are parameters; N_0, N_l, N^*, and ΔL_{clm} are given as

$$N_0 = \frac{C_{oxe} \cdot V_{gsteff}}{q}, \tag{6.182}$$

$$N_l = \frac{C_{oxe} \cdot V_{gsteff}}{q} \cdot \left(1 - \frac{A_{bulk} \cdot V_{dseff}}{V_{gsteff} + 2 \cdot V_t}\right), \tag{6.183}$$

$$N^* = \frac{V_t \cdot (C_{oxe} + C_d + CIT)}{q}, \tag{6.184}$$

and

$$\Delta L_{\text{clm}} = Litl \cdot \ln \left(\frac{\dfrac{V_{\text{DS}} - V_{\text{dseff}}}{Litl} + EM}{\dfrac{2 \cdot VSAT}{\mu_{\text{eff}}}} \right), \tag{6.185}$$

where EM is a model parameter and the expression of $Litl$ has been introduced before. The spectral drain current noise power density in the subthreshold regime is given by

$$S_{\text{id,sub}}(f) = \frac{NOIA \cdot k_{\text{B}} \cdot T \cdot I_{\text{DS}}^2}{f^{Ef} \cdot L_{\text{eff}} \cdot W_{\text{eff}} \cdot N^{*2} \cdot 10^{10}}. \tag{6.186}$$

6.14.2 Channel Thermal Noise Model

Two options for the channel thermal noise are provided in BSIM4. One is the charge-based model from BSIM3v3 and the other is the holistic model. They can be selected by a model parameter **tnoiMod**.

When **tnoiMod** $= 0$, the charge-based thermal noise model is used in the simulation. The noise current is given by

$$\overline{i_d^2} = \frac{4 \cdot K_{\text{B}} \cdot T \cdot \Delta f}{R_{\text{DS}}(V) + \dfrac{L_{\text{eff}}^2}{\mu_{\text{eff}} \cdot |Q_{\text{inv}}|}} \cdot NTNOI \tag{6.187}$$

where $R_{\text{DS}}(V)$ is the bias-dependent LDD source/drain resistance, $NTNOI$ is the model parameter introduced to improve the simulation accuracy, especially for short-channel devices, and Q_{inv} is modeled by

$$Q_{\text{inv}} = W_{\text{active}} \cdot L_{\text{active}} \cdot C_{\text{oxeff}} \cdot \left[V_{\text{gsteff}} - \frac{A_{\text{bulk}} \cdot V_{\text{dseff}}}{2} + \frac{A_{\text{bulk}}^2 \cdot V_{\text{dseff}}^2}{12 \cdot \left(V_{\text{gsteff}} - \dfrac{A_{\text{bulk}} \cdot V_{\text{dseff}}}{2} \right)} \right].$$
$$\tag{6.188}$$

Figure 6.28(a) illustrates the schematic of the thermal noise model with **tnoiMod** $= 0$.

When **tnoiMod** $= 1$, the holistic thermal noise model is used. In this thermal noise model, all the short-channel effects including the velocity saturation effect incorporated

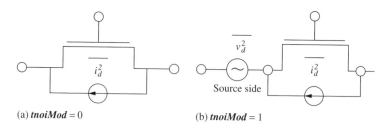

(a) **tnoiMod** $= 0$ (b) **tnoiMod** $= 1$

Figure 6.28 Schematic of the equivalent circuit of the noise model; (a) **tnoiMod** $= 0$ and (b) **tnoiMod** $= 1$

in the *I–V* model are automatically included in the noise calculation. In addition, the amplification of the channel thermal noise through G_m and G_{mbs} as well as the induced gate noise with partial correlation to the channel thermal noise are all captured in the new "noise partition" model. Figure 6.28(b) illustrates the schematic of the thermal noise model with ***tnoiMod*** $= 1$, in which part of the channel thermal noise source is partitioned to the source side.

The noise voltage source partitioned to the source side is given by

$$\overline{v_d^2} = 4 \cdot k_B \cdot T \cdot \theta_{\text{tnoi}} \cdot \frac{V_{\text{dseff}} \cdot \Delta f}{I_{\text{DS}}} \tag{6.189}$$

and

$$\theta_{\text{tnoi}} = 0.37 \cdot \left[1 + TNOIB \cdot L_{\text{eff}} \cdot \left(\frac{\mu_{\text{eff}} \cdot V_{\text{gsteff}}}{2 \cdot VSAT \cdot L_{\text{eff}}} \right)^2 \right]. \tag{6.190}$$

The noise current source in the channel region with gate and body amplification is given by

$$\overline{i_d^2} = \frac{4 \cdot k_B \cdot T \cdot V_{\text{dseff}} \cdot \Delta f}{I_{\text{DS}}} \cdot [G_{\text{DS}} + \beta_{\text{tnoi}}(G_m + G_{mbs})]^2 - \overline{v_d^2} \cdot (G_m + G_{\text{DS}} + G_{mbs})^2 \tag{6.191}$$

and

$$\beta_{\text{tnoi}} = 0.577 \cdot \left[1 + TNOIA \cdot L_{\text{eff}} \cdot \left(\frac{\mu_{\text{eff}} \cdot V_{\text{gsteff}}}{2 \cdot VSAT \cdot L_{\text{eff}}} \right)^2 \right]. \tag{6.192}$$

6.14.3 Other Noise Models

6.14.3.1 *Thermal noise models for parasitic resistances*

BSIM4 calculates the thermal noise contribution from the parasitic resistances at the gate, the drain, the source, and the substrate. The power spectral density of the noise current from the gate resistance is given by

$$S_{\text{it}, R_g} = \frac{4 \cdot k_B \cdot T}{R_g}. \tag{6.193}$$

The power spectral density of the noise current from the drain parasitic resistance is given by

$$S_{\text{it}, R_D} = \frac{4 \cdot k_B \cdot T}{R_D}. \tag{6.194}$$

The power spectral density of the noise current from the source parasitic resistance is given by

$$S_{\text{it}, R_S} = \frac{4 \cdot k_B \cdot T}{R_S}. \tag{6.195}$$

The power spectral density of the noise current from one of the five substrate resistances is given by

$$S_{\text{it}, R_{\text{subx}}} = \frac{4 \cdot k_{\text{B}} \cdot T}{R_{\text{subx}}} \tag{6.196}$$

where R_{subx} can be *RBPS*, or *RBPD*, or *RBSB*, or *RBDB*, or *RBPB*.

6.14.3.2 Shot noise model for gate tunneling current

Shot noise contributed from the gate tunneling current is also modeled in BSIM4. On the basis of the gate tunneling current model discussed earlier, the shot noise spectral density can be calculated by

$$S_{\text{sn}, \text{igtx}} = 2 \cdot q \cdot I_{\text{Gtx}}, \tag{6.197}$$

where I_{Gtx} can be the gate-to-source tunneling current, or the gate-to-drain tunneling current, or the gate-to-substrate tunneling current.

6.15 JUNCTION DIODE MODELS

6.15.1 Junction Diode *I–V* Model

Three junction diode model options are provided in BSIM4. A model parameter ***dioMod*** is introduced to select a specific diode model in the simulation. When ***dioMod*** = 0, a resistance-free diode model is referred. In this model, no current limiting feature, which is modeled in the diode model when ***dioMod*** = 1 and 2, is accounted for. However, depending on whether the model parameter *XJBVS* or *XJBVD* is given in the model file, the junction breakdown feature can be turned on or off in the simulation. The ***dioMod*** = 1 diode model comes from BSIM3v3, in which the current limiting feature was included in the forward-bias region by introducing a current limiting parameter *IJTHSFWD* or *IJTHDFWD*. However, the junction breakdown is not modeled. When ***dioMod*** = 2, the junction diode model accounts for the diode breakdown with current limiting in both forward- and reverse-bias regions. In the following, the diode model equations for the source region are discussed. The equations of the diode model in the drain region are exactly the same except replacing the subscript "s" for the source region with the subscript "d" for the drain region.

When ***dioMod*** = 0, the junction current in the source region can be given by

$$I_{\text{jBS}} = I_{\text{sBS}} \cdot \left[\exp\left(\frac{V_{\text{jBS}}}{NJS \cdot V_{\text{t}}} \right) - 1 \right] \cdot \left[1 + XJBVS \cdot \exp\left(-\frac{BVS + V_{\text{jBS}}}{NJS \cdot V_{\text{t}}} \right) \right] \tag{6.198}$$

where I_{sBS} is the total saturation current from the contributions of both the area component and the periphery component, including both the gate-edge and the isolation-edge sidewall components, and will be given later after discussing all the model options.

When ***dioMod*** = 1, the junction current in the source region can be given by

$$I_{\text{jBS}} = I_{\text{sBS}} \cdot \left[\exp\left(\frac{V_{\text{jBS}}}{NJS \cdot V_{\text{t}}} \right) - 1 \right] \quad V_{\text{jBS}} \leq V_{\text{jBSfd}} \tag{6.199}$$

and

$$I_{jBS} = I_{vjBSfd} - I_{sBS} + \frac{I_{vjSBfd}}{NJS \cdot V_t} \cdot (V_{jBS} - V_{jBSfd}) \quad V_{jBS} > V_{jBSfd} \quad (6.200)$$

where V_{jBSfd} and I_{vjBSfd} are given by

$$I_{vjBSfd} = I_{sBS} \cdot \exp\left(\frac{V_{jBSfd}}{NJS \cdot V_t}\right), \quad (6.201)$$

$$V_{jBSfd} = NJS \cdot V_t \cdot \ln\left(1 + \frac{IJTHSFWD}{I_{sBS}}\right). \quad (6.202)$$

When **dioMod** $= 2$, both current limiting and breakdown are included in the model. The junction current in the source region is given by

$$I_{jBS} = \left[I_{vjBSrev} - I_{sBS} \cdot \frac{XJBVS \cdot (V_{jBS} - V_{jBSrev})}{NJS \cdot V_t} \cdot \exp\left(-\frac{BVS - V_{jBSrev}}{NJS \cdot V_t}\right) \right]$$
$$\cdot \left[\exp\left(\frac{V_{BS}}{NJS \cdot V_t}\right) - 1 \right] \quad V_{jBS} < V_{jBSrev}, \quad (6.203)$$

$$I_{jBS} = I_{sBS} \cdot \left[\exp\left(\frac{V_{jBS}}{NJS \cdot V_t}\right) + XJBVS \cdot \exp\left(-\frac{BVS}{NJS \cdot V_t}\right) - 1 \right.$$
$$\left. - XJBVS \cdot \exp\left(-\frac{BVS + V_{jBS}}{NJS \cdot V_t}\right) \right] \quad V_{jBSrev} \le V_{jBS} < V_{jBSfd}, \quad (6.204)$$

$$I_{jbs} = I_{sbs} \cdot \left[\exp\left(\frac{V_{jbsfd}}{NJS \cdot v_t}\right) + \exp\left(-\frac{BVS}{NJS \cdot v_t}\right) - 1 \right.$$
$$\left. - XJBVS \cdot \exp\left(-\frac{V_{jbsfd}}{NJS \cdot v_t}\right) - \exp\left(-\frac{BVS + V_{jbsfd}}{NJS \cdot v_t}\right) \right] \quad V_{jbs} \ge V_{jbsfd},$$
$$(6.205)$$

where V_{jBSrev} and $I_{vjBSrev}$ are given by

$$V_{jBSrev} = -BVS - \frac{NJS \cdot V_t}{XJBVS} \cdot \ln\left(\frac{IJTHSREV}{I_{sBS}}\right), \quad (6.206)$$

$$I_{vjbsrev} = I_{sbs} \cdot \left[1 + XJBVS \cdot \exp\left(-\frac{BVS + V_{jbs}}{NJS \cdot v_t}\right) \right]. \quad (6.207)$$

In the above, *IJTHSREV* and *IJTHSFWD* are model parameters introduced to limit the current in reverse bias and forward bias.

In all the diode models above, the saturation current I_{sBS} is used and is given in the following:

$$I_{sBS} = A_{seff} \cdot J_{ss}(T) + P_{seff} \cdot J_{ssws}(T) + W_{effcj} \cdot NF \cdot J_{sswgs}(T) \quad (6.208)$$

where A_{seff} and P_{seff} are the effective area and peripheral length of the source region, and they will be discussed in detail in the next section. $J_{ss}(T)$, $J_{ssws}(T)$, and $J_{sswgs}(T)$ are given by

$$J_{ss}(T) = JSS \cdot \exp\left[\frac{\dfrac{E_g(TNOM)}{V_t(TNOM)} - \dfrac{E_g(T)}{V_t(T)} + XTIS \cdot \ln\left(\dfrac{T}{TNOM}\right)}{NJS}\right], \tag{6.209}$$

$$J_{ssws}(T) = JSSWS \cdot \exp\left[\frac{\dfrac{E_g(TNOM)}{V_t(TNOM)} - \dfrac{E_g(T)}{V_t(T)} + XTIS \cdot \ln\left(\dfrac{T}{TNOM}\right)}{NJS}\right], \tag{6.210}$$

and

$$J_{sswgs}(T) = JSSWGS \cdot \exp\left[\frac{\dfrac{E_g(TNOM)}{V_t(TNOM)} - \dfrac{E_g(T)}{V_t(T)} + XTIS \cdot \ln\left(\dfrac{T}{TNOM}\right)}{NJS}\right], \tag{6.211}$$

where E_g is the energy band gap of silicon and is calculated by

$$E_g(T) = 1.16 - \frac{7.02 \cdot 10^{-4} \cdot T^2}{T + 1108}. \tag{6.212}$$

6.15.2 Junction Diode Capacitance Model

Source/drain junction capacitances consist of three components, the bottom junction capacitance, the sidewall junction capacitance along the junction edge at the isolation side, and the sidewall junction capacitance along the junction edge at the gate side. In the following, we discuss these capacitance components, taking only the source junction as an example. The corresponding model equations for the drain junction can be given by replacing the subscripts from "s" to "d".

The total junction capacitance for a junction capacitance at the source side can be described by the following equations:

$$C_{BS} = A_{seff} \cdot C_{jBS} + P_{seff} \cdot C_{jBSsw} + W_{effcj} \cdot N_f \cdot C_{jBSswg} \tag{6.213}$$

where C_{jBS} is the unit-area bottom S/B junction capacitance, C_{jBSsw} is the unit-length S/B junction sidewall capacitance along the junction edge at the isolation side, C_{jBSswg} is the unit-length S/B junction sidewall capacitance along the junction edge at the gate side. A_{seff} and P_{seff} are the model parameters for effective area and periphery length. N_F is the number of fingers for a multifinger device.

The parameter C_{jBS} in the above is calculated by

$$C_{jBS} = C_{JS}(T) \cdot \left(1 - \frac{V_{BS}}{P_{BS}(T)}\right)^{-MJS} \qquad V_{BS} < 0, \tag{6.214}$$

$$C_{jBS} = C_{JS}(T) \cdot \left(1 - MJS \cdot \frac{V_{BS}}{P_{BS}(T)}\right) \qquad V_{BS} \geq 0. \tag{6.215}$$

The parameter C_{jBSws} in the above is calculated by

$$C_{jBSws} = C_{JSWS}(T) \cdot \left(1 - \frac{V_{BS}}{P_{BSWS}(T)}\right)^{-MJSW} \qquad V_{BS} < 0, \qquad (6.216)$$

$$C_{jBSws} = C_{JSWS}(T) \cdot \left(1 - MJSW \cdot \frac{V_{BS}}{P_{BSWS}(T)}\right) \qquad V_{BS} \geq 0. \qquad (6.217)$$

The parameter C_{jBSwsg} in the above is calculated by

$$C_{jBSwsg} = C_{JSWSG}(T) \cdot \left(1 - \frac{V_{BS}}{P_{BSWSG}(T)}\right)^{-MJSWG} \qquad V_{BS} < 0, \qquad (6.218)$$

$$C_{jBSwsg} = C_{JSWSG}(T) \cdot \left(1 - MJSWG \cdot \frac{V_{BS}}{P_{BSWSG}(T)}\right) \qquad V_{BS} \geq 0, \qquad (6.219)$$

where *MJ*, *MJSW*, and *MJSWG* are model parameters. The temperature dependences of the junction capacitances are described by the following:

$$C_{JS}(T) = CJS \cdot [1 + TCJ \cdot (T - TNOM)], \qquad (6.220)$$

$$C_{JSWS}(T) = CJSWS \cdot [1 + TCJSW \cdot (T - TNOM)], \qquad (6.221)$$

$$C_{JSWSG}(T) = CJSWGS \cdot [1 + TCJSWG \cdot (T - TNOM)]. \qquad (6.222)$$

The temperature dependences of the built-in potentials on the source side are described by

$$P_{BS}(T) = PBS - TPB \cdot (T - TNOM), \qquad (6.223)$$

$$P_{BSWS}(T) = PBSWS - TPBSW \cdot (T - TNOM), \qquad (6.224)$$

$$P_{BSWGS}(T) = PBSWGS - TPBSWG \cdot (T - TNOM). \qquad (6.225)$$

6.16 LAYOUT-DEPENDENT PARASITICS MODEL

BSIM4 considers the layout geometry–dependent parasitics model, which can calculate the parasitic geometry information according to the layouts of the devices, such as isolated, shared, or merged source/drain, and multifinger, and so on. Depending on the layout difference, many new model parameters have been introduced to select the model for the corresponding parasitic geometry calculation.

6.16.1 Effective Junction Perimeter and Area

In this section, only the case of the source-side junction is discussed. The same approach is used for the drain-side junction.

If a PS parameter is given in the model file, the effective perimeter of the source junction, P_{seff}, can be calculated by

$$P_{seff} = PS \quad \textbf{perMod} = 0, \tag{6.226}$$

$$P_{seff} = PS - W_{effcj} \cdot NF \quad \textbf{perMod} \neq 0. \tag{6.227}$$

Please note that P_{seff} (not PS) will be used in calculating the junction DC and CV in BSIM4 and it refers to the total source junction periphery length subtracted by total gate-side periphery. Also, the meaning of PS is different depending on value of the **perMod** parameter in the model file. When **perMod** = 0, the value of PS does not include the periphery length at the gate side and hence $P_{seff} = PS$. When **perMod** is not zero, PS follows the traditional definition, which includes the contribution of the periphery length at the gate side. However, the P_{seff} parameter subtracts the contribution of the periphery length at the gate side by using Eq. (6.227).

If AS parameter is given in the model file, the effective area of the source junction, A_{seff}, is calculated by

$$A_{seff} = AS. \tag{6.228}$$

If PS and AS parameters are not given in the model file, both P_{seff} and A_{seff} will be calculated by the parameters N_F, DWJ, **geoMod**, $DMCG$, $DMCI$, $DMDG$, $DMCGT$, and MIN.

It has been a common practice that compact MOSFET models consider only the case of a single-finger (isolated) device, when calculating the source/drain geometry parameters. In real applications, the layout structures are more complex than that. A more common example in analog/RF applications is a multifinger device, in which the source and drain share the same junction area to save the total area of the device and reduce the parasitic capacitance. Another example is a device with merged junctions, which are widely used in digital applications. BSIM4 takes these cases into account, and the calculations of the geometry parameters (and hence the parasitics) in source/drain regions are determined by the model parameter **geoMod** as shown in Table 6.3.

Table 6.3 **geoMod** parameters and the layout options of the source/drain regions

geoMod	End source	End drain	Note
0	Isolated	Isolated	N_F must be odd
1	Isolated	Shared	N_F can be odd or even
2	Shared or none	Isolated	N_F can be odd or even
3	Shared or none	Shared or none	N_F can be odd or even
4	Isolated	Merged	N_F must be odd
5	Shared or none	Merged	N_F can be odd or even
6	Merged	Isolated	N_F must be odd
7	Merged	Shared or none	N_F can be odd or even
8	Merged	Merged	N_F must be odd
9	Isolated/shared or none	Shared or none	N_F must be even
10	Shared or none	Isolated/shared or none	N_F must be even

It is too space-consuming to list the detailed equations for the source/drain geometric parameters for all the cases of the **geoMod**. In the following, we give the corresponding expressions to calculate A_{seff}, A_{deff}, P_{seff}, and P_{deff} for the isolated, merged, and shared cases, respectively, on the basis of which the detailed expression for each **geoMod** case can be derived.

The effective area and periphery parameters of the isolated end source/drain regions can be calculated by

$$AX_{eff} = NX_{end} \cdot (DMCG + DMCI - 2 \cdot DMCGT) \cdot W_{eff,cj}$$
$$+ NX_{int} \cdot 2 \cdot (DMCG - DMCGT) \cdot W_{eff,cj}, \qquad (6.229)$$

$$PX_{eff} = NX_{end} \cdot [2 \cdot (DMCG + DMCI - 2 \cdot DMCGT) + W_{eff,cj}]$$
$$+ NX_{int} \cdot 4 \cdot (DMCG - DMCGT), \qquad (6.230)$$

where AX_{eff} can be AS_{eff} or AD_{eff}, the effective source or drain area; PX_{eff} can be PS_{eff} or PD_{eff}, the effective source or drain perimeter; NX_{end} can be NS_{end} or ND_{end}, the numbers of end source or drain regions; NX_{int} can be NS_{int} or ND_{int}, the numbers of the internal source or drain regions. $DMCG$ is the distance from the center of source/drain contact to the gate edge. $DMCI$ is the distance between the center of the source/drain contact and the isolation edge in the channel-length direction. The definitions of $DMCG$, $DMCI$, and $W_{eff,cj}$ can be found in Figure 6.29. $DMCGT$ is a model parameter to make necessary corrections to $DMCI$ and $DMCG$ is a model parameter to fit the measured data from a specific test structure.

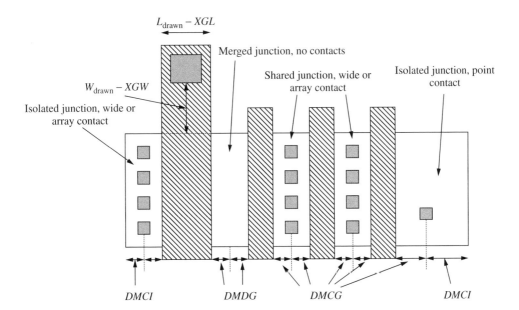

Figure 6.29   Illustration of the definitions of the layout-dependent model parameters

The effective area and periphery parameters of the shared end source/drain regions can be calculated by

$$AX_{\text{eff}} = NX_{\text{end}} \cdot (DMCG - DMCGT) \cdot W_{\text{eff,cj}}$$

$$+ NX_{\text{int}} \cdot 2 \cdot (DMCG - DMCGT) \cdot W_{\text{eff,cj}}, \tag{6.231}$$

$$PX_{\text{eff}} = NX_{\text{end}} \cdot 2 \cdot (DMCG - DMCGT) + NX_{\text{int}} \cdot 4 \cdot (DMCG - DMCGT). \tag{6.232}$$

The effective area and periphery parameters of the merged end source/drain regions can be calculated by

$$AX_{\text{eff}} = NX_{\text{end}} \cdot (DMDG - DMCGT) \cdot W_{\text{eff,cj}}$$

$$+ NX_{\text{int}} \cdot 2 \cdot (DMCG - DMCGT) \cdot W_{\text{eff,cj}}, \tag{6.233}$$

$$PX_{\text{eff}} = NX_{\text{end}} \cdot 2 \cdot (DMDG - DMCGT) + NX_{\text{int}} \cdot 4 \cdot (DMCG - DMCGT), \tag{6.234}$$

where $DMDG$ is the distance between the center of a merged source/drain junction and the gate edge.

The numbers of N_{xend} and N_{xint} depend on the N_F and MIN parameters. If N_F is odd, N_{xend} is 1, and N_{xint} is determined by $(N_F-1)/2$. If N_F is even, the calculation of N_{xend} is different depending on the MIN parameter. If $MIN = 0$, it means that the number of drain contacts is to be minimized. In other words, the outer junction regions of the device with even finger will be considered as source regions. If $MIN = 1$, it means that the number of source contacts is to be minimized, that is, the outer junction regions of the device will be considered as drain regions. The calculation of ND_{end}, NS_{end}, ND_{int}, and NS_{int} for even number of fingers are given in the following:

$$ND_{\text{end}} = \begin{cases} 2 & MIN = 1 \\ 0 & MIN \neq 1 \end{cases}, \tag{6.235}$$

$$NS_{\text{end}} = \begin{cases} 0 & MIN = 1 \\ 2 & MIN \neq 1 \end{cases}, \tag{6.236}$$

$$ND_{\text{int}} = \begin{cases} \dfrac{N_F}{2} - 1 & MIN = 1 \\ \dfrac{N_F}{2} & MIN \neq 1 \end{cases}, \tag{6.237}$$

$$NS_{\text{int}} = \begin{cases} \dfrac{N_F}{2} & MIN = 1 \\ \dfrac{N_F}{2} - 1 & MIN \neq 1 \end{cases}. \tag{6.238}$$

6.16.2 Source/drain Diffusion Resistance Calculation

The calculation of source/drain diffusion resistance in BSIM4 depends also on the layouts discussed above. A model parameter **rgeoMod** is introduced to define the method calculating the source/drain diffusion resistances to account for the fact that the path of current

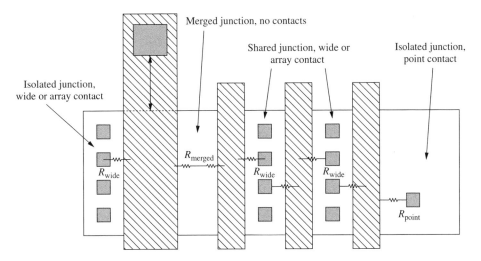

Figure 6.30 Different source/drain resistance components R_{wide}, R_{point}, and R_{merged}, depending on the device layouts

flow at the source/drain and hence the source/drain resistance could be different owing to different layout structures with specific type of contacts – point, wide, or merged as shown in Figure 6.30.

When ***rgeoMod*** $= 0$, no source/drain diffusion resistance R_{sdiff} or R_{ddiff} is generated.

When ***rgeoMod*** $\neq 0$, if the instance parameter *NRS* or *NRD*, for the number of source or drain squares, is given, the source/drain diffusion resistances (R_{sdiff} and R_{ddiff}) are calculated with the following simple equations:

$$R_{\text{sdiff}} = NRS \cdot RSH, \tag{6.239}$$

$$R_{\text{ddiff}} = NRD \cdot RSH, \tag{6.240}$$

where *RSH* is the sheet resistance of the source/drain diffusion region.

When ***rgeoMod*** $\neq 0$ and the instance parameter *NRS* or *NRD* is not given, the source/drain diffusion resistances (R_{sdiff} and R_{ddiff}) are calculated from model parameters *NF*, *DWJ*, ***geoMod***, *DMCG*, *DMCI*, *DMDG* with equations that are a little more complex.

The equations calculating R_{wide} and R_{point} can be given by

$$R_{\text{wide}} = \frac{RSH \cdot (DMCG - DMCGT)}{W_{\text{effcj}}}, \tag{6.241}$$

$$R_{\text{point}} = \frac{RSH \cdot W_{\text{effcj}}}{3 \cdot (DMCG + DMCI - 2DMCGT)}. \tag{6.242}$$

BSIM4 assumes all interjunctions to be shared junctions with wide contacts. So Eq. (6.241) will be used in most of the cases to calculate the source/drain diffusion resistance. The point contact is only used at an end junction.

The diffusion resistance of a merged junction (no contacts) can be calculated with the following equation:

$$R_{\text{merged}} = \frac{RSH \cdot (DMDG - DMCGT)}{W_{\text{effcj}}}. \tag{6.243}$$

Table 6.4 Options for source/drain contacts selected with *rgeoMod*

rgeoMod	Source end contact	Drain enc contact	Intercontact
0	No R_s	No R_d	Wide
1	Wide	Wide	–
2	Wide	Point	Wide
3	Point	Wide	Wide
4	Point	Point	Wide
5	Wide	Merged	Wide
6	Point	Merged	Wide
7	Merged	Wide	Wide
8	Merged	Point	Wide

RgeoMod is the model selector to determine which of the above equations is to be used to calculate the diffusion resistance, depending on whether the contact is a wide one or a point one or a merged one, as defined in Table 6.4.

Owing to the various options given above, the model implementation of the source/drain diffusion resistance becomes sort of complex to cover all the cases. This makes the BSIM4 source/drain series resistance model a little more difficult to use. In some cases, a sub-circuit model with user-defined parasitic components in the intrinsic device model is implemented by bypassing the complex calculations.

REFERENCES

Akers L. A. *et al.* (1982) Threshold voltage models of short, narrow, and small geometry MOSFET, *IEEE Trans. Electron Devices*, **ED-29**, 776–778.

Arora N. D. and Sharama M. (1991) MOSFET substrate current model for circuit simulation, *IEEE Trans. Electron Devices*, **ED-38**, 1392–1398.

Chan M. *et al.* (1994) Relaxation time approach to model the non-quasi-static transient effects in MOSFETs, *Tech. Dig. (IEDM)*, 169–172.

Chen J. *et al.* (1987) Subbreakdown drain leakage current in MOSFETs, *IEEE Electron Device Lett.*, **EDL-8**, 515–517.

Cheng Y. *et al.* (1997a), *BSIM3 version 3 User's Manual*, Memorandum No. UCB/ERL M97/2, University of California, Berkeley.

Cheng Y., Sugii T., Chen K., and Hu C. (1997b) Modeling of small size MOSFETs with reverse short channel and narrow width effects for circuit simulations, *Solid-State Electron.*, **41**(9), 1227–1231.

Cheng Y. *et al.* (1998) A unified MOSFET channel charge model for device modeling in circuit simulation, *IEEE Trans. Computer-Aided Design Integrated Circuits Syst.*, **17**, 641–644.

Cheng Y. and Hu C. (1999) *MOSFET Modeling & BSIM3 User's Guide*, Kluwer Academic publishers, Norwell, MA.

Cheng Y. *et al.* (2002) High-frequency small signal AC and noise modeling of MOSFETs for RF IC design, *IEEE Trans. Electron Devices*, **49**, 400–408.

Enz C. and Cheng Y. (2000) MOS transistor modeling for RF IC design, *IEEE J. Solid-State Circuits*, **35**(2), 248–257.

Frohman-Bentchknowsky D. and Grove A. S. (1969) Conductance of MOS transistor in saturation, *IEEE Trans. Electron Devices*, **ED-16**, 108–113.

King Y. C. *et al.* (1997) AC charge centroid model for quantization of inversion layer in NMOS-FETs, *International Symposium on VLSI Technology, Systems, and Applications*, pp. 245–249.

King Y.-C. *et al.* (1998) DC electrical oxide thickness model for quantization of the inversion layer in MOSFET's, *Semicond. Sci. Technol.*, **13**, 963–966.

Hsu S. T. *et al.* (1991) Physical mechanism of the reverse short channel effect in MOS transitors, *Solid-state Electronics*, **34**(6), 605–608.

Hori A. *et al.* (1993) A 0.1 μm CMOS with a step channel profile formed by ultrahigh vacuum CVD and in situ doped ions, *Tech. Dig. (IEDM)*, 909–911.

Huang J. *et al.* (1992) A physical model for MOSFET output conductance, *Tech. Dig. (IEDM)*, 569–572.

Huang C. L. *et al.* (1993) Effect of polysilicon depletion on MOSFET I-V characteristics, *Electron Lett.*, **29**, 1208, 1209.

Huang J. *et al.* (1994) *BSIM3 Manual (version 2.0)*, University of California, Berkeley, CA.

Liu Z. *et al.* (1993) Threshold voltage model for deep-submicron MOSFETs, *IEEE Trans. Electron Devices*, **ED-40**, 86–98.

Liu W. *et al.* (1999) An efficient and accurate compact model for thin-oxide-MOSFET intrinsic capacitance considering the finite charge layer thickness, *IEEE Trans. Electron Devices*, **46**(5), 1070–1072.

Liu W. *et al.* (2000) *BSIM4 User's Manual*, University of California, Berkeley, CA.

Lu, C.-Y. and Sung J. M. (1989) Reverse short-channel effects on threshold voltage in submicrometer salicide devices, *IEEE Electron Device Lett.*, **10**, 446–448.

Majkusiak B. (1990) Gate tunnel current in an MOS transistors, *IEEE Trans. Electron Devices*, **37**, 1087–1092.

Rafferty C. S. *et al.* (1993) Explanation of reverse short channel effect by defect gradients, *Tech. Dig. (IEDM)*, 311–314.

Sabnis A. G. and Clemens J. T. (1979) Characterization of electron velocity in the inverted <100> Si surface, *Tech. Dig. (IEDM)*, 18–21.

Sodini C. G. *et al.* (1984) The effects of high fields on MOS device and circuit performance, *IEEE Trans. Electron Devices*, **ED-31**, 1386.

Talkhan E. A. *et al.* (1972) Investigation of the effect of drift-field-dependent mobility on MOSFET characteristics, *IEEE Trans. Electron Devices*, **ED-19**, 899.

Troutman R. R. (1979) VLSI Limitation from drain-induced-barrie-lowering, *IEEE Trans. Electron Devices*, **ED-26**, 461.

Viswanathan C. R. *et al.* (1985) Threshold voltage in short channel MOS devices, *IEEE Trans. Electron Devices*, **ED-32**(7), 932–940.

7

The EKV Model

7.1 INTRODUCTION

The EKV model was first published in the original work by Enz, Krummenacher, and Vittoz (Enz *et al*. 1995). The name of the model is constructed from the first letter of each of the authors' last name. Since the introduction, the model has become popular worldwide owing to its simplicity and the relatively few parameters involved. The model has been implemented in many commercial circuit simulators such as Star-Hspice (*www.avanticorp.com*), Eldo (*www.mentor.com*), and AIM-Spice (*www.aimspice.com*). Visit the EKV Web site at *http://legwww.epfl.ch/ekv/* for further information on the availability of the model in other simulators.

In this chapter, we first present a list of model features to give you a quick glimpse of the capabilities of this model. Then, we discuss the backbone of the EKV model that consists of the expressions for the drain current and terminal charges of long-channel MOSFETs. Then extensions to the basic model are discussed. The model equations and parameters presented in this chapter are based on version 2.6 of the model. At the time of writing, version 3.0 of the EKV model was being developed. Unfortunately, not all equations of this new major release had been fixed at that time. Therefore, we are not able to present the model here. However, at the end of this chapter we list new features known at the time of writing (October 2002).

7.2 MODEL FEATURES

As mentioned in the introduction, the EKV model has gained in popularity owing to its simplicity. In addition, the EKV model has many other attractive features, which are listed below.

- Uses the bulk node as the reference for all voltages.
- Symmetric forward/reverse operation of the MOSFET.
- Modeling of weak and moderate inversion to provide accurate predictions of low-voltage, low-current designs.
- Geometry- and process-related dependencies.
- Effects of doping profiles.

Device Modeling for Analog and RF CMOS Circuit Design. T. Ytterdal, Y. Cheng and T. A. Fjeldly
© 2003 John Wiley & Sons, Ltd ISBN: 0-471-49869-6

- Single-expression model that preserves continuity of all derivatives for improved convergence in circuit simulations.
- High-field, short-channel, and narrow width effects including mobility degradation, velocity saturation, channel-length modulation (CLM), source and drain charge-sharing, and reverse short-channel effect (RSCE).
- Modeling of substrate current due to impact ionization.
- Charge-based dynamic model including a first-order non-quasi-static (NQS) description of the transadmittances.
- Thermal and flicker noise modeling.
- Estimation of short-distance geometry- and bias-dependent device mismatch properties.
- Emphasis on accurate modeling of the inversion coefficient in all regions of inversion. This feature is important when the circuit designer is using the g_m/I_D design concept.
- Consistent normalization of device characteristics, in particular for drain current.

Many of the features listed above are discussed in the following sections.

7.3 LONG-CHANNEL DRAIN CURRENT MODEL

The other MOSFET models discussed in this book use the source node as the reference node. The EKV model has adopted a different approach and has assigned the bulk node as the reference node. The argument by the authors in Enz *et al.* (1995) is that the bulk node is chosen as the reference node to exploit the intrinsic symmetry of the MOSFETs. In Figure 7.1 an *n*-channel MOSFET is shown with definitions of the independent voltages and the direction of the drain current I_d.

The top level expression for the drain current is

$$I_d = I_F - I_R \tag{7.1}$$

where I_F and I_R are called the forward and reverse currents, respectively. The expression of I_F and I_R is, in its simplest form, as follows:

$$I_{F(R)} = I_S \ln^2 \left(1 + e^{\dfrac{V_P - V_{sb(db)}}{2V_{th}}} \right). \tag{7.2}$$

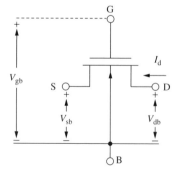

Figure 7.1   Voltage definitions used in the EKV model

This is an interpolating function[1] that tends to a square law for $I_{F(R)} \gg I_S$ (corresponds to strong inversion) and to an exponential for $I_{F(R)} \ll I_S$ (corresponds to weak inversion). Hence, the model describes the weak, moderate, and strong inversion regimes of operation with a single expression. In (7.2), I_S is the specific current, V_P is the pinch-off voltage, and V_{th} is the thermal voltage. We should mention here that in the implementation of the EKV model in computer programs, the semiempirical interpolation used in the original EKV model is replaced with a fully physical charge-based expression (see Bucher *et al.* (1997)).

The expressions for I_S and V_P are given below.

$$I_S = 2n\beta V_{th}^2, \tag{7.3}$$

$$\beta = KP \cdot \frac{W}{L}, \tag{7.4}$$

$$V_P = V'_{gb} - \phi - \gamma \left(\sqrt{V'_{gb} + \left(\frac{\gamma}{2}\right)^2} - \frac{\gamma}{2} \right), \tag{7.5}$$

$$V'_{gb} = V_{gb} - V_{T0} + \phi + \gamma\sqrt{\phi}, \tag{7.6}$$

$$n = 1 + \frac{\gamma}{2\sqrt{V_P + \phi + 4V_{th}}}. \tag{7.7}$$

Here, n is the slope factor, KP is the transconductance parameter, W and L are the effective gate width and length, respectively, ϕ is the bulk Fermi potential, γ is the body effect factor, and V_{T0} is the nominal threshold voltage at zero bias. The quantity V'_{gb} in (7.6) is called the effective gate voltage. The basic set of model parameters required for the long-channel drain current model is listed in Table 7.1.

Optionally, instead of specifying the parameter KP, the oxide thickness TOX and the low-field mobility U0 can be specified and KP will then be calculated. From TOX, the oxide capacitance C_{ox} can be calculated as $C_{ox} = \varepsilon_0 \varepsilon_i / \text{TOX}$, where $\varepsilon_0 (= 8.85418 \times 10^{-12} \text{ F/m})$ is the permittivity in vacuum and ε_i is the dielectric constant of the oxide material. Strictly speaking, the only parameter that must be specified for the charge/transcapacitance model, as well as thermal noise model, is TOX (or C_{ox}).

Note that since all voltages in the EKV model are referred to the bulk node of the transistor, the model parameter value specified for V_{T0} is also referred to the bulk node.

Table 7.1   Long-channel drain current model parameters

Model parameter	SPICE name	Description	Default value	Unit
γ	GAMMA	Body effect parameter	1	$V^{1/2}$
KP	KP	Transconductance parameter	50×10^{-6}	A/V^2
ϕ	PHI	Bulk Fermi potential	0.7	V
V_{T0}	VTO	Nominal threshold voltage	0.5	V

[1] This is not the default interpolation function in the EKV model implementation in AIM-Spice. To use this function, model parameter EKVINT has to be specified with a value different from zero.

7.4 MODELING SECOND-ORDER EFFECTS OF THE DRAIN CURRENT

Second-order effects of the drain current in MOS transistors include high-field effects and short- and narrow-channel effects. In the EKV model, the following second-order effects are included: velocity saturation, mobility degradation, channel-length modulation, charge-sharing, and reverse short-channel effect. In this section we will discuss how the core long-channel drain current model is modified to take into account these effects.

7.4.1 Velocity Saturation and Channel-length Modulation

The effects of velocity saturation and CLM are taken into account in the EKV model by replacing the length L in the expression for β in (7.4) with an equivalent length L_{eq} as follows:

$$\beta = KP \cdot \frac{W}{L_{eq}} \tag{7.8}$$

where L_{eq} is given by

$$L_{eq} = L - \Delta L + \frac{V_{ds} + V_{ip}}{E_{crit}}. \tag{7.9}$$

$$\Delta L = \lambda L_C \ln\left(1 + \frac{V_{ds} - V_{ip}}{L_C E_{crit}}\right), \tag{7.10}$$

$$L_C = \sqrt{\frac{\varepsilon_0 \varepsilon_{si}}{C_{ox}} x_j}, \tag{7.11}$$

$$V_{dsx} = \frac{V_{db} - V_{sb}}{2}, \tag{7.12}$$

$$V_{ip} = \sqrt{V_{DSS}^2 + \Delta V^2} - \sqrt{(V_{dsx} - V_{DSS})^2 + \Delta V^2}, \tag{7.13}$$

$$\Delta V = 4 V_{th} \sqrt{\lambda\left(\sqrt{\frac{I_F}{I_S}} - \frac{V_{DSS}}{V_{th}}\right) + \frac{1}{64}}, \tag{7.14}$$

$$V_{DSS} = V_C\left[\sqrt{\frac{1}{4} + \frac{V_{th}}{V_C}\sqrt{\frac{I_F}{I_S}}} - \frac{1}{2}\right], \tag{7.15}$$

$$V_C = E_{crit} L. \tag{7.16}$$

Most of the parameters used in the equations above are model parameters and are listed in Table 7.2. In addition, ε_{si} is the relative permittivity of silicon (=11.7).

In the equation for L_{eq} in (7.9), the second term on the right side represents a reduction in the effective channel length due to CLM, while the last term on the right side represents an increase in the effective channel length due to velocity saturation. This approach of modeling the effect of velocity saturation is based on the work by Arora *et al.* (1994).

Table 7.2 Parameters used for modeling second-order effects in the drain current

Model parameter	SPICE name	Description	Default value	Unit
C_{ox}	COX	Oxide capacitance	7×10^{-4}	F/m^2
E_{crit}	UCRIT	Longitudinal critical electric field	2×10^6	V/m
λ	LAMBDA	Channel-length modulation coefficient	0.5	–
L_η	LETA	Coefficient for short-channel effect	0.1	–
L_K	LK	Characteristic length for RSCE	0.29×10^{-6}	m
Q_0	Q0	Peak charge density for RSCE	0	C/m^2
θ	THETA	Mobility reduction coefficient	0	V^{-1}
W_η	WETA	Coefficient for narrow-channel effect	0.25	–
x_j	XJ	Junction depth	0.1×10^{-6}	m

7.4.2 Mobility Degradation due to Vertical Electric Field

There are two models available in EKV for taking into account the reduction in the mobility of carriers caused by the vertical electric field – one simple and one more advanced. Only the simple model is discussed here. For a description of the advanced model, we refer to the EKV user manual by Bucher *et al.* (1997). If the model parameter E0 is not specified, the simple model is used. In the simple model the equation for β is modified as follows:

$$\beta = KP \cdot \frac{W}{L_{\text{eq}}} \frac{1}{1 + \theta V_{\text{P}}}, \tag{7.17}$$

where θ is a model parameter called the mobility reduction coefficient.

7.4.3 Effects of Charge-sharing

At short gate lengths, the depletion widths associated with the source–channel and drain–channel junctions may represent a significant fraction of the total gate length. This depletion charge is shared with the depletion charges of the contact junctions and causes a reduction in the total depletion charge. The macroscopic manifestation of this effect is a reduction in the effective threshold voltage as the gate length is decreased. In addition, since the depletion widths are voltage-dependent, the effective threshold voltage also becomes dependent on the drain-source voltage. This effect is called drain-induced barrier lowering (DIBL) in many texts.

The effects of charge-sharing are included in the EKV model by introducing an effective body effect parameter γ' as follows:

$$\gamma' = \gamma - \frac{\varepsilon_{\text{si}}}{C_{\text{ox}}} \left[\frac{L_\eta}{L} \left(\sqrt{V_{\text{SB}}} + \sqrt{V_{\text{DB}}} \right) - \frac{3W_\eta}{W} \sqrt{V_{\text{P}} + \phi} \right]. \tag{7.18}$$

Here, L_η and W_η are the short-channel-effect and the narrow-channel-effect coefficients, respectively. They are both model parameters and are listed in Table 7.2. The effective body effect parameter γ' is now used wherever γ appears in the equations above. Notice that γ' decreases with decreasing L and increasing V_{sb} and V_{db} as is expected from the discussion above.

For analog circuit designers, the most important implication of charge-sharing is that the output resistance of the transistors drops drastically. This is illustrated in the SPICE example presented in Section 7.5.

7.4.4 Reverse Short-channel Effect (RSCE)

The reverse short-channel effect is caused by nonuniform doping along the channel. The effect is that the threshold voltage is larger at the edges of the channel than in the middle. Hence, as the channel length is decreased, the effective threshold is increased. In the EKV model, the RSCE is taken into account by adding an extra term in the expression for the effective gate voltage V'_{gb}. The modified version of (7.6) then becomes

$$V'_{gb} = V_{gb} - V_{T0} + V_{RSCE} + \phi + \gamma\sqrt{\phi} \qquad (7.19)$$

where

$$V_{RSCE} = \frac{2Q_0}{C_{ox}} \frac{1}{\left[1 + \frac{1}{2}\left(\xi + \sqrt{\xi^2 + C_\varepsilon}\right)\right]^2}, \qquad (7.20)$$

$$\xi = C_A\left(10\frac{L}{L_K} - 1\right), \qquad (7.21)$$

$$C_A = 0.028; \quad C_\varepsilon = 4(22 \times 10^{-3})^2. \qquad (7.22)$$

In the equations above, L_K and Q_0 are model parameters and are listed in Table 7.2.

With the discussion of the reverse short-channel effect, we end this section on second-order effects and the drain current model. The EKV model also includes another contribution to the drain current, which is the impact ionization current. However, the equations implemented in the EKV model are the same as the equations discussed in Chapter 1 and are not repeated here.

Before we continue with the dynamic model, we take a break from the listing of equations and present a SPICE example of how second-order effects affect important small-signal characteristics of the transistors.

7.5 SPICE EXAMPLE: THE EFFECT OF CHARGE-SHARING

As mentioned in Section 7.4.3, the most important implication of charge-sharing for analog circuit designers is that the output resistance of the transistors drops drastically. In this SPICE example we will use the AIM-Spice implementation of the EKV model to

```
The effect of charge-sharing
* Circuit description
m1 d g 0 0 mn l=0.5u w=10u
vgs g 0 dc 0.7
vds dd 0 dc 2.5
vid dd d dc 0
* Simulation command
.dc vds 0 5 0.01
.plot dc i(vid)
* MOS model definition
.model mn nmos level=23 cox=3.45m xj=0.15u
+ vto=0.6 gamma=0.71 phi=0.97 kp=150u e0=88e6
+ ucrit=4.5e6 dl=-0.05u dw=-0.02u lambda=0.23
+ ibn=1.0 iba=200e6 ibb=350e6
+ weta=0.05 q0=280u lk=0.5u rsh=510
+ leta=0.28
```

Figure 7.2   The SPICE netlist used in the charge-sharing example

illustrate how pronounced this effect can be. The SPICE input file used in this example is shown in Figure 7.2.

The model parameters are the same as those published on the EKV home page. This parameter set is said to reflect a typical 0.5-μm CMOS process. We use a minimum length transistor supported by this parameter set that is 0.5 μm. As we note from the value of the `vto` parameter, the nominal threshold voltage of this process is 0.6 V. Furthermore, we intend to bias the transistor only 100 mV above the nominal threshold voltage. We have chosen to use a low gate voltage because the effect of charge-sharing is less for high gate voltage overdrives. Hence, we have specified a gate-source voltage of 0.7 V.

We want to run a DC simulation sweeping the drain-source voltage from zero to 5 V. The corresponding simulator command is shown in the line starting with `.dc` in Figure 7.2. In the following line we request that the simulator plot the current through the voltage source `vid`, which is used as an ampere meter to measure the drain current of the transistor. In many SPICE simulators, including AIM-Spice, the drain current can be plotted directly without having to insert an extra voltage source. However, we have chosen to use the syntax and capabilities of standard Berkeley SPICE.

Well, let us take a look at the results. In Figure 7.3 we have plotted the simulated drain current. The thick and thin line represent the drain current with the charge-sharing effect enabled and disabled, respectively. To disable the charge-sharing effect, the model parameters `leta` and `weta` are set to zero in Figure 7.2.

We note from Figure 7.3 that there is quite a large difference in the two curves. In the saturation region, both the current level and the slope are quite different. Since we are most interested in the output resistance, we also plot the inverse of the slope of the two curves in Figure 7.3. This is shown in Figure 7.4, where again the thick and thin line correspond to enabled and disabled charge-sharing effects, respectively.

Notice that in the saturation region, the output resistance is reduced by a factor of about 10 because of charge-sharing. In many amplifier topologies this means the same factor of reduction in the gain.

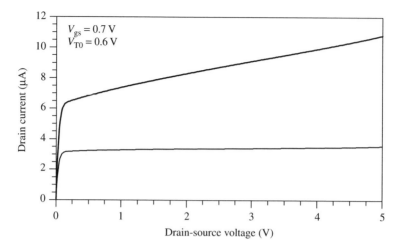

Figure 7.3   Simulated drain current of the circuit shown in Figure 7.2 with charge-sharing enabled (thick line) and disabled (thin line)

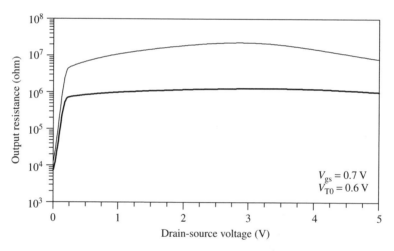

Figure 7.4   Simulated output resistance of the circuit shown in Figure 7.2 with charge-sharing enabled (thick line) and disabled (thin line)

7.6 MODELING OF CHARGE STORAGE EFFECTS

The equations listed in the previous sections are only good for modeling the DC characteristics of a MOS transistor. To be able to predict the dynamic performance of MOSFETs, a model for the capacitances or stored charges is also required. In the EKV model, the effects of charge storage within the intrinsic MOS transistor are implemented using a quasi-static charge-based approach with a first-order extension to NQS operation in AC analysis. The charge storage model is based on equations for the charge stored at each of the four terminals of the transistor. By forcing the sum of the terminal charges to be equal to zero, charge conservation is assured.

The starting point for the charge storage model is the slope factor for charge calculations n_q, which is given by

$$n_q = 1 + \frac{\gamma}{2\sqrt{V_P + \phi}}. \tag{7.23}$$

To split the total inversion charge between the source and the drain terminals, empirical fitting is done as follows:

$$x_f = \sqrt{\frac{1}{4} + \frac{I_F}{I_S}}, \tag{7.24}$$

$$x_r = \sqrt{\frac{1}{4} + \frac{I_R}{I_S}}. \tag{7.25}$$

The normalized drain and source terminal charges q_D and q_S are now given by

$$q_D = -n_q \left(\frac{4}{15} \frac{3x_r^3 + 6x_r^2 x_f + 4x_r x_f^2 + 2x_f^3}{(x_f + x_r)^2} - \frac{1}{2} \right), \tag{7.26}$$

$$q_S = -n_q \left(\frac{4}{15} \frac{3x_f^3 + 6x_f^2 x_r + 4x_f x_r^2 + 2x_r^3}{(x_f + x_r)^2} - \frac{1}{2} \right). \tag{7.27}$$

The total normalized inversion charge is the sum of q_D and q_S:

$$q_I = q_D + q_S = -n_q \left(\frac{4}{3} \frac{x_f^2 + x_f x_r + x_r^2}{x_f + x_r} - 1 \right). \tag{7.28}$$

The bulk charge is given by

$$q_B = \begin{cases} -\gamma \sqrt{V_P + \phi} \dfrac{1}{V_{th}} - \left(\dfrac{n_q - 1}{n_q} \right) q_I & \text{for } V'_{GB} > 0 \\ -\dfrac{V'_{GB}}{V_{th}} & \text{for } V'_{GB} \leq 0 \end{cases}. \tag{7.29}$$

The conservation of charge is assured by forcing the normalized gate charge to be the negative sum of the other charges:

$$q_G = -q_I - q_B. \tag{7.30}$$

The total terminal charges can now be written in terms of their normalized counterparts as follows:

$$Q_B = LWC_{ox} V_{th} q_B, \tag{7.31}$$

$$Q_D = LWC_{ox} V_{th} q_D, \tag{7.32}$$

$$Q_G = LWC_{ox} V_{th} q_G, \tag{7.33}$$

$$Q_I = LWC_{ox} V_{th} q_I, \tag{7.34}$$

$$Q_S = LWC_{ox} V_{th} q_S. \tag{7.35}$$

The intrinsic capacitances can now be obtained from the expressions of the terminal charges as follows:

$$C_{xy} = \pm \frac{\partial Q_x}{\partial V_y} \tag{7.36}$$

where $x, y = B, D, G, S$ and the positive sign is chosen when $x = y$, otherwise the negative sign is used. Taking the different derivatives, the result is simple, analytical, and continuous expressions for all transcapacitances in terms of x_f, x_r, the pinch-off voltage and the slope factor, and derivatives thereof, from weak to strong inversion and nonsaturation to saturation. A simpler model using the five intrinsic capacitances can be obtained when neglecting the slight dependence on the slope factor, resulting in the following simple expressions:

$$C_{gs} = \frac{2}{3} C_{ox} LW \left[1 - \frac{x_r^2 - x_r + \frac{1}{2} x_f}{(x_r + x_f)^2} \right], \tag{7.37}$$

$$C_{gd} = \frac{2}{3} C_{ox} LW \left[1 - \frac{x_f^2 - x_f + \frac{1}{2} x_r}{(x_r + x_f)^2} \right], \tag{7.38}$$

$$C_{gb} = C_{ox} LW \left(\frac{n_q - 1}{n_q} \right) \left(1 - \frac{C_{gs} + C_{gd}}{C_{ox} LW} \right), \tag{7.39}$$

$$C_{sb} = (n_q - 1) C_{gs}, \tag{7.40}$$

$$C_{db} = (n_q - 1) C_{gd}. \tag{7.41}$$

The EKV model also includes extrinsic capacitances. However, the equations implemented in the EKV model are the same as the equations discussed in Chapter 1 and are not repeated here.

7.7 NON-QUASI-STATIC MODELING

As discussed in Chapter 1, for very high frequencies, comparable to the inverse carrier transport time of the channel, the MOSFET models should consider the distributed nature of the channel charge. The EKV model includes a first-order NQS model for small-signal AC analysis. The expression of the NQS drain current is obtained from the quasi-static value of the drain current, which is then first-order low-pass filtered according to

$$I_d(s) = \frac{I_{d \text{ quasi-static}}}{1 + \text{NQS} \cdot s \cdot \tau}. \tag{7.42}$$

Here, NQS is a model parameter allowing the NQS model to be disabled and τ is a characteristic time constant given by

$$\tau = \tau_0 \frac{4}{15} \frac{x_f^2 + 3 x_f x_r + x_r^2}{(x_f + x_r)^3}, \tag{7.43}$$

where τ_0 is the intrinsic time constant, which is given by

$$\tau_0 = \frac{L^2}{2\mu V_{\text{th}}}, \tag{7.44}$$

where μ is the mobility of the carriers.

7.8 THE NOISE MODEL

In the EKV model, the inherent noise of MOS transistors is modeled by a current source INDS between intrinsic source and drain. It is composed of a thermal noise component and a flicker noise component and has the following power spectral density (PSD):

$$S_{\text{INDS}} = S_{\text{thermal}} + S_{\text{flicker}}, \tag{7.45}$$

where the thermal noise component is given by

$$S_{\text{thermal}} = 4k_{\text{B}}T\beta|q_{\text{I}}|. \tag{7.46}$$

Here, k_{B} is the Boltzmann constant, T is the absolute temperature in Kelvin, and q_{I} is the normalized inversion charge given by (7.28). Note that (7.46) is valid in all regimes of operation including below threshold and for small V_{ds}.

The flicker noise PSD is given by

$$S_{\text{flicker}} = \frac{\text{KF} \cdot g_{\text{m}}}{WLC_{\text{ox}} f^{\text{AF}}}, \tag{7.47}$$

where KF and AF are model parameters and f is the frequency of operation.

7.9 TEMPERATURE EFFECTS

The modeling of temperature effects as well as default values may be implementation-dependent. Therefore, please consult the documentation of your simulator.

The temperature appears explicitly in exponential terms through the thermal voltage, which is given by

$$V_{\text{th}} = \frac{k_{\text{B}}T}{q}, \tag{7.48}$$

where q is the electron charge. The temperature dependency of the band gap of silicon is modeled according to

$$E_{\text{g}} = 1.16 - 0.000702 \frac{T^2}{T + 1108}. \tag{7.49}$$

The nominal threshold voltage V_{T0} is modeled as decreasing linearly with the temperature as follows:

$$V_{\text{T0}}(T) = V_{\text{T0}} - \text{TCV}(T - T_{\text{ref}}), \tag{7.50}$$

where TCV is a model parameter and T_{ref} is the reference temperature at which model parameters are extracted. Other relevant model parameters are made temperature-dependent by the following equations:

$$KP(T) = KP \left(\frac{T}{T_{\text{ref}}} \right)^{\text{BEX}}, \tag{7.51}$$

$$E_{\text{crit}}(T) = E_{\text{crit}} \left(\frac{T}{T_{\text{ref}}} \right)^{\text{UCEX}}, \tag{7.52}$$

$$\phi(T) = \phi \frac{T}{T_{\text{ref}}} - 3V_{\text{th}} \ln \left(\frac{T}{T_{\text{ref}}} \right) - E_{\text{g}}(T_{\text{ref}}) \frac{T}{T_{\text{ref}}} + E_{\text{g}}(T). \tag{7.53}$$

Here, BEX and UCEX are model parameters.

7.10 VERSION 3.0 OF THE EKV MODEL

As mentioned in the introduction of this chapter, a new major release of the EKV model is right around the corner. This upcoming release has been labeled version 3.0 and provides many improvements compared to version 2.6 that was described earlier. Several papers have been published that present the different new features available. These papers are Sallese *et al.* (2000), Sallese *et al.* (2000b), Bucher *et al.* (2001), Porret *et al.* (2001), Sallese *et al.* (2001), Bucher *et al.* (2002), Bucher *et al.* (2002a), Bucher *et al.* (2002b), Enz (2002), Enz *et al.* (2002b), and Martin *et al.* (2002). The improvements to version 3.0 presented in these papers are as follows:

- Extension of the charge-based model so that all of its aspects from static to dynamic operation, including NQS modeling as well as noise, are handled within the same coherent framework.
- Inclusion of important aspects in very deep submicron CMOS related to the use of high doping in the channel region and the polysilicon gate. This leads to effects such as poly-gate depletion and quantization effects in the channel.
- Description of effects caused by nonuniformities both in the vertical and the longitudinal directions.
- Effects that are a result of nonideal field distributions in short- and narrow-channel devices such as, for example, DIBL.
- Modeling of vertical field–dependent mobility and velocity saturation effects are entirely handled within the framework of the charge model.
- Verification of the model for accurate description of high-frequency and RF operation of MOS transistors.

According to the authors of the new version, the release is scheduled for early 2003.

REFERENCES

Arora N. D., Rios R., Huang C., and Raol K. (1994) PCIM: a physically based continuous short-channel IGFET model for circuit simulation, *IEEE Trans. Electron Devices*, **ED-41**(6), 988–997.

Bucher M. *et al.* (1997) The EPFL-EKV MOSFET Model Equations for Simulation, ReVision II, *EKV Version 2.6 Users Manual*, can be downloaded from http://legwww.epfl.ch/ekv/model.html.

Bucher M., Sallese J., and Lallement C. (2001) Accounting for quantum effects and polysilicon depletion in an analytical design-oriented MOSFET model, in *Simulation of Semiconductor Processes and Devices 2001*, Tsoukalas D., Tsamis C., eds, Springer Verlag, pp. 296–299.

Bucher M. *et al.* (2002) The EKV 3.0 Compact MOS transistor model: accounting for deep-submicron aspects, Presented at the *Workshop on Compact Modeling (WCM), MSM 2002*, San Juan, Puerto Rico, pp. 670–673.

Bucher M. *et al.* (2002a) EKV 3.0: an analog design-oriented MOS transistor model, *Proc. 9th Int. Conf. on Mixed Design (MIXDES 2002)*, Wroclaw, Poland.

Bucher M. *et al.* (2002b) Analysis of transconductances at all levels of inversion in deep submicron CMOS, *Proc. 9th IEEE Int. Conf. on Electronics, Circuits and Systems (ICECS 2002)*, Vol. III, Dubrovnik, Croatia, pp. 1183–1186.

Enz C., Krummenacher F., and Vittoz E. A. (1995) An analytical MOS transistor model valid in all regions of operation and dedicated to low-voltage and low-current applications, *Analog Integrated Circuits and Signal Processing*, Vol. 8, Kluwer Academic Publishers, Boston, MA, pp. 83–114.

Enz C. (2002) An MOS transistor model for RF IC design valid in all regions of operation. *IEEE Trans. Microwave Theory Tech.*, **50**(1), 342–359.

Enz C. *et al.* (2002b) The foundations of the EKV MOS transistor charge based model, *Workshop on Compact Models-MSM 2002*, Puerto Rico, April 2002, pp. 666–669.

Martin P., Bucher M., and Enz C. (2002) MOSFET Modeling and Parameter Extraction for Low Temperature Analog Circuit Design. *5th European Workshop on Low Temperature Electronics (WOLTE-5), Grenoble, France, June 2002 – Journal de Physique IV, No. 12*, 2002, Les Editions de Physique, Les Ulis, France, Pr3 51–56.

Porret A.-S., Sallese J.-M., and Enz C. (2001) A compact non quasi-static extension of a charge-based MOS model, *IEEE Trans. Electron Devices*, **48**(8), 1647–1654.

Sallese J., Bucher M., and Lallement C. (2000) Improved analytical modeling of polysilicon depletion in MOSFETs for circuit simulation, *Solid-State Electron.*, **44**(6), 905–912.

Sallese J. and Porret A.-S. (2000b) A novel approach to non-quasi-static model of the MOS transistor valid in all modes of operation, *Solid-State Electron.*, **44**(6), 887–894.

Sallese J. *et al.* (2001) Advancements in DC and RF MOSFET modeling with the EPFL-EKV charge-based model, *Proc. 8th Int. Conf. on Mixed-Signal Design (MIXDES 2001)*, 2001.

8
Other MOSFET Models

8.1 INTRODUCTION

In this chapter, we describe two other MOSFET models that are used by design engineers. The first one is the MOS Model 9 that was developed by Philips engineers and became a standard among many designers and foundries in Europe by the mid-nineties. The second model presented is the MOSA1 model developed by two of the authors of this book (Fjeldly and Ytterdal) together with Michael Shur. Both models were developed in the early 1990s.

The objective of this chapter is to provide a reference of the two models for the circuit designers to use in a design situation. Thus, a detailed discussion of the model equations and background physics are not given.

8.2 MOS MODEL 9

The MOS Model 9 (MM9) was developed on the basis of the ideas presented in the book by deGraaff and Klaassen (1990) and was first published in Velghe *et al.* (1994).

The main characteristics of this model that ensure proper operations in the analog domain are the following:

- The consistency of current and charge descriptions by using the same carrier-density and electrical-field expressions.
- Accurate prediction of the transition from weak to strong inversion.
- Accurate prediction of the transition from linear to saturation region.
- Continuity in all charges and current expressions and their derivatives.
- A reduced parameter set to describe an individual transistor.

The model has been implemented in many commercial circuit simulators such as Star-Hspice (*http://www.avanticorp.com/*) and Eldo (*http://www.mentor.com/*). Visit the MM9 Web site at *http://www-us2.semiconductors.philips.com/Philips_Models/* for further information on the availability of the model.

Device Modeling for Analog and RF CMOS Circuit Design. T. Ytterdal, Y. Cheng and T. A. Fjeldly
© 2003 John Wiley & Sons, Ltd ISBN: 0-471-49869-6

Very recently, in June 2002, a new MOS model 11 (MM11) was released by Philips. The new release is a successor of MM9 and was especially developed to give not only an accurate description of currents and charges and their first-order derivatives (i.e., transconductance, conductance, capacitances) but also of the higher-order derivatives to be able to provide accurate description of electrical distortion behavior as discussed in Chapter 5 of this book. However, since the model has not yet gained widespread implementation in commercially available circuit simulators, we have decided to discuss the MM9 here, more specifically, version 903 of MM9.

8.2.1 The Drain Current Model

The basic model parameters of the drain-source DC current model in MM9 are listed in Table 8.1.

Table 8.1 Basic model parameters of the MM9 model

Parameter	Description	Unit
A1	Factor for the weak-avalanche current	–
A2	Exponent for the weak-avalanche current	V
A3	Factor of the drain-source voltage above which the weak avalanche occurs	–
ALP	Channel-length modulation parameter	–
BET	Gain factor for the reference transistor	A/V^2
COL	Gate overlap capacitance per unit width	F/m
ETADS	Substrate bias dependency of the static feedback effect	–
ETAGAM	Substrate bias dependence of DIBL	–
ETAM	Exponent of the back-bias dependence of m	–
GAM1	Coefficient for the drain-induced threshold shift for large gate drive	$V^{(1-\eta_{DS})}$
GAMOO	DIBL coefficient	–
K	High-back-bias body factor	$V^{1/2}$
KO	Low-back-bias body factor	$V^{1/2}$
M	Number of devices in parallel	–
MO	Parameter for the subthreshold ideality factor model	–
NFA	First coefficient of the flicker noise	$1/(Vm^4)$
NFB	Second coefficient of the flicker noise	$1/(Vm^2)$
NFC	Third coefficient of the flicker noise	1/V
NT	Coefficient of the thermal noise	J
PHIB	Surface potential for strong inversion	V
THE1	Coefficient of the mobility reduction due to the gate-induced field	1/V
THE2	Coefficient of the mobility reduction due to the back-bias	$1/V^{1/2}$
THE3	Coefficient of the mobility reduction due to the lateral field	1/V
TOX	Gate oxide thickness	m
VP	Characteristic voltage of the channel-length modulation	V
VSBT	Limiting voltage of the V_{sb} dependence of m and γ_0	V
VSBX	Transition voltage for the dual-k-factor model	V
VTO	Nominal threshold voltage with no substrate or drain bias	V
ZET1	Weak inversion correction factor	–

Note: DIBL; drain-induced barrier lowering.

Before we start listing the expressions that build up the drain-source current model in MM9, we should briefly discuss the geometry scaling principle applied in the model. Take a look at the description of the fifth parameter in Table 8.1. It states that the parameter BET is the gain factor of the reference transistor. What is the reference transistor? In MM9 the reference transistor is usually a long and wide square transistor and all the model parameters listed in Table 8.1 are extracted on the basis of measurements of this transistor. Therefore, in many implementations of the MM9 model in circuit simulators, all the model parameter names include an "R" at the end to indicate that the model parameter is for the reference transistor. For example, the parameter THE1 is called THE1R. In this chapter, the effective length and width of the reference transistor is denoted as $L_{\mathrm{eff,ref}}$ and $W_{\mathrm{eff,ref}}$, respectively. When the expressions are applied to an actual transistor, most of the model parameters are modified using geometry scaling expressions. This feature is discussed in Section 8.2.2.

In the MM9 model the top level drain-source current expression is given by

$$I_{\mathrm{ds}} = \beta G_3 \frac{V_{\mathrm{gt3}} V_{\mathrm{ds1}} - \left(\dfrac{1 + \delta_1}{2}\right) V_{\mathrm{ds1}}^2}{\{1 + \theta_1 V_{\mathrm{gt1}} + \theta_2 (u_{\mathrm{s}} - u_{\mathrm{s0}})\}(1 + \theta_3 V_{\mathrm{ds1}})}. \tag{8.1}$$

Here, the terms in the denominator make up the mobility model that includes mobility reduction due to the vertical field (the term containing θ_1), the bulk-source voltage (the term containing θ_2), and the lateral field (the term containing θ_3). In the MM9 model the source and drain parasitic resistances are absorbed into the theta values as follows:

$$\theta_1 = \theta_{1\mathrm{T}} + (R_{\mathrm{s}} + R_{\mathrm{d}})\beta, \tag{8.2}$$

$$\theta_3 = \theta_{3\mathrm{T}} - R_{\mathrm{d}}\beta - \frac{\mathrm{THE1}}{2}. \tag{8.3}$$

Here, R_{d} and R_{s} are the drain and source parasitic series resistances, respectively. Furthermore, $\theta_{1\mathrm{T}}$ and $\theta_{3\mathrm{T}}$ are the temperature- and geometry-updated versions of the model parameters THE1 and THE3, respectively (see (8.48) and (8.50)).

The effects of channel-length modulation and drain-induced barrier lowering (DIBL) are take into account by the MM9 model through the term G_3 in (8.1). This term consists of an interpolation expression, which interpolates between the subthreshold and the strong inversion regimes of operation by the following expressions:

$$G_3 = \frac{\zeta(1 - e^{-V_{\mathrm{ds}}/V_{\mathrm{th}}}) + G_1 G_2}{\zeta_1^{-1} + G_1}, \tag{8.4}$$

$$G_2 = 1 + \alpha \ln\left(1 + \frac{V_{\mathrm{ds}} - V_{\mathrm{ds1}}}{V_P}\right), \tag{8.5}$$

and

$$G_1 = e^{V_{\mathrm{gt2}}/(2m V_{\mathrm{th}})}, \tag{8.6}$$

where V_{ds} is the drain-source voltage and V_{th} is the thermal voltage.

To assure the continuity of the higher-order derivatives at the saturation voltage, the following smoothing function is used for V_{ds1}:

$$V_{ds1} = \text{hyp}_5(V_{ds}, V_{dss1}, \varepsilon_3), \tag{8.7}$$

$$\text{hyp}_5(x, x_0, \varepsilon) = x_0 - \text{hyp}_1\left(x_0 - x - \frac{\varepsilon^2}{x_0}, \varepsilon\right), \tag{8.8}$$

$$\text{hyp}_1(x, \varepsilon) = \frac{1}{2}\left[x + \sqrt{x^2 + 4\varepsilon}\right], \tag{8.9}$$

$$V_{dss1} = \frac{V_{gt3}}{1 + \delta_1} \frac{2}{1 + \sqrt{1 + \frac{2\theta_3 V_{gt3}}{1 + \delta_1}}}, \tag{8.10}$$

$$\varepsilon_3 = 0.3\frac{V_{dss1}}{1 + V_{dss1}}, \tag{8.11}$$

$$\delta_1 = \frac{0.3}{u_s}\left(k + \frac{(k_0 - k)V_{sbx}^2}{V_{sbx}^2 + (0.1V_{gt1} + V_{sb})^2}\right), \tag{8.12}$$

$$V_{gt3} = \begin{cases} 2m V_{th} \ln(1 + G_1), & V_{gt2} < V_{gta} \\ V_{gt2} + 10^{-8} & V_{gt2} \geq V_{gta} \end{cases}, \tag{8.13}$$

$$V_{gta} = 2m V_{th} \cdot 37. \tag{8.14}$$

In (8.13), m is the subthreshold ideality factor, which ideally is 1.0. However, as discussed in Chapter 1, in real CMOS technologies, the capacitance of the depletion region under the channel makes the ideality factor larger than 1.0. In the MM9 model the sensitivity of the ideality factor on the source-bulk voltage is governed by

$$m = 1 + m_0\left(\frac{u_{s0}}{u_{s1}}\right)^{\eta_0}, \tag{8.15}$$

$$u_{s0} = \sqrt{\phi_B}, \tag{8.16}$$

$$u_{s1} = \text{hyp}_2(u_s, u_{st}, 0.01), \tag{8.17}$$

$$\text{hyp}_2(x, x_0, \varepsilon) = x_0 - \text{hyp}_1(x - x_0, \varepsilon), \tag{8.18}$$

$$u_s = \sqrt{h_1}, \tag{8.19}$$

$$u_{st} = \sqrt{V_{sbt} + \phi_B}, \tag{8.20}$$

$$h_1 = \text{hyp}_1(V_{sb} + 0.5\phi_B, 0.01) + 0.5\phi_B. \tag{8.21}$$

The threshold voltage model of the MM9 model takes into account all important effects present in modern MOS transistors, which include substrate bias effects, DIBL, and static feedback.

From the expressions listed above, we see that the MM9 model defines two gate voltage overdrives, V_{gt1} and V_{gt2} (V_{gt3} is not really a gate voltage overdrive). It comes as no surprise that also two threshold voltages are involved. The first gate voltage overdrive is defined as

$$V_{gt1} = \text{hyp}_1(V_{gs} - V_{T1}, 5 \cdot 10^{-4}), \tag{8.22}$$

where the threshold voltage in this case is given by

$$V_{T1} = V_{T0} + \Delta V_{T0}. \tag{8.23}$$

Here,

$$\Delta V_{T0} = k \left\{ \sqrt{\text{hyp}_4(V_{sb}, V_{sbx}, 0.1) + \left(\frac{k}{k_0} u_{sx}\right)^2} - \frac{k}{k_0} u_{sx} \right\}$$

$$+ k_0 \left\{ \sqrt{h_1 - \text{hyp}_4(V_{sb}, V_{sbx}, 0.1)} - u_{s0} \right\}, \tag{8.24}$$

$$\text{hyp}_4(x, x_0, \varepsilon) = \text{hyp}_1(x - x_0, \varepsilon) - \text{hyp}_1(-x_0, \varepsilon), \tag{8.25}$$

$$u_{sx} = \sqrt{V_{sbx} + \phi_B}. \tag{8.26}$$

From (8.23) and (8.24) we note that this first version of the threshold voltage contains only the effect of substrate bias.

The second gate voltage overdrive V_{gt2} uses a second threshold voltage that contains all the different effects listed above. V_{gt2} is defined as follows:

$$V_{gt2} = V_{gs} - V_{T2}, \tag{8.27}$$

$$V_{T2} = V_{T1} + \Delta V_{T1}, \tag{8.28}$$

$$\Delta V_{T1} = -\gamma_0 \frac{V_{gtx}^2}{V_{gtx}^2 + V_{gt1}^2} V_{ds} - \gamma_1 \frac{V_{gt1}^2}{V_{gtx}^2 + V_{gt1}^2} V_{ds}^{ETADS}, \tag{8.29}$$

$$V_{gtx} = \frac{\sqrt{2}}{2}, \tag{8.30}$$

$$\gamma_0 = \gamma_{00} \left(\frac{u_{s1}}{u_{s0}}\right)^{n_\gamma}. \tag{8.31}$$

The total drain current is the sum of I_{ds}, the weak-avalanche current, and the leakage current of the drain-bulk diode. In most implementations of the MM9 model in circuit simulators, a dedicated model called JUNCAP (Velghe 1995) is available for describing the bulk-drain and bulk-source diodes. This model is not discussed here. The weak-avalanche current I_{dba} in MM9 is given by

$$I_{dba} = \begin{cases} 0, & V_{ds} \le V_{dsa} \\ I_{ds} a_1 e^{\{-a_2/(V_{ds} - V_{dsa})\}}, & V_{ds} > V_{dsa} \end{cases}, \tag{8.32}$$

where $V_{dsa} = a_3 V_{dss1}$.

8.2.2 Temperature and Geometry Dependencies

The expressions for the effective channel length and width used in the MM9 model are

$$L_{eff} = L_{drawn} + \text{LVAR} - 2 \cdot \text{LAP} \tag{8.33}$$

and

$$W_{\text{eff}} = W_{\text{drawn}} + \text{WVAR} - 2 \cdot \text{WOT}, \tag{8.34}$$

where L_{drawn} and W_{drawn} are gate length and width used on the layout. Furthermore, LVAR, LAP, WVAR, and WOT are model parameters that are listed in Table 8.2.

Table 8.2 Model parameters for geometry and temperature dependencies

Parameter	Description	Unit
ETAALP	Coefficient of the length dependence of α	–
ETABET	Exponent of the temperature dependence of the gain factor	–
ETAZET	Exponent of the length dependence of ζ_1	–
LAP	Lateral diffusion on each side of the gate	m
LVAR	Difference between the actual and the drawn gate length	m
SLA1	Coefficient of the length dependence of a_1	m
SLA2	Coefficient of the length dependence of a_2	Vm
SLA3	Coefficient of the length dependence of a_3	m
SLALP	Coefficient of the length dependence of α	$m^{\eta\alpha}$
SLGAM1	Coefficient of the length dependence of γ_1	$V^{(1-\eta_{DS})}m$
SLGAMOO	Coefficient of the length dependence of γ_{00}	$m^{1/2}$
SLK	Coefficient of the length dependence of k	$V^{1/2}m$
SLKO	Coefficient of the length dependence of k_0	$V^{1/2}m$
SLMO	Coefficient of the length dependence of m_0	m
SLTHE1	Coefficient of the length dependence of θ_1	m/V
SLTHE2	Coefficient of the length dependence of θ_2	$m/(V)^{1/2}$
SLTHE3	Coefficient of the length dependence of θ_3	m/V
SLVBST	Coefficient of the length dependence of V_{sbt}	Vm
SLVBSX	Coefficient of the length dependence of V_{sbx}	Vm
SLVTO	Coefficient of the length dependence of V_{T0}	Vm
SLVTO2	Second coefficient of the length dependence of V_{T0}	Vm^2
SLZET1	Coefficient of the length dependence of ζ_1	$m^{\eta\zeta}$
STA1	Coefficient of the temperature dependence of a_1	1/K
STLTHE1	Coefficient of the temperature dependence of the length dependence of θ_1	m/(VK)
STLTHE2	Coefficient of the temperature dependence of the length dependence of θ_2	$m/(V^{1/2}K)$
STLTHE3	Coefficient of the temperature dependence of the length dependence of θ_3	m/(VK)
STMO	Coefficient of the temperature dependence of m_0	1/K
STTHE1	Coefficient of the temperature dependence of θ_1	1/(VK)
STTHE2	Coefficient of the temperature dependence of θ_2	$1/(V^{1/2}K)$
STTHE3	Coefficient of the temperature dependence of θ_3	1/(VK)
STVTO	Coefficient of the temperature dependence of V_{T0}	V/K
SWA1	Coefficient of the width dependence of a_1	m
SWA2	Coefficient of the width dependence of a_2	Vm
SWA3	Coefficient of the width dependence of a_3	m
SWALP	Coefficient of the width dependence of α	m

Table 8.2   (*continued*)

Parameter	Description	Unit
SWGAM1	Coefficient of the width dependence of γ_1	$V^{(1-\eta_{DS})}m$
SWK	Coefficient of the width dependence of k	$V^{1/2}m$
SWKO	Coefficient of the width dependence of k_0	$V^{1/2}m$
SWTHE1	Coefficient of the width dependence of θ_1	m/V
SWTHE2	Coefficient of the width dependence of θ_2	$m/(V)^{1/2}$
SWTHE3	Coefficient of the width dependence of θ_3	m/V
SWVSBT	Coefficient of the width dependence of V_{sbt}	Vm
SWVSBX	Coefficient of the width dependence of V_{sbx}	Vm
SWVTO	Coefficient of the width dependence of V_{TO}	Vm
WOT	Channel-length reduction per side due to the lateral diffusion of the source/drain dopant ions	m
WVAR	Difference between the actual and the drawn gate width	m

These effective quantities are then used together with model parameters to include geometry and temperature dependencies of the model parameters as follows:

$$\beta = \text{BET}\frac{W_{\text{eff}}}{L_{\text{eff}}}\left(\frac{T_{\text{nom}}}{T}\right)^{\text{ETABET}}, \tag{8.35}$$

where ETABET is a model parameter and T_{nom} is the temperature at which the model parameters were extracted. All model parameters related to geometry and temperature dependency are listed in Table 8.2.

The surface potential is updated versus temperature as follows:

$$\phi_B = \text{PHIB} + \frac{\text{PHIB} - 1.13 - 2.5 \cdot 10^{-4}T_{\text{nom}}}{300}(T - T_{\text{nom}}). \tag{8.36}$$

The threshold voltage at zero-bias V_{TO} depends both on geometry and temperature according to the following expression:

$$
\begin{aligned}
V_{TO} = {} & \text{VTO} + \text{STVTO}(T - T_{\text{nom}}) + \text{SLVTO}\left(\frac{1}{L_{\text{eff}}} - \frac{1}{L_{\text{eff,ref}}}\right), \\
& + \text{SLVTO2}\left(\frac{1}{L_{\text{eff}}^2} - \frac{1}{L_{\text{eff,ref}}^2}\right) + \text{SWVTO}\left(\frac{1}{W_{\text{eff}}} - \frac{1}{W_{\text{eff,ref}}}\right).
\end{aligned}
\tag{8.37}
$$

The channel-length modulation effect and the weak inversion correction parameter are made dependent on the geometry by applying the following expressions:

$$\alpha = \text{ALP} + \text{SLALP}\left(\frac{1}{L_{\text{eff}}^{\text{ETAALP}}} - \frac{1}{L_{\text{eff,ref}}^{\text{ETAALP}}}\right) + \text{SWALP}\left(\frac{1}{W_{\text{eff}}} - \frac{1}{W_{\text{eff,ref}}}\right), \tag{8.38}$$

$$V_P = \text{VP}\left(\frac{L_{\text{eff}}}{L_{\text{eff,ref}}}\right), \tag{8.39}$$

and

$$\zeta_1 = \text{ZET1} + \text{SLZET1}\left(\frac{1}{L_{\text{eff}}^{\text{ETAZET}}} - \frac{1}{L_{\text{eff,ref}}^{\text{ETAZET}}}\right). \tag{8.40}$$

The length dependency of the DIBL parameter γ_{00} is quadratic and is given by

$$\gamma_{00} = \text{GAMOO} + \text{SLGAMOO}\left(\frac{1}{L_{\text{eff}}^2} - \frac{1}{L_{\text{eff,ref}}^2}\right). \tag{8.41}$$

The subthreshold slope parameter MO has both temperature and length dependencies that are modeled as follows:

$$m_0 = \text{MO} \cdot \text{STMO} \cdot (T - T_{\text{nom}}) + \text{SLMO}\left(\frac{1}{L_{\text{eff}}^{1/2}} - \frac{1}{L_{\text{eff,ref}}^{1/2}}\right). \tag{8.42}$$

The body effect parameters in the threshold voltage model depend on the geometry of the actual devices as follows:

$$k_0 = \text{KO} + \text{SLKO}\left(\frac{1}{L_{\text{eff}}} - \frac{1}{L_{\text{eff,ref}}}\right) + \text{SWKO}\left(\frac{1}{W_{\text{eff}}} - \frac{1}{W_{\text{eff,ref}}}\right), \tag{8.43}$$

$$k = \text{K} + \text{SLK}\left(\frac{1}{L_{\text{eff}}} - \frac{1}{L_{\text{eff,ref}}}\right) + \text{SWK}\left(\frac{1}{W_{\text{eff}}} - \frac{1}{W_{\text{eff,ref}}}\right), \tag{8.44}$$

and

$$V_{\text{sbx}} = \text{VSBX} + \text{SLVSBX}\left(\frac{1}{L_{\text{eff}}} - \frac{1}{L_{\text{eff,ref}}}\right) + \text{SWVBSX}\left(\frac{1}{W_{\text{eff}}} - \frac{1}{W_{\text{eff,ref}}}\right). \tag{8.45}$$

Also, the drain-bias dependency of the threshold voltage varies with the geometry of the device. The dependency is described by the following two expressions:

$$V_{\text{sbt}} = \text{VSBT} + \text{SLVSBT}\left(\frac{1}{L_{\text{eff}}} - \frac{1}{L_{\text{eff,ref}}}\right) \tag{8.46}$$

and

$$\gamma_1 = \text{GAM1} + \text{SLGAM1}\left(\frac{1}{L_{\text{eff}}} - \frac{1}{L_{\text{eff,ref}}}\right) + \text{SWGAM1}\left(\frac{1}{W_{\text{eff}}} - \frac{1}{W_{\text{eff,ref}}}\right). \tag{8.47}$$

The parameters of the mobility model are updated according to the operating temperature and geometry as follows:

$$\theta_{1T} = \text{THE1} + \text{STTHE1}(T - T_{\text{nom}}) + [\text{SLTHE1} + \text{STLTHE1}(T - T_{\text{nom}})]$$
$$\cdot \left(\frac{1}{L_{\text{eff}}} - \frac{1}{L_{\text{eff,ref}}}\right) + \text{SWTHE1}\left(\frac{1}{W_{\text{eff}}} - \frac{1}{W_{\text{eff,ref}}}\right), \tag{8.48}$$

$$\theta_2 = \text{THE2} + \text{STTHE2}(T - T_{\text{nom}}) + [\text{SLTHE2} + \text{STLTHE2}(T - T_{\text{nom}})]$$
$$\cdot \left(\frac{1}{L_{\text{eff}}} - \frac{1}{L_{\text{eff,ref}}}\right) + \text{SWTHE2}\left(\frac{1}{W_{\text{eff}}} - \frac{1}{W_{\text{eff,ref}}}\right), \tag{8.49}$$

and

$$\theta_{3T} = \text{THE3} + \text{STTHE3}(T - T_{\text{nom}}) + [\text{SLTHE3} + \text{STLTHE3}(T - T_{\text{nom}})]$$
$$\cdot \left(\frac{1}{L_{\text{eff}}} - \frac{1}{L_{\text{eff,ref}}} \right) + \text{SWTHE3} \left(\frac{1}{W_{\text{eff}}} - \frac{1}{W_{\text{eff,ref}}} \right). \tag{8.50}$$

The parameters for the weak-avalanche current model is updated versus temperature and geometry according to

$$a_1 = \text{A1} + \text{STA1}(T - T_{\text{nom}}) + \text{SLA1} \left(\frac{1}{L_{\text{eff}}} - \frac{1}{L_{\text{eff,ref}}} \right) + \text{SWA1} \left(\frac{1}{W_{\text{eff}}} - \frac{1}{W_{\text{eff,ref}}} \right), \tag{8.51}$$

$$a_2 = \text{A2} + \text{SLA2} \left(\frac{1}{L_{\text{eff}}} - \frac{1}{L_{\text{eff,ref}}} \right) + \text{SWA2} \left(\frac{1}{W_{\text{eff}}} - \frac{1}{W_{\text{eff,ref}}} \right), \tag{8.52}$$

and

$$a_3 = \text{A3} + \text{SLA3} \left(\frac{1}{L_{\text{eff}}} - \frac{1}{L_{\text{eff,ref}}} \right) + \text{SWA3} \left(\frac{1}{W_{\text{eff}}} - \frac{1}{W_{\text{eff,ref}}} \right). \tag{8.53}$$

8.2.3 The Intrinsic Charge Storage Model

The charge storage model for the intrinsic capacitances in MM9 is charge-based to guarantee charge conservation. The expressions for the terminal charges in an n-channel transistor are as follows (to obtain the expressions for p-channel devices change the signs):

$$Q_{\text{D}} = -C_{\text{ox}} \left[\frac{1}{2} V_{\text{gt3}} + \Delta 2 \cdot V_{\text{ds2}} \left(\frac{1}{12} F_{\text{j}} + \frac{1}{60} F_{\text{j}}^2 - \frac{1}{3} \right) \right], \tag{8.54}$$

$$Q_{\text{S}} = -C_{\text{ox}} \left[\frac{1}{2} V_{\text{gt3}} + \Delta 2 \cdot V_{\text{ds2}} \left(\frac{1}{12} F_{\text{j}} - \frac{1}{60} F_{\text{j}}^2 - \frac{1}{6} \right) \right], \tag{8.55}$$

$$Q_{\text{B}} = \tfrac{1}{2}(Q_{\text{BS}} + Q_{\text{BD}}), \tag{8.56}$$

$$Q_{\text{G}} = -(Q_{\text{D}} + Q_{\text{S}} + Q_{\text{B}}), \tag{8.57}$$

$$Q_{\text{BD}} = \begin{cases} C_{\text{ox}} \, \text{hyp}_3(V_{\text{gb}} - V_{\text{FB}}, V_{\text{ds2}} + V_{\text{sb}} + V_{\text{T1d}} - V_{\text{FB}}, 0.03), & V_{\text{gb}} < V_{\text{FB}} \\ -C_{\text{ox}}k_0 \left\{ -\frac{k_0}{2} + \sqrt{\frac{k_0^2}{4} + \text{hyp}_3(V_{\text{gb}} - V_{\text{FB}}, V_{\text{ds2}} + V_{\text{sb}} + V_{\text{T1d}} - V_{\text{FB}}, 0.03)} \right\}, & V_{\text{gb}} \geq V_{\text{FB}} \end{cases} \tag{8.58}$$

$$Q_{\text{BS}} = \begin{cases} C_{\text{ox}} \, \text{hyp}_3(V_{\text{gb}} - V_{\text{FB}}, V_{\text{sb}} + V_{\text{T1}} - V_{\text{FB}}, 0.03), & V_{\text{gb}} < V_{\text{FB}} \\ -C_{\text{ox}}k_0 \left\{ -\frac{k_0}{2} + \sqrt{\frac{k_0^2}{4} + \text{hyp}_3(V_{\text{gb}} - V_{\text{FB}}, V_{\text{sb}} + V_{\text{T1}} - V_{\text{FB}}, 0.03)} \right\}, & V_{\text{gb}} \geq V_{\text{FB}}. \end{cases} \tag{8.59}$$

In the equations above, V_{FB} is the flat-band voltage, which is given by

$$V_{FB} = V_{T0} - \phi_B - k_0\sqrt{\phi_B} \qquad (8.60)$$

and

$$C_{ox} = \varepsilon_{ox}\frac{W_{eff}L_{eff}}{TOX}M, \qquad (8.61)$$

where ε_{ox} is the permittivity of the oxide (absolute value). Furthermore,

$$F_j = \frac{(1+\delta_2)(1+\theta_3 V_{ds2})V_{ds2}}{2V_{gt3} - (1+\delta_2)V_{ds2}}, \qquad (8.62)$$

$$V_{ds2} = hyp_5(V_{ds}, V_{dss2}, \varepsilon_7), \qquad (8.63)$$

$$V_{dss2} = \frac{V_{gt3}}{1+\delta_2}\frac{2}{1+\sqrt{1+\dfrac{2\theta_3 V_{gt3}}{1+\delta_2}}}, \qquad (8.64)$$

$$\Delta 2 = \frac{\partial V_{gt3}}{\partial V_{sb}} + \frac{\partial V_{gt3}}{\partial V_{gs}} + \frac{\partial V_{gt3}}{\partial V_{ds}}, \qquad (8.65)$$

$$\delta_2 = \frac{\partial V_{T2}}{\partial V_{sb}} - \frac{\partial V_{T2}}{\partial V_{gs}} - \frac{\partial V_{T2}}{\partial V_{ds}}, \qquad (8.66)$$

$$V_{T1d} = V_{T0} + \Delta V_{T0d}, \qquad (8.67)$$

$$\Delta V_{T0d} = k\left\{\sqrt{hyp_4(V_{db}, V_{sbx}, 0.1) + \left(\frac{k}{k_0}u_{sx}\right)^2} - \frac{k}{k_0}u_{sx}\right\}$$

$$+ k_0\left\{\sqrt{h_2 - hyp_4(V_{db}, V_{sbx}, 0.1)} - u_{s0}\right\}, \qquad (8.68)$$

$$h_2 = hyp_1(V_{db} + 0.5\phi_B, 0.01) + 0.5\phi_B, \qquad (8.69)$$

$$\varepsilon_7 = 0.1\frac{V_{dss2}}{1+V_{dss2}}. \qquad (8.70)$$

From the expressions for the terminal charges listed above, the intrinsic capacitances can be obtained by taking the corresponding derivatives of the terminal charges with respect to the voltages. If one is interested in the Meyer capacitances (see Section 1.4.4), they can be calculated as follows:

$$C_{gs\_i} = \left.\frac{\partial Q_G}{\partial V_{gs}}\right|_{V_{gd}, V_{gb}}, \qquad (8.71)$$

$$C_{gd\_i} = \left.\frac{\partial Q_G}{\partial V_{gs}}\right|_{V_{gs}, V_{gb}}, \qquad (8.72)$$

and

$$C_{gd\_i} = \left.\frac{\partial Q_G}{\partial V_{gb}}\right|_{V_{gs}, V_{gd}}. \qquad (8.73)$$

Here, we have used the subscript "i" to indicate intrinsic capacitances. To obtain the total capacitances we have to include the overlap capacitances as shown below.

$$C_{gs} = C_{gs\_i} + W_{eff}M \cdot \text{COL.} \tag{8.74}$$

$$C_{gd} = C_{gd\_i} + W_{eff}M \cdot \text{COL.} \tag{8.75}$$

Capacitances related to the bulk-drain and bulk-source *pn*-junctions are also part of the equivalent circuit of MM9. In most implementations of MM9 in circuit simulators, a dedicated model called JUNCAP (see Velghe 1995) is available for describing the bulk-drain and bulk-source capacitances. We refer to this document for a description of junction capacitances.

8.2.4 The Noise Model

The total noise output of a transistor consists of a thermal and a flicker noise part that create fluctuations in the channel current. Because of the capacitive coupling between gate and channel regions, current fluctuations are also induced in the gate current. These two aspects are covered in MM9 by assigning two correlated noise-current sources, one connected between drain and source i_{nth} and the other one between gate and source i_{ng}, and an uncorrelated noise-current source between drain and source i_{nfl}. The correlated current sources are directional. This is illustrated in Figure 8.1.

The expressions for the different noise sources in the MM9 model are now presented. The contribution from the thermal noise has the following spectral density function:

$$S_{i_{nth}} = N_T h_6, \tag{8.76}$$

where

$$N_T = \frac{T}{T_{nom}} \text{NT}, \tag{8.77}$$

$$h_6 = \begin{cases} h_3 h_4, & h_4 < h_5 \\ h_3 h_5, & h_4 \geq h_5 \end{cases}, \tag{8.78}$$

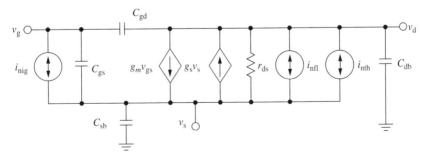

Figure 8.1 Noise-current sources inserted into the small-signal equivalent circuit of a MOS transistor

$$h_5 = \frac{V_{dss1}}{2V_{th}}, \tag{8.79}$$

$$h_4 = 1 + \theta_3 V_{ds1} + \frac{1}{3}F_I^2, \tag{8.80}$$

$$h_3 = \beta G_3 \frac{V_{gt3} - \left(\frac{1+\delta_1}{2}\right)V_{ds1}}{\{1 + \theta_1 V_{gt1} + \theta_2(u_s - u_{s0})\}(1 + \theta_3 V_{ds1})}, \tag{8.81}$$

$$F_j = \frac{(1+\delta_1)(1+\theta_3 V_{ds1})V_{ds1}}{2V_{gt3} - (1+\delta_1)V_{ds1}}. \tag{8.82}$$

The spectral density function used to model the effects of flicker noise is given as

$$S_{i_{nfl}} = \frac{S_{si}S_{wi}}{S_{si} + S_{wi}}, \tag{8.83}$$

which is a parallel combination of the noise spectral density in weak inversion S_{wi} and in strong inversion S_{si}. The noise spectral density due to flicker noise in weak inversion is written as

$$S_{wi} = N_{FA}\frac{V_{th}I_{ds}^2}{f N''^2}, \tag{8.84}$$

where

$$N_{FA} = \frac{W_{eff,ref}L_{eff,ref}}{W_{eff}L_{eff}}\frac{NFA}{M} \tag{8.85}$$

and

$$N'' = \frac{\varepsilon_{ox}}{qTOX}V_{th}(m_0 + 1). \tag{8.86}$$

In strong inversion, the flicker noise spectral density is described by

$$\begin{aligned}
S_{si} = &\frac{1}{f}\frac{V_{th}q^2 I_{ds}\beta \cdot TOX^2}{\varepsilon_{ox}^2\{1 + \theta_1 V_{gt1} + \theta_2(u_s - u_{s0})\}}\\
&\cdot\left[N_{FA}\ln\frac{N_0 + N''}{N_L + N''} + N_{FB}(N_0 - N_L) + \frac{1}{2}N_{FC}(N_0^2 - N_L^2)\right]\\
&+ \frac{V_{th}I_{ds}^2}{f}\frac{G_2 - 1}{G_2}\left\{\frac{N_{FA} + N_{FB}N_L + N_{FC}N_L^2}{(N_L + N'')^2}\right\}, \tag{8.87}
\end{aligned}$$

where

$$N_{FB} = \frac{W_{eff,ref}L_{eff,ref}}{W_{eff}L_{eff}}\frac{NFB}{M}, \tag{8.88}$$

$$N_{FC} = \frac{W_{eff,ref}L_{eff,ref}}{W_{eff}L_{eff}}\frac{NFC}{M}, \tag{8.89}$$

$$N_0 = \frac{\varepsilon_{ox}}{q \cdot TOX}V_{gt3}. \tag{8.90}$$

and

$$N_L = \frac{\varepsilon_{ox}}{q \cdot TOX}(V_{gt3} - V_{ds1}).$$ (8.91)

The spectral density of the final noise source i_{nig} in Figure 8.1 is expressed as

$$S_{i_{nig}} = N_T \frac{(2\pi f C_{ox})^2}{3g_m} \frac{1}{1 + 0.075\left(\dfrac{2\pi f C_{ox}}{g_m}\right)^2}.$$ (8.92)

8.3 THE MOSA1 MODEL

The MOSA1 model is based on our unified charge control model (UCCM) and the universal FET model (see, for example, Shur *et al.* (1992); Lee *et al.* (1993); Fjeldly *et al.* (1998)). These models exhibit improved accuracy over the basic MOSFET models in several areas that will be explained later. Most importantly, both subthreshold and above-threshold currents are accounted for by using UCCM for the mobile channel charge density. The model also accounts for velocity saturation in the channel, the channel-length modulation (CLM), the threshold voltage shift due to drain bias–induced lowering of the injection barrier between the source and the channel (DIBL), the effects of bulk charge, the bias-dependent average low-field mobility, and self-heating. Another important feature of the model is the utilization of a single continuous expression for the drain current, which is valid in all regimes of operation, effectively removing discontinuities in all derivatives, thereby improving the convergence properties of the model. The model parameters, such as the average low-field mobility, the saturation velocity, the source and drain resistances, are extractable from experimental data using a direct extraction method described in Shur *et al.* (1992).

Clearly, the subthreshold current is very important since it has consequences for the bias and logic levels needed to achieve a satisfactory *off*-state in digital operations. Hence, it affects the power dissipation in logic circuits. Likewise, the holding time in dynamic memory circuits is controlled by the magnitude of the subthreshold current.

To correctly model the subthreshold operation of MOSFETs, we need a charge control model for this regime. Also, in order to avoid convergence problems when using the model in circuit simulators, it is preferable to use a UCCM that covers both the above- and below-threshold regimes in one continuous expression. One such model is the unified charge control model (UCCM), which is described below (see also Section 1.5).

8.3.1 The Unified Charge Control Model

The standard Metal Insulator Semiconductor (MIS) charge control model described in Chapter 1 postulates that the interface inversion charge of electrons, qn_s, is proportional to the applied voltage swing, $V_{gt} = V_{gs} - V_T$, where V_{gs} is the gate-source voltage and V_T is the threshold voltage. This model is an adequate description of the strong inversion regime of the MIS capacitor, but fails for voltages near and below V_T (i.e., in the weak inversion and depletion regime). Byun *et al.* (1992) proposed a new UCCM, which was applied to HFETs, MOSFETs, polysilicon thin film transistors (TFTs), and other FET devices (see also Moon *et al.* (1990); Moon *et al.* (1991); Park *et al.* (1991); Park (1992)). According to this model, the electronic inversion charge in the FET channel is related to the applied voltage swing as follows:

$$V_{GT} - \alpha V_F = \eta V_{th} \ln \left(\frac{n_s}{n_o} \right) + a(n_s - n_o). \tag{8.93}$$

Here V_F is the quasi-Fermi potential measured relative to the Fermi potential at the source side of the channel, α is the bulk effect parameter, n_o is the sheet density of carriers at threshold, η is the ideality factor in the subthreshold regime, V_{th} is the thermal voltage, and $a \approx q/c_i$, $c_i = \varepsilon_i/d_i$ is the effective gate capacitance per unit area, ε_i is the dielectric permeability of the gate insulator, and d_i is the thickness of the gate insulator (including a quantum correction to the effective insulator thickness).

We note that in strong inversion, the linear term will dominate on the right-hand side of (8.93) and the drift current will dominate the charge transport in the channel. Hence, according to the theory of charge transport, V_F can usually be replaced by the channel potential V. Therefore, the UCCM reduces to a form of the simple parallel plate capacitor model above threshold.

In order to understand the origin of the bulk effect parameter α, we consider the expression for the above-threshold inversion sheet charge density in a MOSFET given by

$$q n_s(x) = c_i[V_{GS} - V_{TX} - V(x)], \tag{8.94}$$

where, assuming a constant substrate doping N_a,

$$V_{TX} = V_{FB} + 2\varphi_b + \sqrt{2\varepsilon_s q N_a[2\varphi_b + V(x) - V_{BS}]}/c_i. \tag{8.95}$$

Here V_{TX} may be regarded as a generalization of the threshold voltage V_T of (1.18) in Chapter 1, where the dependence of the depletion charge density on the local channel potential V is included. Linearizing (8.95) with respect to V, we obtain $V_{TX} \approx V_T + (\alpha - 1)V$ where $\alpha = 1 + \frac{1}{2}\sqrt{2\varepsilon_s q N_a/(2\varphi_b - V_{BS})}/c_i$. Hence, the strong inversion charge control equation, including the effect of the variation in the depletion charge along the channel, can be written as $q n_s \approx c_i(V_{GT} - \alpha V)$. In practical modeling, α may be regarded as an adjustable parameter with a value close to unity.

Below threshold, the linear term on the right-hand side of (8.93) can be neglected. This allows us to estimate the subthreshold drain-source current as follows in terms of the gradient of the Fermi potential:

$$I_{ds} = q n_s W \mu_n \frac{dV_F}{dx} \approx -\frac{q \eta W D_n}{\alpha} \frac{dn_s}{dx}. \tag{8.96}$$

Here we used quasi-equilibrium electron statistics that relates the electron density to the Fermi potential. We also used the Einstein relation, $D_n = \mu_n V_{th}$, to relate the mobility to the diffusion coefficient. The derivative of n_s with respect to lateral position x in (8.96) indicates that the subthreshold current is a pure diffusion current. However, for short channels, the lateral variation of the channel potential may be significant, as indicated in Figure 1.23, giving rise to a drift component in the transport mechanism. Moreover, the injection of charge carriers at the source and drain contacts is dominated by thermionic emission. In fact, for very short channels, the latter may be the dominant overall transport mechanism, similar to what takes place in a Shottky barrier.

Finally, we note that an approximate, analytical solution of the UCCM equation (8.93) can be obtained from that of the MOS capacitor in (1.38) by replacing V_{GT} by $V_{GT} - \alpha V_F$, that is,

$$n_s = 2n_o \ln \left[1 + \frac{1}{2} \exp \left(\frac{V_{GT} - \alpha V_F}{\eta V_{th}} \right) \right]. \tag{8.97}$$

8.3.2 Unified MOSFET $I-V$ Model

The basic analytical models discussed in Section 1.4 need improvements in several areas, especially for application to modern MOSFET technology. This issue is addressed in the unified drain current model presented here. Most importantly, the subthreshold regime is considered in terms of the unified charge control modeling approach. Furthermore, the universal model also accounts for series drain and source resistances, velocity saturation in the channel, gate-bias dependence of mobility, impact ionization, channel-length modulation (CLM), and for the threshold voltage shift caused by DIBL. In order to avoid convergence problems and excessive computation time in circuit simulations, continuous (unified) expressions are used for the current–voltage and capacitance–voltage characteristics, covering all regimes of operation.

Another important feature of the model is the use of relatively few, physically based parameters, all of which are extractable from experimental data. Details of the parameter extraction procedures for the unified MOSFET model are found in Lee *et al.* (1993).

In order to establish a continuous expression for the $I-V$ characteristics, valid in all regimes of operation, we start by expressing the drain current in each regime and then make smooth transitions between them. Moreover, we want the model to be extrinsic, that is, to include the effects of the parasitic source and drain resistances, R_s and R_d (note that the corresponding intrinsic model is recovered simply by setting all parasitic resistances to zero). These resistances can be accounted for by relating the intrinsic gate–source and drain–source voltages, V_{GS} and V_{GD}, considered so far to their extrinsic (measured) counterparts, V_{gs} and V_{gd}, by including the voltage drops across the series resistances:

$$V_{GS} = V_{gs} - R_s I_d, \tag{8.98}$$

$$V_{DS} = V_{ds} - (R_s + R_d) I_d. \tag{8.99}$$

Note that by applying (8.98) in the saturation current expression (1.54) for the simple velocity saturation model and by solving with respect to I_{sat}, we obtain the extrinsic expression

$$I_{sat} = \frac{\beta V_{gt}^2}{1 + \beta R_s V_{gt} + \sqrt{1 + 2\beta R_s V_{gt} + (V_{gt}/V_L)^2}}, \tag{8.100}$$

where $V_{gt} = V_{gs} - V_T$ is the extrinsic gate voltage overdrive and $\beta = W\mu_n \varepsilon_i / d_i L$ is the transconductance parameter. The extrinsic saturation voltage V_{sat} can likewise be expressed as

$$V_{sat} = V_{gt} + \left(R_d - \frac{1}{\beta V_L} \right) I_{sat}. \tag{8.101}$$

Similar analytical expressions can also be obtained for the velocity saturation model based on other velocity-field expressions such as (1.52) by Sodini (for n-channel MOSFETs) and (1.53) with $m = 1$ (for p-channel MOSFETs).

For drain-source voltages in the linear regime, well below the saturation voltage, the drain-source current can be written as

$$I_{ds} \approx g_{chi} V_{DS} \approx g_{ch} V_{ds}, \tag{8.102}$$

where g_{chi} and g_{ch} are the intrinsic and the extrinsic channel conductances in the linear region, respectively. The linear conductances are related by

$$g_{ch} = \frac{g_{chi}}{1 + g_{chi}(R_s + R_d)}. \tag{8.103}$$

In deriving (8.103), we assumed that the drain current is so small that the transconductance is much less than the channel conductance. The linear intrinsic channel conductance can be written as, using (8.97) for n_s and setting $V_F = 0$,

$$g_{chi} = \frac{q n_s W \mu_n}{L} \approx \frac{2 W q \mu_n n_o}{L} \ln \left[1 + \frac{1}{2} \exp \left(\frac{V_{gt}}{\eta V_{th}} \right) \right]. \tag{8.104}$$

We note that near threshold and in the subthreshold regime, the channel resistance will normally be much larger than the parasitic source and drain resistances, allowing us to make the approximations $V_{ds} \approx V_{DS}$ and $V_{gs} \approx V_{GS}$, and $g_{ch} \approx g_{chi}$.

The next task is to bridge the transition between the linear and the saturation regimes with one single expression. For this purpose, we have proposed the following extrinsic, universal interpolation expression (Shur *et al.* 1992):

$$I_{ds} = \frac{g_{ch} V_{ds} (1 + \lambda V_{ds})}{[1 + (g_{ch} V_{ds}/I_{sat})^m]^{1/m}}. \tag{8.105}$$

Here, m is a parameter that determines the shape of the characteristics in the knee region. The factor $(1 + \lambda V_{ds})$ is used for describing the finite output conductance in saturation, mainly caused by CLM, that is, the effect of the finite extent of the saturated region of the channel (see Figure 1.12). We note that (8.105) has the correct asymptotic behavior in the linear regime, in agreement with (8.102). In addition, when $\lambda = 0$, I_d asymptotically approaches I_{sat} in saturation, as required.

In Figure 8.2, we illustrate the link between a typical experimental MOSFET $I-V$ characteristic and the saturation voltage I_{sat}, the linear extrinsic channel conductance g_{ch}, and the finite output conductance in saturation g_{chs}. When CLM is the cause of the latter, we have $g_{chs} \approx \lambda I_{sat}$ in deep saturation. This result can be used to determine λ as part of the device characterization. A more detailed discussion of the effect of channel-length modulation is given in Section 1.5.2.

In the subthreshold regime, we can calculate the saturation current from (8.96), making use of the subthreshold approximation of the UCCM expression, that is, $n_s \approx n_o$

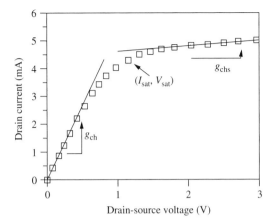

Figure 8.2 Typical experimental FET I–V characteristic with finite output conductance in saturation

$\exp[(V_{gt} - \alpha V_F)/\eta V_{th}]$. Straightforward integration over the channel length, corresponding to a variation in V_F from 0 to V_{ds}, gives the following idealized expression for the subthreshold current:

$$I_{sub} \approx \frac{q n_o \eta D_n W}{\alpha L} \exp\left(\frac{V_{gt}}{\eta V_{th}}\right)\left[1 - \exp\left(-\frac{\alpha V_{ds}}{\eta V_{th}}\right)\right]. \tag{8.106}$$

From this expression, the subthreshold saturation current is obtained for $V_{ds} > 2V_{th}$, where the term in the square bracket approaches unity, causing the dependence on V_{ds} to vanish. However, as indicated earlier, the inclusion of short-channel effects in the model will restore some degree of drain-bias dependence in the saturated subthreshold regime.

A remaining task is now to join the above-threshold and the subthreshold current expressions into one unified model expression. We first note that in strong inversion we can write $\beta V_{gt} = g_{chi}$ since $\beta = W\mu_n C_i/L$ and $q n_s = C_i V_{gt}$ at small V_{ds}. Hence, all occurrences of the combination βV_{gt} in (8.100) can be substituted by g_{chi}. Next, we replace all remaining occurrences of V_{gt} by the effective gate voltage overdrive V_{gte} that coincides with V_{gt} well above threshold and equals $2V_{th}$ below threshold. This converts (8.100) to the following unified form that approaches the correct limiting behavior above and below threshold:

$$I_{sat} = \frac{g_{chi} V_{gte}}{1 + g_{chi} R_s + \sqrt{1 + 2g_{chi} R_s + (V_{gte}/V_L)^2}}. \tag{8.107}$$

Here, V_{gte} is the effective gate voltage overdrive, which can be written as (see Section 1.5.3).

$$V_{gte} = V_{th}\left[1 + \frac{V_{gt}}{2V_{th}} + \sqrt{\delta^2 + \left(\frac{V_{gt}}{2V_{th}} - 1\right)^2}\right], \tag{8.108}$$

where δ determines the width of the transition region at threshold ($V_{gt} = 0$). Typically, $\delta = 3$ is a good choice.

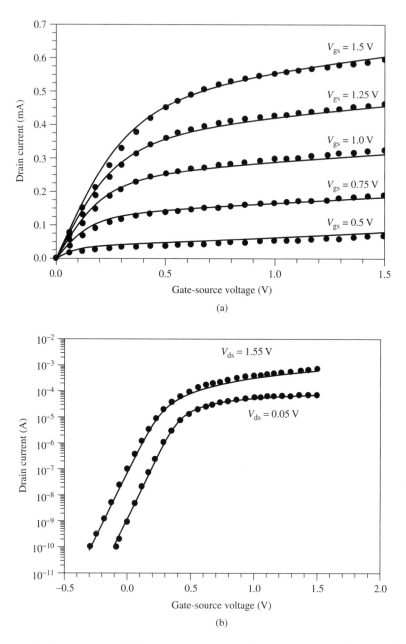

Figure 8.3 Experimental (symbols) and modeled (a) above-threshold and (b) subthreshold characteristics for a deep submicrometer nMOS with effective gate length $L = 0.09\,\mu\text{m}$. Device parameters: $W = 1\,\mu\text{m}$, $d_i = 3.5\,\text{nm}$, $\mu_n = 0.026\,\text{m}^2/\text{Vs}$, $v_s = 6 \times 10^4\,\text{m/s}$, $m = 2.2$, $R_s = R_d = 200\,\Omega$, $\lambda = 0.142/\text{V}$, $\eta = 1.7$, $V_{T0} = 0.335\,\text{V}$, $\sigma_0 = 0.11$, $V_\sigma = 0.2\,\text{V}$, $V_{\sigma t} = 0.18\,\text{V}$. Substrate-source bias $V_{bs} = 0\,\text{V}$. Reproduced from Ytterdal T., Shur M., and Fjeldly T. A. (1994) Sub-0.1 μm MOSFET modeling and circuit simulation, *Electron. Lett.*, **30**, 1545, 1546 and Mii Y. *et al.* (1994) Experimental high performance sub-0.1 μm channel nMOSFET's, *IEEE Electron Devices Lett.*, **EDL-15**, 28–30

Hence, using (8.107) for the saturation current and (8.103) for the channel conductance, the $I-V$ expression of (8.105) becomes truly unified, with the correct limiting behavior both above and below threshold, and with a continuous transition between all regimes of operation. Additional nonideal effects related to short channels and high fields, to be discussed next, can now be added to this model.

The quality of the present universal MOSFET model is illustrated in Figure 8.3 for a deep submicrometer n-channel MOSFET. Figure 8.3(a) shows the above-threshold $I-V$ characteristics and Figure 8.3(b) shows the subthreshold transfer characteristics in a semilogarithmic plot. As can be seen from the figures, our model reproduces quite accurately the experimental data in the entire range of bias voltages, over several decades of current variation. The parameters used in the model calculations were obtained from the parameter extraction procedure described by Shur *et al.* (1992) (see also Lee *et al.* (1993)).

8.3.3 Unified $C-V$ Model

The charge storage model used in the MOSA1 model is the unified Meyer $C-V$ model described in Section 1.5.3.1 and is not further discussed here.

REFERENCES

Byun Y. H., Shur M., Hack M., and Lee K. (1992) New analytical poly-silicon thin film transistor model for CAD and parameter characterization, *Solid-State Electron.*, **35**(5), 655–663.

deGraaff H. and Klaassen F. (1990) *Compact Transistor Modeling for Circuit Design*, Springer-Verlag, Berlin.

Fjeldly T. A., Ytterdal T., and Shur M. (1998) *Introduction to Device Modeling and Circuit Simulation*, John Wiley & Sons, New York.

Lee K., Shur M., Fjeldly T. A., and Ytterdal T. (1993) *Semiconductor Device Modeling for VLSI*, Prentice Hall, Englewood Cliffs, NJ.

Mii Y. *et al.* (1994) Experimental high performance sub-0.1 μm channel nMOSFET's, *IEEE Electron Devices Lett.*, **EDL-15**, 28–30.

Moon B., Byun Y., Lee K., and Shur M. (1990) New continuous heterostructure field effect transistor model and unified parameter extraction technique, *IEEE Trans. Electron Devices*, **ED-37**(4), 908–918.

Moon B. *et al.* (1991) Analytical model for p-Channel MOSFETs, *IEEE Trans. Electron Devices*, **ED-38**(12), 2632–2646.

Park C. K. *et al.* (1991) A unified charge control model for long channel n-MOSFETs, *IEEE Trans. Electron Devices*, **ED-38**(2), 399–406.

Park C. K. (1992) *A Unified Current-Voltage Modeling for Deep Submicron CMOS FETs*, Ph.D. thesis, Korea Advanced Institute of Science and Technology, Korea.

Shur M., Fjeldly T. A., Ytterdal T., and Lee K. (1992) Unified MOSFET model, *Solid-State Electron.*, **35**(12), 1795–1802.

Velghe R., Klaassen D., and Klaassen F. (1994) *MOS Model 9*, Unclassified Report NL-UR 003/94, Philips Electronics N.V., Eindhoven, The Netherlands.

Velghe R. (1995) *JUNCAP*. Unclassified Report NL-UR 028/95, Philips Electronics N.V.

Ytterdal T., Shur M., and Fjeldly T. A. (1994) Sub-0.1 μm MOSFET modeling and circuit simulation, *Electron. Lett.*, **30**, 1545, 1546.

9
Bipolar Transistors in CMOS Technologies

9.1 INTRODUCTION

As most analog designers have experienced, bipolar junction transistors (BJTs) can be realized in CMOS technologies. In most cases the bipolar action is unwanted and causes problems for the designers. This is why the BJTs in CMOS technologies are often referred to as parasitic devices. However, in some analog functions, the exponential relationship between the emitter current and the base-emitter voltage of bipolar transistors is needed. Probably the most important applications are PTAT[1] current sources and band-gap voltage reference (BVR) circuits.

In this chapter, we will describe how BJTs are realized in CMOS technologies and discuss important parameters of this device.

9.2 DEVICE STRUCTURE

In an n-well CMOS technology, a vertical BJT structure is formed by placing a region of p^+ diffusion within an n-type well as shown in Figure 9.1. As indicated by the BJT symbol included in the figure, the device formed is a PNP transistor.

Since the width of the base is determined by the depth of the well, which is relatively large, the current gain β of the BJT is quite low, usually between 5 and 10. Also, notice from the figure that the substrate itself realizes the collector terminal. Hence, the transistor is not very suited for use in amplifiers.

We should also mention that some foundries refer to this device as an n-well junction diode in which the emitter corresponds to the anode of the diode and the well contact is the cathode.

9.3 MODELING THE PARASITIC BJT

As mentioned in the introduction, for the most important applications of the parasitic BJT, the only interesting characteristic is the emitter current versus the base-emitter voltage.

---

[1] PTAT: Proportional to absolute temperature.

Device Modeling for Analog and RF CMOS Circuit Design.   T. Ytterdal, Y. Cheng and T. A. Fjeldly
© 2003 John Wiley & Sons, Ltd   ISBN: 0-471-49869-6

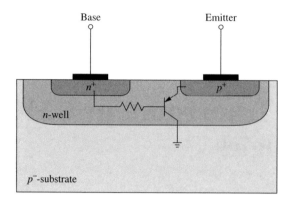

Figure 9.1   A BJT structure realized in an n-well CMOS technology

Therefore, in this chapter we have chosen to discuss only the modeling of the base-emitter pn-junction and we assume that the base terminal is grounded. We treat the device as a diode where the anode is represented by the emitter terminal, as shown in Figure 9.1, and the cathode is connected to ground.

The symbol and the definitions of the voltage across the device and the current through the device are shown in Figure 9.2.

Before we present the model for this device, we would like to discuss its characteristics. In Figure 9.3 we have plotted the emitter current I_e versus the emitter voltage V_e using a logarithmic scale on the vertical axis. The plot is based on simulation results of a parasitic BJT from a commercially available 0.5-μm CMOS technology. We notice from the figure that for current levels from about 1 pA to 0.1 mA the device exhibits the wanted exponential relationship between the applied voltage and the current through the device since we have a straight line in a semilogarithmic plot. In this region, the characteristics resemble that of an ideal diode and we can apply the well-known diode equation. However, at current levels below 1 pA and above 0.1 mA, the characteristics deviate considerably from a straight line. Next we will discuss how to model these characteristics to take into account the nonideal effects.

When designing circuits that are based on the assumption of exponential characteristics, the designer should generate a plot similar to the one shown in Figure 9.3 and identify the range in which the assumption holds. Then the designer should bias the device within this range.

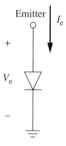

Figure 9.2   Symbol and definition of voltages and currents for the parasitic pnp BJT device

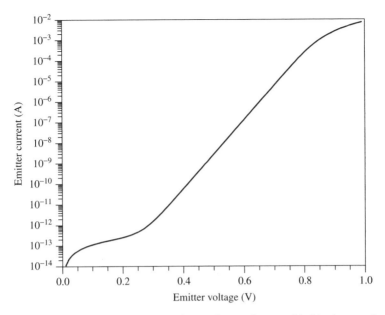

Figure 9.3   Emitter current versus emitter voltage of a parasitic bipolar transistor

We start the description of the model by presenting the ideal diode equation and then show how to extend it to account for the nonideal effects. We will not include the background physics for the derivation of the ideal diode equation. Interested readers are referred to other texts such as Fjeldly *et al.* (1998).

9.3.1 The Ideal Diode Equation

The ideal diode equation can be written as

$$I_e = I_s[e^{V_e/(nV_{th})} - 1], \qquad (9.1)$$

where I_s is the junction saturation current and n is the ideality factor. Furthermore, V_{th} is the thermal voltage given by $k_B T/q$, where k_B is the Boltzmann constant ($= 1.38066 \times 10^{-23}$ J/K), T is the absolute temperature in kelvin, and q is the electronic charge ($= 1.60218 \times 10^{-19}$ C). The junction saturation current I_s can be expressed as

$$I_s = Aq \left(\frac{D_p n_{po}}{L_p} + \frac{D_n p_{no}}{L_n} \right), \qquad (9.2)$$

where A is the area of the device, D_p is the diffusion constant for holes, n_{po} is the minority carrier concentration of electrons on the p-side, L_p is the hole diffusion length, D_n is the electron diffusion constant, p_{no} is the minority carrier concentration of holes on the n-side, and L_n is the electron diffusion length. The minority carrier concentrations are given as the ratio of the square of the intrinsic carrier concentration n_i and the doping

Table 9.1   Silicon diode parameters at room temperature

Parameter	Description	Value
D_n	Electron diffusion constant	$37.5 \, \text{cm}^2/\text{s}$
D_p	Hole diffusion constant	$13 \, \text{cm}^2/\text{s}$
L_n	Electron diffusion length	$\sim 0.08 \, \text{cm}^a$
L_p	Hole diffusion length	$\sim 0.03 \, \text{cm}^a$
n_i	Intrinsic carrier concentration	$1.02 \times 10^{10}/\text{cm}^{-3}$

[a] Valid only for low doping concentrations

concentrations as follows:

$$n_{po} = \frac{n_i^2}{N_a},$$
(9.3)

$$p_{no} = \frac{n_i^2}{N_d}.$$
(9.4)

Here, N_a and N_d are the concentration of dopants on the p- and n-sides, respectively. In Table 9.1, approximate parameter values are given. The ideality factor n varies from 1.0 to about 1.5.

9.3.2 Nonideal Effects

The ideal diode equation described by (9.1) to (9.4) is in reasonable agreement with experimental data for only a limited range of currents. Outside this range, many nonideal effects come into play such as carrier generation/recombination in the depletion region, series resistance, and high-level injection. Here follows a brief discussion of some important nonideal mechanisms that are relevant for the applications we are interested in.

9.3.2.1 Series resistance

In real devices, the parasitic series resistance R_s of the device contacts and of the semiconductor quasi-neutral regions may play an important role, especially at high forward bias. Then the diode equation for the total current has to be modified according to

$$I_e = I_s[e^{(V_e - I_e R_s)/(n V_{th})} - 1] = I_s[e^{V_{ei}/(n V_{th})} - 1],$$
(9.5)

where V_{ei} is the intrinsic voltage. An approximate analytical solution of I_e versus V_e for this case was discussed by Fjeldly *et al.* (1991). We note that at sufficiently large forward bias, most of the applied voltage will fall across the series resistance and the $I_e - V_e$ characteristic approaches a linear form with a slope determined by R_s.

9.3.2.2 High injection

At large forward bias, when the voltage across the intrinsic diode approaches the built-in potential, the junction barrier is severely reduced and the assumption of a low-level injection (where the densities of minority carriers are small compared to those of the majority carriers) is no longer valid. Instead, the injected minority carrier concentrations may become comparable to those of the majority carriers, and we enter the so-called high-injection regime. It can be shown that in the high-injection regime the denominator of the argument to the exponential function in (9.1) is changed to $2nV_{th}$. The result is that the rate of increase of the current with respect to the voltage is reduced.

The effect of the series resistance and high-injection effects are visible in the $I-V$ characteristics in Figure 9.3 in the high-current region when the current level is approaching 1 mA. In this region we notice that the current levels off and becomes less than what it would have been if we had continued on the trace of the ideal diode equation. The initial leveling off is caused by high-injection effects and eventually the series resistance comes into play resulting in a linear increase in the current with respect to the applied voltage.

9.3.2.3 Generation/recombination in the depletion region

Generation and recombination processes are taking place in the depletion region (neglected in the ideal diode) seeking to restore equilibrium. These processes give rise to current contributions that may play an important or even dominant role. These contributions depend on the concentration, distribution, and energy levels of traps in the depletion region. Traps associated with various impurities and defects are always present in any semiconductor material. The simplest model describing the generation and recombination currents is based on the assumption that we have only one type of dominant trap, uniformly distributed in the device. In reality, traps may be nonuniformly distributed, and more than one type of trap may be involved. By applying these assumptions, it can be shown that the generation/recombination current can be modeled as

$$I_{gr} = I_{sr}\left(1 - \frac{V_{ei}}{V_j}\right)^m [e^{V_{ei}/(n_r V_{th})} - 1] \tag{9.6}$$

where I_{sr} is a saturation current parameter related to the generation/recombination process and n_r is the ideality factor for the generation/recombination current ($\sim$2). Note that the first bracket on the right-hand side of (9.6) gives the scaling of the depletion layer width with the potential V_{ei} for an arbitrary grading of the junction doping profile, specified by the grading parameter m (1/2 for an abrupt junction and 1/3 for a linearly graded junction). V_j is the contact (built-in) potential of the junction ($\sim$0.7 V for silicon at room temperature). The generation/recombination current I_{gr} is added to the ideal diode equation in (9.1).

The effect of generation/recombination currents is visible in the $I-V$ characteristics in Figure 9.3 in the low-current region. Notice that the deviation from a straight line in this region causes the total current to become higher than the ideal current. This effect is caused by generation/recombination currents.

REFERENCES

Fjeldly T. A., Moon B., and Shur M. (1991) Analytical solution of generalized diode equation, *IEEE Trans. Electron Devices*, **38**(8), 1976, 1977.

Fjeldly T. A., Ytterdal T., and Shur M. (1998) *Introduction to Device Modeling and Circuit Simulation*, John Wiley & Sons, New York.

10
Modeling of Passive Devices

10.1 INTRODUCTION

Passive devices such as resistors, capacitors, and inductors find wide usage in analog and RF CMOS integrated circuits. In this chapter, we will discuss the different structures utilized to implement the different passive devices and describe how to model them.

10.2 RESISTORS

Analog circuits usually include resistors in a variety of applications to implement functions such as current limiting and voltage division. Fortunately, most processes offer a choice of several different resistor structures utilizing different materials. Some structures are available in standard digital processes, while others require specific analog options that add extra cost to the fabrication of the circuit. In this section we will describe different resistor structures and discuss how they are modeled. However, first we will briefly discuss resistivity, sheet resistance, contact resistance, and resistor layout.

The resistance of a resistor can be calculated from its dimensions and resistivity. In Figure 10.1, we show a simple resistor structure constructed from a slab of a homogeneous material and perfectly conductive contacts.

Using the symbols defined in the figure, the resistance of the structure shown can be written as

$$R = \rho \frac{L}{Wt} \tag{10.1}$$

where ρ is the resistivity of the material. For integrated CMOS circuits the resistors are either diffusions or depositions that can be treated as films of constant thickness. Therefore, it is common to combine resistivity and thickness into a single term, called the sheet resistance $R_{\mathrm{sh}} = \rho/t$. Using this definition of the sheet resistance, (10.1) can be written as

$$R = R_{\mathrm{sh}} \frac{L}{W}. \tag{10.2}$$

In (10.2), the ratio L/W is often called the number of resistance squares. Note that (10.1) and (10.2) are valid only for low frequencies in which the skin effect is not present. As

Device Modeling for Analog and RF CMOS Circuit Design.   T. Ytterdal, Y. Cheng and T. A. Fjeldly
© 2003 John Wiley & Sons, Ltd   ISBN: 0-471-49869-6

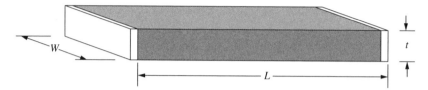

Figure 10.1   A simple resistor structure consisting of a slab of a homogeneous material contacted by perfectly conductive terminals

the frequency is increased, the current is not distributed uniformly throughout the cross section of the resistor. Instead, the current is concentrated at the surface of the material. To account for this effect, we modify the expression for R_{sh} to include the skin effect as follows:

$$R_{sh} = \frac{\rho}{\delta(1 - e^{-t/\delta})},$$ (10.3)

where

$$\delta = \sqrt{\frac{\rho}{\omega\mu_0}}.$$ (10.4)

Here, δ, ω_0, and μ_0 represent the skin depth, the radial frequency, and the permeability in vacuum, respectively.

In the above discussion, we have assumed that the contacts at both ends of the resistor have zero resistivity. Since real contacts have a finite resistance, we have to include terms in (10.2) that account for the resistance of the contacts. A typical layout of a resistor consisting of a body and two contact heads is shown in Figure 10.2. Here L_b and W_b are the length and the width of the body, respectively, and W_c and W_o are the width of the contact and the overlap of the head, respectively.

Two different mechanisms give rise to resistance in the contacts. The first mechanism is caused by a potential barrier that exists between the metallization and the material of the resistor. The second mechanism is due to the current spreading or crowding as it enters the contact. If $W_c + 2W_o$ is greater than W_b, the current spreads out and the resulting resistance is slightly decreased. The total resistance of the resistor structure in Figure 10.2 is given by

$$R = R_b + 2R_h,$$ (10.5)

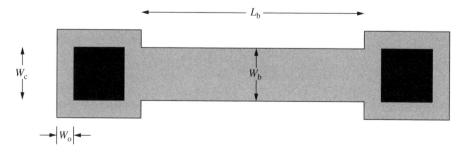

Figure 10.2   The layout of a typical resistor consisting of the resistor body and two resistor heads

where R_b and R_h are the resistances of the body and the head, respectively, and is given by

$$R_b = R_{sh} \frac{L_b}{W_b} \qquad (10.6)$$

and

$$R_h = R_c + R_{sh} \left(\frac{W_o}{W_b} + \Delta_\square \right). \qquad (10.7)$$

Here, R_c is the contact resistance and $\Delta_\square$ is an adjustment of the number of resistance squares caused by current spreading or crowding. If $W_c + 2W_o > W_b$, $\Delta_\square$ is negative, otherwise it is positive. The following approximate formula for $\Delta_\square$ was proposed by Ting and Chen (1971):

$$\Delta_\square = \frac{1}{\pi} \left[\frac{1}{k} \ln \left(\frac{k+1}{k-1} \right) + \ln \left(\frac{k^2 - 1}{k^2} \right) \right], \qquad (10.8)$$

where

$$k = \frac{W_b}{W_b - W_c}. \qquad (10.9)$$

This formula is strictly valid only for $W_c \gg W_b - W_c$.

The contact resistance R_c in (10.7) results from the presence of a potential barrier between the resistance material and the metal. Murrmann and Widmann (1969) showed that the contact resistance could be written as

$$R_c = \frac{\sqrt{R_{sh} \rho_c}}{W_c} \coth \left(L_c \sqrt{\frac{R_c}{\rho_c}} \right) \qquad (10.10)$$

where ρ_c is the specific contact resistance and L_c is the length of the contact ($L_c = W_c$ if square contacts are used). In Table 10.1, we list contact resistances for two common contact systems.

This finalizes our general discussion of resistors. In the following sections, we will discuss the most important resistors types available in CMOS processes in terms of sheet resistance and equivalent circuits including parasitics.

10.2.1 Well Resistors

In digital CMOS processes, where there is no high sheet resistance poly available, we may use one or more wells to implement resistors. In conventional CMOS processes,

Table 10.1 Typical contact resistances for two common contact systems (Hastings 2001)

Contact system	Sheet resistance R_{sh} (Ω/sq)	R_c ($\Omega\mu m^2$)
Al-Cu-Si	160	750
Al-Cu/Ti-W/PtSi	160	1250

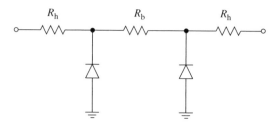

Figure 10.3   Equivalent circuit of a well resistor

there is usually only one type of well available, either n-well or p-well. However, in state-of-the-art scaled down processes (as of this writing, 0.18-μm CMOS and newer), it becomes more and more common to have several well types.

The most important feature of well resistors is the relative high sheet resistance, on the order of 1 to 10 kilo-ohms per square. The disadvantages of well resistors are high temperature coefficients (TCs) (may be as high as 6000 ppm/°C), voltage dependency, and large parasitic capacitances to ground since the well is located close to the substrate. A 1π equivalent circuit of a well resistor valid up to moderate frequencies ($\sim$100 MHz) is shown in Figure 10.3. Note that the parasitic capacitors are implicitly included through the reverse-biased diodes. By using diodes in the equivalent circuit, leakage currents to ground are also accounted for in the equivalent circuit, which is important at elevated temperatures. If the resistor is long, a 1π equivalent circuit may not be accurate enough. In such cases the equivalent circuit in Figure 10.3 can be easily extended to 3π and 5π equivalents.

As a final note on well resistors, we would like to point out an important issue. When laying out well resistors, remember that the width should be at least twice as large as the depth of the well. Otherwise the resistor does not achieve full junction depth and the sheet resistance becomes much higher than the values reported by the foundry.

10.2.2 Metal Resistors

When small resistances are desired, maybe the best choice is to use one of the metal layers. Sheet resistance of the different metal layers is typically in the range 20 to 40 mΩ/sq. The advantages of metal resistors are low parasitic coupling to the substrate, low-voltage dependency, and low TCs.

10.2.3 Diffused Resistors

Diffused resistors can be realized in a CMOS process by making contacts to each side of an implanted region (the same type of region that is used for the drain and source of MOS transistors). Resistors made this way exhibit sheet resistances in the range 20 to 50 Ω/sq if silicide blocks are used. Hence, using this resistor type may add extra cost to the manufacturing (check the documentation of the process you are using). This resistor type is not often used, since most CMOS processes offer poly resistors that have equal or greater sheet resistances (see next section).

10.2.4 Poly Resistors

Polysilicon is used in all modern CMOS processes for producing the gate of the MOS transistors. After deposition, the poly is heavily doped to improve conductivity for obtaining high-speed operation of the MOS transistors. The sheet resistance of heavily doped poly lies in the range of 1 to 20 Ω/sq. At en extra cost, a mask can be manufactured that stops heavy poly doping at regions where resistors are desired. As a result, lightly doped poly can be produced with sheet resistance varying betwcen 20 to 1000 Ω/sq. The TC of poly resistors can have both positive and negative values depending on the doping density and the type of doping atoms used. In general, the absolute value of the TC increases with the sheet resistance.

Similar to well resistors, the equivalent circuit of poly resistors also includes parasitic capacitances to ground. However, since the poly layer is located further from the substrate compared to wells, the capacitance per unit resistor area is smaller for poly resistors compared to well resistors. Poly resistors may have unit-area parasitic capacitances on the order of 0.1 fF/μm^2. A 1π equivalent circuit of a poly resistor valid up to moderate frequencies is shown in Figure 10.4.

In RF CMOS circuits, only poly and metal resistors are used since the other resistor types are located too close to the substrate, which causes the substrate loss to be larger than necessary. At gigahertz (GHz) frequencies, the equivalent circuits discussed so far are not accurate enough. Effects that need to be included are the substrate loss and the parasitic series inductance. An equivalent circuit valid at frequencies up to about 10 GHz is shown in Figure 10.5.

The only intended element in Figure 10.5 is the ideal resistor R. All the other elements are considered parasitics. The self-inductance of the metal or the poly resistor is modeled by inductor L_p. Parasitics due to the substrate are commonly modeled in terms of C_p, C_s, and R_s. The lateral dimensions of the structure are on the order of hundreds of micrometers, which is much larger than the oxide thickness and comparable to the substrate thickness. Thus, the substrate capacitance and resistance are approximately proportional to the area occupied by the resistor and can be estimated by

$$C_p = \frac{1}{2}L_b W_b \frac{\varepsilon_0 \varepsilon_{ox}}{t_{ox}}, \tag{10.11}$$

$$R_{sub} = R_{sub0} L_b / W_b, \tag{10.12}$$

$$C_s = \frac{1}{2}L_b W_b C_{sub}, \tag{10.13}$$

$$R_s = \frac{2}{L_b W_b G_{sub}}. \tag{10.14}$$

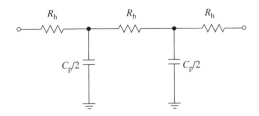

Figure 10.4 Equivalent circuit of a poly resistor

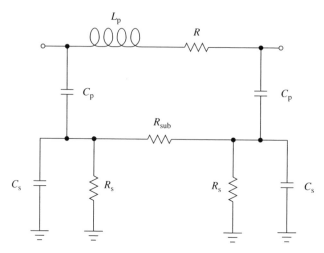

Figure 10.5 Equivalent circuit of a poly resistor that is valid at gigahertz frequencies

Here $\varepsilon_0 (= 8.85418 \times 10^{-12}\,\text{F/m})$ is the permittivity in vacuum, and ε_{ox} and t_{ox} are the dielectric constant and the thickness of the field oxide, respectively. Furthermore, R_{sub0}, C_{sub}, and G_{sub} are technology parameters that must be provided by the foundry. Equations (10.11)–(10.14) are valid only if the body of the resistor is much larger than the heads.

10.3 CAPACITORS

Capacitors have become ubiquitous in analog-integrated circuits particularly owing to the switched capacitor technique for realization of analog-to-digital and digital-to-analog data converters and discrete-time filters. Other applications include continuous-time filters, RF building blocs, and for compensation in feedback amplifiers.

Farad is the SI unit for capacitance. In the context of integrated circuits, the range of values is between several femtofarads (fF) to about 100 picofarads (pF). Capacitors used in integrated circuits are parallel-plate capacitors in which a slab of an insulating material called the *dielectric* is sandwiched between two conductive plates. The capacitance value of a parallel-plate capacitor is probably known to the reader, but is repeated here for your convenience:

$$C = \frac{A\varepsilon_0\varepsilon_i}{t_i}, \tag{10.15}$$

where A is the area of the plates, ε_i and t_i are the relative permittivity (or dielectric constant) and thickness of the insulator, respectively. The dielectric constants of the two most common insulators used in CMOS are shown in Table 10.2.

The expression in (10.15) slightly underestimates the capacitance since not all field lines go through the insulator. Some of the field lines go through the air and is called the fringing field, which increases the apparent area of the plates. This effect is proportional to the thickness of the insulator. In integrated circuits it is usually neglected since the vertical dimensions are usually much smaller than the dimension of the plates.

Table 10.2 Relative permittivities of insulators used in CMOS technologies

Material	Relative permittivity
Silicon dioxide (SiO$_2$)	3.9
Silicon nitride (Si$_3$N$_4$)	7.5

The available capacitor structures in CMOS technologies are poly-insulator-poly (or simply poly-poly) capacitors, metal-insulator-metal (MIM) capacitors, MOSFET capacitors, and junction capacitors. These different capacitor types are discussed in the following sections.

10.3.1 Poly–poly Capacitors

As the name indicates, both plates of a poly-insulator-poly (or simply poly-poly) capacitor are made of deposited polysilicon that is doped to keep the resistivity low. The bottom plate is usually implemented using the same layer as the poly gate of MOSFETs. The other plate must be supported by a second poly layer. There are extra processing steps involved in poly-poly capacitors since the insulator is unique to this structure. Hence, if you want poly-poly capacitors you would have to pay extra for the fabrication of your circuits. An example vertical cross section of a poly-poly structure is shown in Figure 10.6.

We note from the figure that the bottom plate is placed on top of field oxide to shield the plate from the substrate. Of course, the shielding is not perfect and there is a parasitic capacitance connected between the bottom plate and the substrate. This parasitic capacitance, which is labeled C_{p1} in Figure 10.6, can be up to 20% of the intended capacitance (labeled C). The parasitic capacitance labeled C_{p2} is much smaller than C_{p1} since the distance between the metal layer and the substrate is large.

An equivalent circuit valid up to moderate frequencies can be extracted from the vertical cross section in Figure 10.6 and is presented in Figure 10.7. At gigahertz frequencies the lossy substrate has to be accounted for in the equivalent circuit using an approach similar to that was used for resistors in Section 10.2.4.

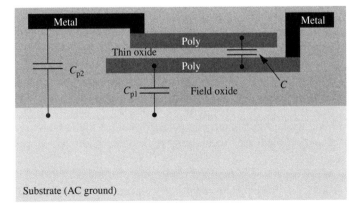

Figure 10.6 Vertical cross section of a poly-poly capacitor structure

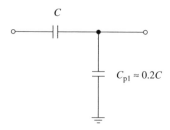

Figure 10.7 Equivalent circuit of poly-poly capacitors valid up to moderate frequencies

The capacitance per unit area of poly-poly capacitors changes from one technology generation to the next because both vertical and lateral dimensions are scaled. Hence, it is not possible to provide a number that can be universally applied. For example, at the 0.5-μm technology node, the capacitance per unit area was around 1 fF/μm^2. You would have to refer to the documentation of your current process for accurate numbers.

There is a voltage modulation of poly-poly capacitors caused by poly-depletion since the conductivity of the polysilicon plates is finite. Usually, the voltage dependency is modeled using a second-order polynomial description as follows:

$$C = C_0(1 + k_1 V + k_2 V^2), \qquad (10.16)$$

where C_0 is the capacitance at zero applied voltage, k_1 and k_2 are the first- and second-order coefficients of the voltage dependency, and V is the voltage across the capacitor. The coefficients k_1 and k_2 must be supplied by the foundry or extracted from measured data. If the linearity of poly-poly capacitors is not sufficient, one can instead use metal plates. This type of capacitors is discussed in the following section.

10.3.2 Metal–insulator–metal Capacitors

Usually, when referring to metal-insulator-metal (MIM) capacitors, we assume that structures involving thin insulators are provided by dedicated process steps. This is a relatively novel structure that made its introduction in recent technology generations. However, the principle of using metal layers as the plate material has been utilized by design houses for a long time to cut cost since the double poly option could be skipped. The drawback of such capacitors is the low capacitance per unit area (typically 0.05 fF/μm^2) since the oxide used between metal layers is quite thick. Such capacitors will not be further discussed here. Instead we will focus on MIM capacitors since they are rapidly becoming very popular owing to their high linearity and high unit-area capacitance. In modern state-of-the-art processes, MIM capacitors are replacing poly-poly capacitors owing to their improved linearity and mismatch characteristics since the conductivity of the metal plates is higher than that for the corresponding polysilicon plate, which reduces the effect of depletion.

Usually, the MIM capacitor is realized using one of the conventional metal layers as the bottom plate and a dedicated thin metal layer placed between two conventional metal layers to realize the top plate. The insulator is made of a thin dielectric with a thickness of about 40 nm.

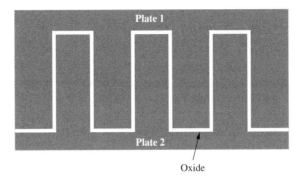

Figure 10.8   Top view of a lateral flux capacitor utilizing a single metal layer

The area consumed by capacitors tends to be large. Therefore, the search for more area-effective structures is on and considerable amount of research has been conducted to exploit new structures both in the vertical and in the lateral dimension. In the vertical dimension, the area efficiency can be improved by using more than two metal layers. With the continuous downscaling of the minimum line width in CMOS technologies, the lateral field has been exploited to realize capacitor structures such as the lateral flux capacitance and quasi-fractal capacitors with enhanced capacitance per unit area (see, for example, Akcasu (1993); Samavati *et al.* (1998); Stolmeijer and Greenlaw (1999)). The top view of an example lateral flux capacitor structure realized using a single metal layer is shown in Figure 10.8. As a final remark on MIM capacitors, we should mention that the equivalent circuits that are used for poly-poly capacitors can also be used for MIM capacitors.

10.3.3 MOSFET Capacitors

MOS transistors can be used as capacitors. Their selling point is the high unit-area capacitance due to the thin oxide. For example, in 0.13-μm CMOS processes the unit-area capacitance is about $10\,fF/\mu m^2$. If the transistors are placed in isolated wells, floating capacitors with both negative and positive applied voltages can be realized. The connection of the transistor in such a configuration is shown in Figure 10.9. One of the most important applications of such capacitors is frequency compensation in feedback systems since the accuracy of the capacitance value is not important; it just has to be larger than a given value. There is only one issue to be aware of when using such a capacitor as a frequency compensation element, and that is the dip in the capacitance around the threshold voltage of the transistor as shown in Figure 10.10.

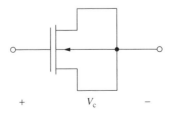

Figure 10.9   MOS transistor in an isolated well configured as a floating capacitor

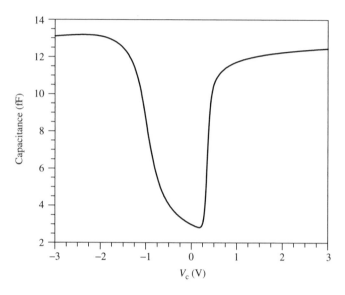

Figure 10.10   Capacitance-voltage characteristics of a MOS transistor configured as in Figure 10.9

Another important application of the capacitance element shown in Figure 10.9 is as a variable capacitor (varactor) in RF circuits as a frequency-tuning device. In this type of application, we want to be in the dip region in Figure 10.10 where the capacitance is most sensitive to the voltage across it. In varactor structures the gate structure and the contacts can be placed over both *n*- and *p*-type wells to create both conventional MOS transistors and accumulation mode transistors.

10.3.4 Junction Capacitors

Another type of capacitor available to designers of CMOS circuits is the junction capacitor that utilizes the depletion region surrounding a reverse-biased *pn*-junction as the insulator. An example structure is shown in Figure 10.11 where the capacitor is formed by placing an n^+ diffusion region inside a *p*-type well. The permittivity of silicon is about three times higher than that of silicon dioxide, so the capacitance per unit area is quite high. For example, in 0.13-μm CMOS processes it is on the order of 2 fF/μm^2.

We note from Figure 10.11 that the effective area to be used when estimating the capacitance value of the structure consists of two parts, the bottom and the sidewalls of the n^+ diffusion region. The area of the bottom region is usually approximated by the length multiplied by the width of the n^+ diffusion region. An approximate, and most often used, estimate of the area of the sidewalls is the depth of the junction, which is a technology parameter, multiplied by the perimeter of the n^+ diffusion region times a constant to account for the nonrectangular shape. Usually, the foundries provide numbers for two unit capacitors and the total capacitance at zero bias can be calculated as follows:

$$C_{total0} = C_{j0} A + C_{jsw0} P, \tag{10.17}$$

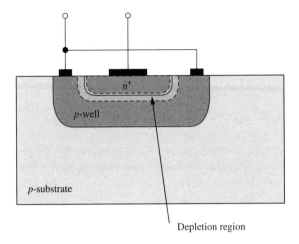

Figure 10.11   A junction capacitance formed by an n^+ diffusion region inside a p-type well

where A and P are the area and the perimeter of the n^+ diffusion region, respectively, and C_{j0} and C_{jsw0} are the capacitance at zero bias per unit area and perimeter of the of the n^+ diffusion region, respectively.

Since we have denoted the above parameters at zero bias, you have probably guessed that the junction capacitance is voltage-dependent. To illustrate this, we have plotted in Figure 10.12 the junction capacitance versus applied reverse voltage for a commercially available 0.13-μm CMOS technology. We note from the figure that the capacitance is maximum at zero voltage and has decreased by about 40% at a reverse bias of 2 V. The capacitance decreases with the square root of the applied voltage because the width of the depletion region increases at this rate.

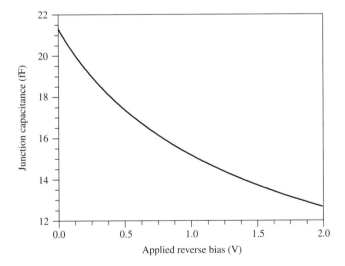

Figure 10.12   Junction capacitance versus applied reverse voltage for a commercially available 0.13-μm CMOS technology

Table 10.3   Diode junction parameters in SPICE

Parameter	Description	Default value
CJ0	Zero-bias junction capacitance	0 F
CJSW	Zero-bias sidewall capacitance	0 F
MJ	Grading coefficient	0.5
MJSW	Grading coefficient of the sidewalls	0.33

The voltage dependency illustrated in Figure 10.12 is the most serious drawback of the junction capacitance. Another problem, that is particularly troublesome at elevated temperatures, is the conductance that exists in parallel with the capacitor that causes the capacitor to leak. This effect is dominated by generation/recombination currents in the depletion region.

Usually, a junction diode is used to model junction capacitors, where the key model parameters in SPICE are listed in Table 10.3.

10.4 INDUCTORS

Interest in integrated spiral inductors in CMOS processes has surged with the recent growing demand for CMOS RF communication circuits. This is because inductors form the core to a successful integrated radio transmitter and receiver (transceiver). While the integration of resistors and capacitors in CMOS is a well understood process, the inclusion of inductors is still a challenging endeavor. In recent years a great amount of work has been published on the analysis and modeling of on-chip spiral inductors (see, for example, Lutz *et al.* (1999); Yue and Wong (2000); Kythakyapuzha and Kuhn (2001)).

The spiral inductor is the most common structure for implementing on-chip inductors in standard CMOS processes. A spiral structure is laid out using two or more of the top metal layers. The top metal layers are used since this will minimize the resistive coupling to the substrate to reduce the loss as much as possible. In Figure 10.13, we show a typical layout of an inductor using two metal layers. More than two metal layers can be utilized to increase the Q factor (Lutz *et al.* 1999).

The inductance value of spiral inductors is a complicated function of geometry and is usually estimated from two-dimensional electromagnetic (EM) simulations. For a crude estimate (within $\pm 30\%$), the following equation can be used (Lee 1998):

$$L = \mu_0 n^2 r \approx 1.2 \cdot 10^{-6} n^2 r, \tag{10.18}$$

where n and r represent the number of turns and the radius of the spiral, respectively.

The key to accurate modeling of the spiral inductor at high frequencies is the ability to identify the important parasitics. Fortunately, in most cases the foundries do this job for us providing an equivalent circuit and the values for each circuit element. The values for the circuit elements are usually extracted from predefined structures. Hence, if the designers wish to modify the structures to optimize the inductor for either a special frequency range or an application, the characterization task has to be performed again.

A commonly used equivalent circuit of an integrated spiral inductor is shown in Figure 10.14 (see, for example, Yue and Wong (2000)).

Figure 10.13   Typical layout of a spiral inductor using two metal layers

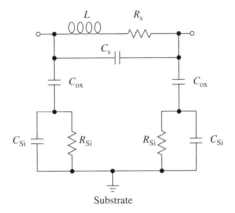

Figure 10.14   Equivalent circuit of an integrated spiral inductor on a silicon substrate

Note that since inductors are inserted in a circuit to store magnetic energy, both capacitors and resistors are considered parasitics, resistors because they introduce loss (decreasing the Q value) and capacitors because they store electric energy and provide coupling to the lossy substrate. Thus, the only intended element in Figure 10.14 is the ideal inductor L.

The resistance of the metal spiral is modeled by the resistor R_s in series with the inductor. The resistance value can be calculated using (10.2) and (10.3).

The capacitor denoted as C_s in Figure 10.14 models the capacitive coupling between the input and the output ports of the spiral inductor. By inspecting the physical layout of the inductor in Figure 10.13, we notice that both the overlap between the spiral and the underpass and cross-talk between adjacent turns contributes to C_s. In most practical inductors, adjacent turns are on almost the same potential and the total capacitance is dominated by the overlap capacitances. Hence, it is sufficient to model C_s as a sum of overlap capacitances as follows:

$$C_s = n W_s^2 \frac{\varepsilon_0 \varepsilon_{ox}}{t_{ox\_m}}, \tag{10.19}$$

where W_s is the spiral line width, ε_{ox} is the dielectric constant of the oxide, and t_{ox_m} is the vertical distance between the spiral and the underpass.

Parasitics due to the substrate are commonly modeled in terms of C_{ox}, C_{Si}, and R_{Si} (see, for example, Hughes and White (1975)). For spiral inductors on silicon, the lateral dimensions of the spiral structure are on the order of hundreds of micrometers, which is much larger than the oxide thickness and comparable to the substrate thickness. Thus, the substrate capacitance and resistance are approximately proportional to the area occupied by the spiral and can be estimated by

$$C_{ox} = \frac{1}{2} L_s W_s \frac{\varepsilon_0 \varepsilon_{ox}}{t_{ox}}, \tag{10.20}$$

$$C_{Si} = \frac{1}{2} L_s W_s C_{sub}, \tag{10.21}$$

$$R_{Si} = \frac{2}{L_s W_s G_{sub}}. \tag{10.22}$$

Here, C_{sub} and G_{sub} are technology parameters usually provided by the foundry.

REFERENCES

Akcasu O. E. (1993) *High Capacitance Structures in a Semiconductor Device*, U.S. Patent 5,208,725.

Hastings A. (2001) *The Art of Analog Layout*, Prentice Hall, Upper Saddle River, NJ.

Hughes G. W. and White R. M. (1975) Microwave properties of nonlinear MIS and Schottky-barrier microstrip, *IEEE Trans. Electron Devices*, **22**(10), 945–955.

Kythakyapuzha S. R. and Kuhn W. B. (2001) Modeling of inductors and transformers, *Proc. Radio Frequency Integrated Circuits (RFIC) Symposium*, pp. 283–286.

Lee T. H. (1998) *The design of CMOS Radio Frequency Integrated Circuits*, Oxford Press, Cambridge.

Lutz R. D. *et al.* (1999) Modeling and analysis of multilevel spiral inductors for RF ICs, *1999 IEEE MTT-S International Microwave Symposium Digest*, Vol. 1, pp. 43–46.

Murrmann H. and Widmann D. (1969) Current crowding on metal to planar devices, *IEEE Trans. Electron Devices*, **16**(12), 1022–1024.

Samavati H. *et al.* (1998) Fractal capacitors, *IEEE J. Solid-State Circuits*, **33**, 2035–2041.

Stolmeijer A. and Greenlaw D. C. (1999) *High Quality Capacitor for Submicrometer Integrated Circuits*, U.S. Patent 5,939,766.

Ting C. Y. and Chen C. Y. (1971) A study of the contacts of a diffused resistor, *Solid-State Electron.*, **14**, 434.

Yue C. P. and Wong S. S. (2000) Physical modeling of spiral inductors on silicon. *IEEE Trans. Electron Devices*, **47**(3), 560–568.

11

Effects and Modeling of Process Variation and Device Mismatch

11.1 INTRODUCTION

As CMOS IC fabrication technology becomes more and more advanced, the control of process variation and manufacturing uncertainty becomes more and more critical. The circuit yield loss caused by the process and device parameter variation has been more pronounced than before. Because CMOS technology demonstrates a promising future for low-power and low-cost analog and mixed-signal applications, designers are currently utilizing advanced deep-submicrometer technologies to achieve the goal of low-voltage and high-performance integration. However, owing to difficulty in controlling the variation in different fabrication steps such as gate oxide growth, channel and source/drain implants, photolithography, etching, and so on in modern technologies, the variation in device parameters will become larger compared with older technology. Designers need physical, predictive, and accurate statistical models to describe the device (and hence the circuit) parameter variation caused by process variations.

Typically, the variations of device/circuit characteristics are divided into two categories, interdie variation and intradie variation/mismatch. The interdie variation describes the die-to-die, wafer-to-wafer, or lot-to-lot process variability. In other words, the same variation is assumed for the devices in the same circuit, so the interdie device variation has little influence on the circuit behavior of some analog circuits such as a current mirror with a constant current bias as long as all transistors can still be biased in the saturation region. The intradie device variation and mismatch describes the die-/wafer-level process variability caused by the variation of some process parameters such as oxide thickness and doping profile across the die/wafer. This is sometimes called local process variation and mismatch (LPVM) (Pronath *et al.* (2000); Zanella *et al.* (1999)). In this case, devices in the same circuit may have different variations in electrical parameters caused by LPVM. For digital circuits, the influence of interdie variations on the circuit performance is important. So most circuit simulators with statistical modeling capability for digital applications ignore the intradie variations when simulating the circuit behavior with considerations of process and device variation, which does not cause problems because the device variation caused by the interdie variability is indeed much larger than that caused by intradie variability.

Device Modeling for Analog and RF CMOS Circuit Design.   T. Ytterdal, Y. Cheng and T. A. Fjeldly
© 2003 John Wiley & Sons, Ltd   ISBN: 0-471-49869-6

The mismatching of devices in the same circuits has little influence on the circuit behavior. However, for analog circuits, not only interdie process variation but also intradie process variation and device mismatch influence significantly the variations of the circuit behaviors (Pelgrom *et al.* (1998); Groon *et al.* (2001); Tarim *et al.* (2000)). Statistical modeling of devices with intradie process variation and mismatch is needed in designing analog circuits to statistically and accurately simulate/predict the analog circuit performance.

Together with the MOS transistors, other devices such as resistors and capacitors used in analog circuit design are also sensitive to the LPVM. It is important to model the LPVM, which influences the electrical behavior of resistors, capacitors, and MOSFETs for analog/RF applications.

11.2 THE INFLUENCE OF PROCESS VARIATION AND DEVICE MISMATCH

11.2.1 The Influence of LPVM on Resistors

Typical CMOS and BICMOS technologies offer several different resistors, such as diffusion n^+/p^+ resistors, n^+/p^+ poly resistors, and n-well resistor. Depending on the applications, designers select different resistors in circuit design. Since different materials have various electrical characteristics and temperature variations, the effects of process variation will be different for different types of resistors. Many factors in the fabrication of a resistor such as the fluctuations of the film thickness, doping concentration, doping profile, and the annealing conditions as well as the dimension variation caused by the photolithographic inaccuracies and nonuniform etch rates can vary the value of a resistor. Poly resistors, which consist of a strip of poly deposited on top of field oxide in LOCOS isolation technology or in shallow trench isolation technology, are widely used in analog circuit design. We take it as an example to discuss the influence of the process variation.

The process control in the deposited polysilicon film thickness is one of the key factors to vary the poly resistor value. Typically a modern CMOS process allows 10 to 15% variation in polysilicon thickness, which results in a similar amount of change in resistance. Further, the doping level determined by various implants and annealing steps in the process will contribute a 10 to 20% variation. The overall effects from the variation in polysilicon thickness and doping will result in a sheet resistance variation of 15 to 25%. In addition, the line-width control, a measure of dimensional variation caused by photolithography and processing, is also critical in determining the resistance variation. Most modern processes can have line-width control of poly dimension less than ±10% of their minimum feature size. However, one should be aware that an extremely narrow poly resistor might have increased variation in the resistance owing to the growth of individual grains across the entire width of the resistor. When considering poly resistor etching, which easily causes a large variation in resistance value depending on the geometry of the resistor, typically, the overall (wafer-to-wafer and lot-to-lot) sheet resistance variation of poly resistors in modern IC technologies is in the range of 20 to 30% (Lane and Wrixon (1989)). The variation on a single wafer (die-to-die) is smaller. Figure 11.1 shows a 3-sigma variation of unsalicided poly resistances measured from a single wafer in a 0.18-μm CMOS technology.

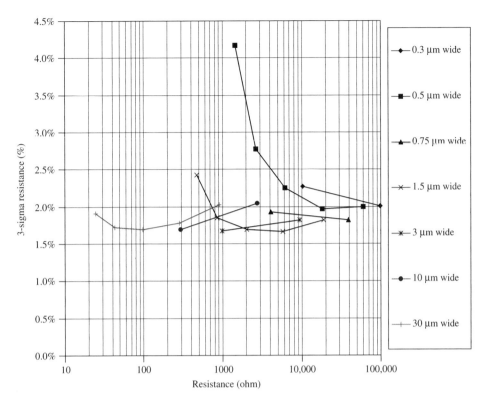

Figure 11.1 Measured data for characterizing the variation of the poly resistor on a single wafer. Reproduced from Cheng Y. (2002) The influence and modeling of process variation and device mismatch for analog/RF circuit design, *Proceedings of the Fourth IEEE International Caracas Conference on Devices, Circuits and Systems*, pp. 282–289

As mentioned above, a poly resistor should always reside on the top of the field oxide, which not only reduces the parasitic capacitance between the resistor and the substrate but also ensures that oxide steps do not cause unexpected resistance variations. In addition, it should be noted that the oxide isolating the resistor will provide the isolation both electrically and thermally. A poly resistor that dissipates sufficient power will experience permanent resistance variations due to self-induced annealing. Extreme power dissipation will melt or crack polysilicon long before diffused resistors of similar dimensions suffer damage.

As discussed above, integrated resistors display significant variations in the sheet resistance due to the influence of the variation of some process and device parameters in the fabrication. However, this does not prevent the resistors from being used in various analog circuit designs as long as the device matching properties are within the range the designs require. For analog circuit applications, the matching property of resistors is as important as, or more important than, the variation of the absolute value of the resistance. The understanding and modeling of the mismatching of resistors is very important for analog and RF applications.

Besides the factors causing the variation in absolute resistance, such as the fluctuations in dimensions, doping, and film thickness, more factors related to the fabrication process and material could influence the matching behavior of the resistors. Typically two parameters, mean (M) and standard deviation (σ), are used to describe the matching characteristics of the devices defined in Eqs. (11.1) and (11.2) (Box *et al.* (1978)):

$$M = \frac{\sum\limits_{i=1}^{N} \delta_i}{N},$$
(11.1)

$$\sigma = \sqrt{\frac{\sum\limits_{i=1}^{N} (\delta_i - M)^2}{N - 1}},$$
(11.2)

where δ_i is the parameter value of the *i*th sample unit. The mean is a measure of the systematic mismatch between the matched devices, caused by mechanisms that influence all of the samples in the same way. The standard deviation describes random mismatch caused by statistical fluctuations in process parameters or material properties. Once the mean and standard deviation have been determined using Eqs. (11.1) and (11.2), they can be used to predict worst-case mismatches according to so called 3-sigma or 6-sigma mismatch.

During the processing, many factors can result in fluctuations of the resistance of the resistor and all these fluctuations can be categorized into two groups: one in which the fluctuations occurring in the whole device are scaled with the device area, called area fluctuations, and the other in which fluctuations take place only along the edges of the device and are therefore scaled with the periphery, called peripheral fluctuations. For a matched resistor pair with width W and resistance R, the standard deviation of the random mismatch between the resistors is (Shyu *et al.* (1982))

$$\sigma = \frac{\sqrt{f_a + \dfrac{f_p}{W}}}{W\sqrt{R}}$$
(11.3)

where f_a and f_p are constants describing the contributions of area and periphery fluctuations, respectively.

Figure 11.2 shows the measured data of 3-sigma deviation versus resistance (geometry) from a group of unsalicided resistors. It shows that the resistors with narrow widths have poorer matching characteristics, which can be explained by Eq. (11.3). In circuit applications, to achieve required matching, resistors with width (at least 2–3 times) wider than minimum width should be used. Also, as shown in Figure 11.2, resistors with higher resistance (longer length) at fixed width exhibit larger mismatching. To achieve the desired matching, it has been a common practice that a resistor with long length (for high resistance) is broken into shorter resistors in series.

11.2.2 The Influence of LPVM on Capacitors

Current CMOS technology provides various capacitance options, such as poly-to-poly capacitors, metal-to-metal capacitors, MOS capacitors, and junction capacitors. The more

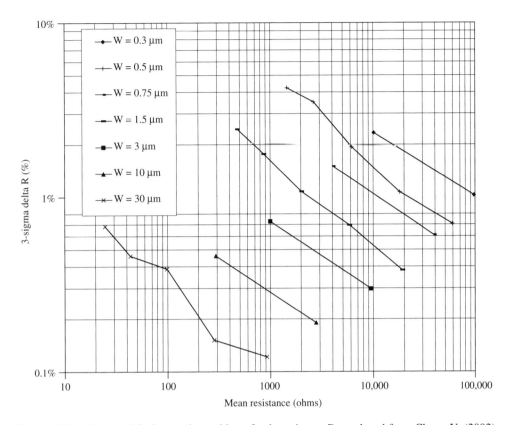

Figure 11.2   Measured 3-sigma mismatching of poly resistors. Reproduced from Cheng Y. (2002)
The influence and modeling of process variation and device mismatch for analog/RF circuit design,
Proceedings of the Fourth IEEE International Caracas Conference on Devices, Circuits and Systems,
pp. 282–289

popular ones used in the design are metal-to-metal capacitors, junction capacitors, and
MOS capacitors. It has been known that the integrated capacitors show significant variability due to the process variation. For example, both the capacitance value and matching
property of a metal-insulator-metal(MIM) capacitor are sensitive to the variation in the
thickness of dielectric and geometry of metal plates. For a MOSFET or a MOS capacitor,
the capacitance values are strongly dependent on the changes in oxide thickness and doping profile in the channel besides the variation in geometries. Typically a modern CMOS
process can maintain a variation in some integrated capacitors such as MOS capacitors
within ±20%. However, the variation in MIM capacitors can be larger than 20%, especially the one called parasitic capacitor or lateral capacitor, distinguishing it from the
typical (vertical) MIM capacitors, because this kind of capacitor is dependent not only
on the thickness of the dielectric and the geometries but also on the permittivity of the
dielectric film that varies with the composition and the growth condition of the dielectric.
Similarly, a ±15–20% variation in junction capacitors should be expected in a modern
process, and the variation becomes larger in capacitors with large area.

Similar to the resistors, the matching behavior of capacitors depends on the random
mismatch due to periphery and area fluctuations (without including the fluctuation of the

oxide thickness) with a standard deviation (Shyu *et al.* (1982)),

$$\sigma = \frac{\sqrt{f_a + \dfrac{f_p}{C}}}{\sqrt{C}} \qquad (11.4)$$

where f_a and f_p are factors describing the influence of the area and periphery fluctuations, respectively. A more complex formula to account for the random oxide and mobility effects has also been reported by Shyu *et al.* (1984).

As given in Eq. (11.4), the contribution of the periphery components decreases as the area (capacitance) increases. For very large capacitors, the area components dominate and the random mismatch becomes inversely proportional to $\sqrt{C}$. Figure 11.3 shows the mismatching characteristics of the MIM capacitor. Even though the mismatching of such kinds of MIM capacitors is larger than that of other capacitors such as regular MIM capacitors, the trend of mismatching versus capacitance approximately follows that represented by Eq. (11.4).

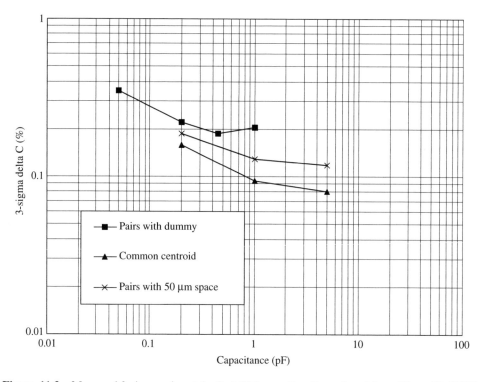

Figure 11.3   Measured 3-sigma mismatch of a MIM capacitor. Reproduced from Cheng Y. (2002) The influence and modeling of process variation and device mismatch for analog/RF circuit design, *Proceedings of the Fourth IEEE International Caracas Conference on Devices, Circuits and Systems*, pp. 282–289

11.2.3 The Influence of LPVM on MOS Transistors

MOSFETs are the most complex components in a CMOS technology fabrication. The variations in many process parameters can result in the variations in device characteristics. The more important ones are the variation in oxide thickness, doping concentration, and profiles in both the channel region and the source/drain region, and device channel length/width. For 0.1-μm or less advanced technology, the variations in doping and annealing condition for the polysilicon gate and in silicon-oxide interface characteristics should be taken into account because they will influence significantly some electrical parameters such as the traps/interface charges, effective oxide thickness, mobility, and so on.

As MOSFETs attract more and more analog/RF applications, the matching characteristics are crucial in achieving precise analog/RF circuit design. Owing to the fast development of very large-scale integration (VLSI) technologies, the minimum feature sizes of MOSFETs in several advanced technology generations, such as $0.13\,\mu$m and less, have been scaled to an unbelievable geometry range, which is reaching the physical limit for the devices to operate properly according to the classic device physics theory. In this case, many factors, which are negligible in influencing the device matching behavior of devices with large feature sizes ($0.5\,\mu$m and above), are now becoming important. Advanced device mismatching models based on the device characterization of modern technology are needed to include new factors that are more pronounced in devices in $0.13\,\mu$m and less technologies since analog IC applications, such as D/A converters and reference sources, would like to have approaches of reducing the layout area without degrading the device matching.

Depending on the applications, different device factors are used to describe mismatching behavior. The one used widely is $\Delta I_d/I_d$, which typically includes two portions, threshold voltage mismatch and current factor mismatch as given in Eq. (11.5):

$$\frac{\Delta I_d}{I_d} = -\frac{G_m}{I_d}\Delta V_{th} - \beta\Delta\frac{1}{\beta} \tag{11.5}$$

where V_{th}, β, I_d, and G_m are the threshold voltage, current factor, drain current, and transconductance of the device, respectively. For analog applications, other factors for describing the mismatching behavior of small-signal parameters such as G_m and G_{ds} are also introduced (Thewes *et al.* (2000)). Furthermore, the influence of mismatching on RF behavior such as the distortion characteristics has been reported recently (Lee *et al.* (2000)).

Figures 11.4 and 11.5 show the three-signal mismatching characteristics measured from a 0.18-μm CMOS technology. The data in Figure 11.4 are measured by monitoring ΔV_{gs}, representing the input offset voltage for a differential pair. ΔV_{gs} is determined by applying the same gate and drain voltages to both transistors and then varying the gate voltage of one of the transistors until the drain current is equal to the other transistor. The data in Figure 11.5 are measured by monitoring the ΔI_d, representing the error between the currents in a simple current mirror. The ΔI_d in Figure 11.5 is defined as the percent difference in I_d normalized by I_d.

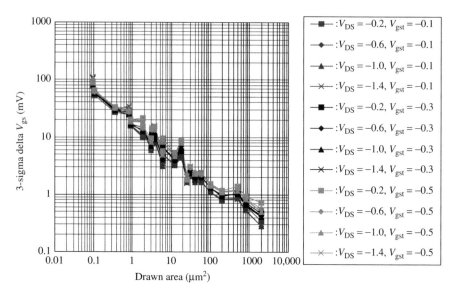

Figure 11.4   Measured data for 3-sigma mismatching of ΔV_{gs} in PMOSFETs. Reproduced from Cheng Y. (2002) The influence and modeling of process variation and device mismatch for analog/RF circuit design, *Proceedings of the Fourth IEEE International Caracas Conference on Devices, Circuits and Systems*, pp. 282–289

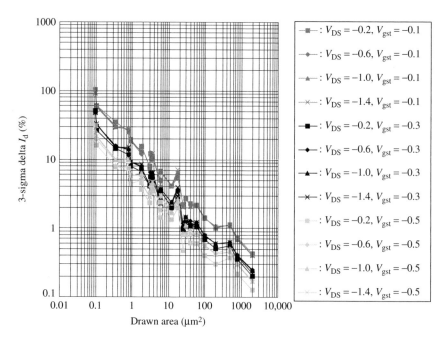

Figure 11.5   Measured data for 3-sigma mismatching of ΔI_d in PMOSFETs. Reproduced from Cheng Y. (2002) The influence and modeling of process variation and device mismatch for analog/RF circuit design, *Proceedings of the Fourth IEEE International Caracas Conference on Devices, Circuits and Systems*, pp. 282–289

11.3 MODELING OF DEVICE MISMATCH FOR ANALOG/RF APPLICATIONS

In the above section, we have reviewed the process variation and device mismatch and their influence on the device characteristics. Next we discuss the statistical modeling of the LPVM of these components.

11.3.1 Modeling of Mismatching of Resistors

The following equation is typically used to model a (polysilicon) resistor:

$$R = R_{sh} \frac{L}{W + \Delta W} + \frac{R_e}{W + \Delta W} \tag{11.6}$$

where R_{sh} is the sheet resistance of the poly resistor, R_e is the end resistance coefficient, W and L are resistor width and length, and ΔW is the resistor width offset.

The correlations between standard deviations (σ) of the model parameters and the standard deviation of the resistance are given in the following:

$$\sigma_R{}^2 = \sigma_{R_{sh}}{}^2 \left[\frac{\delta R}{\delta R_{sh}} \right]^2 + \sigma_{R_e}{}^2 \left[\frac{\delta R}{\delta R_e} \right]^2 + \sigma_{\Delta W}{}^2 \left[\frac{\delta R}{\delta \Delta W} \right]^2, \tag{11.7}$$

$$\sigma_R{}^2 = \sigma_{R_{sh}}{}^2 \frac{L^2}{(W + \Delta W)^2} + \sigma_{R_e}{}^2 \frac{1}{(W + \Delta W)^2} + \sigma_{\Delta W}{}^2 \left[\frac{L \cdot R_{sh}}{(W + \Delta W)^2} + \frac{R_e}{(W + \Delta W)^2} \right]^2. \tag{11.8}$$

To define the resistor matching, based on Eq. (11.7), we have

$$\sigma_{\Delta R/R}{}^2 = \sigma_{R_{sh}}{}^2 \left[\frac{L}{(L \cdot R_{sh} + R_e)} \right]^2 + \sigma_{R_e}{}^2 \left[\frac{1}{(L \cdot R_{sh} + R_e)} \right]^2 + \sigma_{\Delta W}{}^2 \left[\frac{1}{(W + \Delta W)} \right]^2, \tag{11.9}$$

$$\sigma_{R_{sh}} = \frac{A_{R_{sh}}}{(W \cdot L)^{1/2}}, \tag{11.10}$$

$$\sigma_{R_e} = A_{R_e}, \tag{11.11}$$

$$\sigma_{\Delta W} = \frac{A_{\Delta W}}{W^{1/\sqrt{2}}}. \tag{11.12}$$

Figure 11.6 shows the comparison result between the above model and the measured data of 3-sigma mismatching of unsalicided polysilicon resistors with different geometries. The maximum error percentage between the model and the data is less than 3%.

11.3.2 Mismatching Model of Capacitors

A simple MIM capacitor matching model, considering the most important sources of capacitance variation, is given in the following equations:

$$\sigma_{\Delta_c/C}{}^2 = \sigma_p^2 + \sigma_a^2 + \sigma_d^2, \tag{11.13}$$

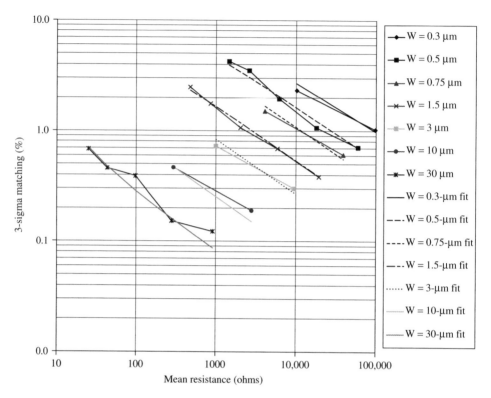

Figure 11.6 Comparison between the model and the measured data for 3-sigma mismatching of resistors. Reproduced from Cheng Y. (2002) The influence and modeling of process variation and device mismatch for analog/RF circuit design, *Proceedings of the Fourth IEEE International Caracas Conference on Devices, Circuits and Systems*, pp. 282–289

$$\sigma_p = \frac{f_p}{C^{3/4}}, \tag{11.14}$$

$$\sigma_a = \frac{f_a}{C^{1/2}}, \tag{11.15}$$

$$\sigma_d = f_d \cdot d, \tag{11.16}$$

where f_p, f_a, and f_d are constants describing the influence of periphery, area, and distance fluctuations.

The periphery component (the first term in Eq. (11.13)) models the effect of edge roughness, and it is most significant for small capacitors, which have a relatively large amount of edge capacitance. The area component (the second term in Eq. (11.13)) models the effect of short-range dielectric thickness variations, and it is most significant for moderate size capacitors. The distance component models the effect of global dielectric thickness variations across the wafer, and it becomes significant for large capacitors or widely spaced capacitors.

Figure 11.7 presents the comparison results between the model and the data for MIM capacitors in a 0.18-μm CMOS process. In Figure 11.7, the distance, that is, the "space"

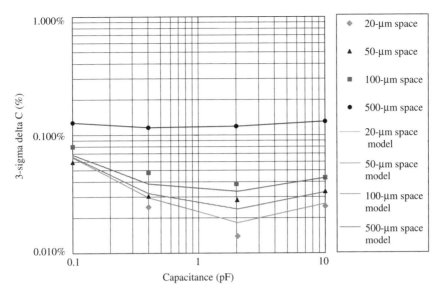

Figure 11.7 Comparison between the model and the measured data for 3-sigma mismatching of MIM capacitors. Reproduced from Cheng Y. (2002) The influence and modeling of process variation and device mismatch for analog/RF circuit design, *Proceedings of the Fourth IEEE International Caracas Conference on Devices, Circuits and Systems*, pp. 282–289

in the figure, is defined from the center of one capacitor to the center of the other capacitor. It shows that the simple model given by Eqs. (11.3) to (11.6) can predict the measured data well.

11.3.3 Mismatching Models of MOSFETs

11.3.3.1 Simple model

The following simple first-order model has been used to the model ΔI_d results in saturation region:

$$\Delta_{I_d} = \Delta_\beta \left[\frac{\delta I_d}{\delta \beta} \right] + \Delta_{V_t} \left[\frac{\delta I_d}{\delta V_{th}} \right], \tag{11.17}$$

$$\sigma_{\Delta I_d / I_d}{}^2 = \frac{\sigma_\beta^2}{\beta^2} + \sigma_{V_t}^2 \left[\frac{2}{V_{gs} - V_{th}} \right]^2, \tag{11.18}$$

$$\sigma_\beta = \frac{f_\beta}{\sqrt{W \cdot L}}, \tag{11.19}$$

$$\sigma_{V_{th}} = \frac{f_{V_{th}}}{\sqrt{W \cdot L}}, \tag{11.20}$$

where $I_d = \beta(V_{gs} - V_{th})^2$ and $\beta = [(\mu_o \cdot C_{ox})/2] \cdot [W/L]$, μ_o is the mobility, C_{ox} is the unit-area oxide capacitance, W is the channel width, and L is the channel length. $\sigma_{v_{th}}$ and

σ_β are standard deviations of V_{th} and β parameters. f_a and f_p are constants describing the influence of area and periphery fluctuations.

Note that it is better to express the β dependence as a function of σ_β/β in percentage, so the absolute value of β is not important and the $V_{gs} - V_{th}$ term can be expressed as V_{gst}. The equation can be rewritten in the following way, which is independent of the model parameters:

$$\sigma_{\Delta I_d/I_d}{}^2 = \left(\frac{\sigma_{\beta\%}}{100}\right)^2 + \sigma_{V_t}{}^2 \left[\frac{2}{V_{gst}}\right]^2, \tag{11.21}$$

$$\sigma_{\beta\%} = \frac{f_{\beta\%}}{\sqrt{W \cdot L}}, \tag{11.22}$$

$$\sigma_{V_{th}} = \frac{f_{V_{th}}}{\sqrt{W \cdot L}}. \tag{11.23}$$

Figure 11.8 gives the comparison between prediction of the simple model and the measured data of 3-sigma mismatching of the devices of 10-μm channel width and 0.18-μm channel length at various bias conditions. The model describes the trend of the 3-sigma

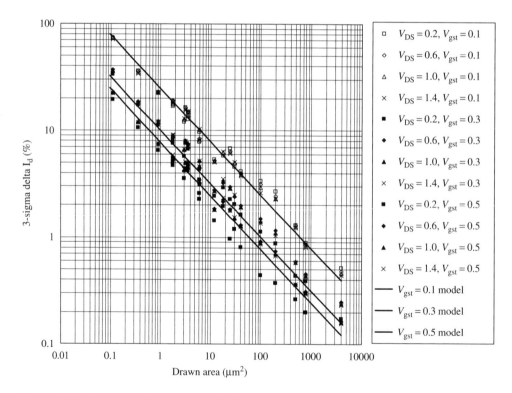

Figure 11.8 Comparison between the simple model and the measured 3-sigma mismatching of MOSFETs. Reproduced from Cheng Y. (2002) The influence and modeling of process variation and device mismatch for analog/RF circuit design, *Proceedings of the Fourth IEEE International Caracas Conference on Devices, Circuits and Systems*, pp. 282–289

mismatching of the device reasonably well by using the extracted values of $\sigma_{\beta\%}$ and $\sigma_{v_{th}}$ from the measured data.

11.3.3.2 A physical model

The simple model discussed above is based on two parameters, used widely by designers, V_{th} and β. However, these two parameters are not independent process variables and depend on many physical process parameters, such as oxide thickness, flat band, channel-length offset, channel width offset, mobility, and doping concentrations in the channel, source/drain, and substrate. By using V_{th} and β directly instead of using these fundamental process parameters, the correlations between V_{th} and β will not be accounted for physically in the model, so the simulated results may not predict correctly the device mismatching behavior, depending on the variation defined for V_{th} and β. Also, V_{gs} mismatch is affected by these independent process parameters instead of just the V_{th} mismatch, which is only a small contribution to the V_{gs} mismatch in some cases. So a mismatching model based on the independent process variables is desirable.

Recently, approaches based on the independent variables to model the mismatching have been reported (Zhang and Liou (2001); Oelun *et al.* (2000); Drennan and McAndrew (1999)). Since the SPICE models such as BSIM3 are implemented in simulators in a way such that the model equations, some of which include expressions with replaced correlations for the independent process variables by fitting parameters, have been hard-coded, it is difficult to reuse the correlations in statistical and mismatch simulation. So a subcircuit approach is adopted (Cheng (2002)), on one hand, to account for the important correlations of the totally independent process variables in the statistical/mismatching model, and on the other hand, to include the important components in RF applications

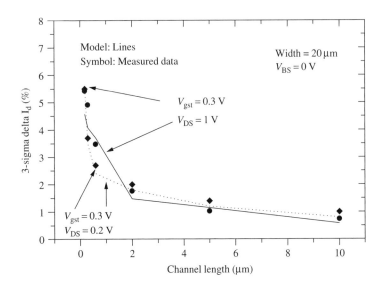

Figure 11.9 Model versus data for devices with different channel lengths. Reproduced from Cheng Y. (2002) The influence and modeling of process variation and device mismatch for analog/RF circuit design, *Proceedings of the Fourth IEEE International Caracas Conference on Devices, Circuits and Systems*, pp. 282–289

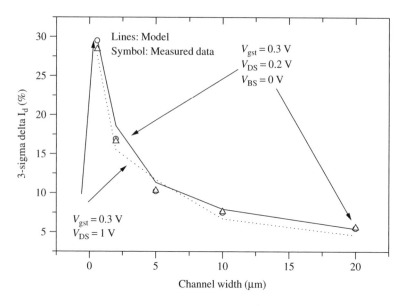

Figure 11.10 Measured data and model for devices with different channel widths. Reproduced from Cheng Y. (2002) The influence and modeling of process variation and device mismatch for analog/RF circuit design, *Proceedings of the Fourth IEEE International Caracas Conference on Devices, Circuits and Systems*, pp. 282–289

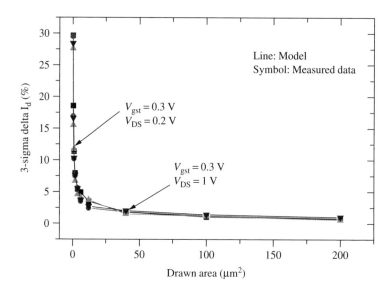

Figure 11.11 Measured data and model for devices with different area ($W \times L$). Reproduced from Cheng Y. (2002) The influence and modeling of process variation and device mismatch for analog/RF circuit design, *Proceedings of the Fourth IEEE International Caracas Conference on Devices, Circuits and Systems*, pp. 282–289

such as gate resistance, substrate resistance, and so on. Both the influence of process variation and device mismatching has been considered in the model. For example, the parameters such as oxide thickness (T_{ox}), doping concentrations in the channel (N_{ch}), in pocket region (N_{pocket}), in source/drain region (N_{sd}), in low-doped drain (LDD) region (N_{ldd}), in the gate (N_{gate}), in the substrate (N_{sub}), channel-length offset (ΔL), and channel width offset (ΔW) are selected as independent process variables. Some model parameters (in BSIM3v3) such as V_{th0}, K_1, K_2, R_{dsw}, C_{j0}, C_{jsw}, C_{jswg}, μ_0, D_{rout}, N_{lx}, D_{vt1}, and so on are modeled as dependent variables linked to these independent parameters by the derived correlations in the subcircuit model.

Figure 11.9 gives the comparison of the model and the data of devices with different channel lengths at two different bias conditions (in both linear and saturation). Similarly, Figures 11.10 and 11.11 show the comparison of the 3-sigma ΔI_d versus channel and area between the model and the data, respectively. To predict the influence of the process variation and device match on the device behavior, both the selection of independent process/device variables and the creation of the correlations linking the independent variables and the device model parameters are important. For modern technology ($0.13\,\mu m$ or less), additional physical effects, becoming significant in advanced technology, should be considered in establishing the correlations. When considering most important physical effects and the correlations between the independent process variables, the model can predict the device mismatch behavior well for devices with different geometry at various bias conditions.

REFERENCES

Box G. *et al.* (1978) *Statistics for Experiments*, John Wiley & Sons, New York.

Cheng Y. (2002) The influence and modeling of process variation and device mismatch for analog/RF circuit design, *Proceedings of the Fourth IEEE International Caracas Conference on Devices, Circuits and Systems*, pp. 282–289.

Drennan P. G. and McAndrew C. C. (1999) A comprehensive MOSFET mismatch model, *Proc. IEEE Int. Electron Device Meeting*, pp. 167–170.

Groon J. A. *et al.* (2001) A simple characterization method for MOS transistor mismatching in deep-submicron technology, *Proceedings of the 2001 International Conference on Microelectronic Test Structures*, pp. 213–218.

Lane W. A. and Wrixon G. T. (1989) The design of thin polysilicon resistors for analog IC applications, *IEEE Trans. Electron Devices*, **36**(4), 738–744.

Lee S. *et al.* (2000) An efficient statistical model using electrical tests for GHz CMOS devices, *5th International Workshop on Statistical Metrology*, pp. 72–75.

Oelun J. *et al.* (2000) A physical approach to mismatch modeling and parameter correlations, *Proceedings of the 2000 IEEE International Symposium on Circuits and Systems*, Vol. 4, Geneva., pp. 377–380.

Pelgrom M. *et al.* (1998) Transistor matching in analog CMOS applications, *Tech. Dig. (IEDM)*, 915–918.

Pronath M. *et al.* (2000) WiCkeD: analog circuit synthesis incorporating mismatch, *Proceedings of the IEEE 2000 Custom Integrated Circuits Conference*, pp. 511–514.

Shyu J. B. *et al.* (1982) Random Errors in MOS capacitors, *IEEE J. Solid-State Circuits*, **sc-17**(6), 1070–1076.

Shyu J. B. *et al.* (1984) Random error effects in matched MOS capacitors and current sources, *IEEE Journal of Solid-State Circuits*, **19**(6), 948–956.

Tarim T. B. *et al.* (2000) Application of a statistical design methodology to low voltage analog MOS integrated circuits, *The 2000 IEEE International Symposium on Circuits and Systems*, Vol. 4, Geneva, pp. 117–120.

Thewes R. *et al.* (2000) Mismatch of MOSFET small signal parameters under analog operation, *IEEE Electron Device Lett.*, **21**(12), 552, 553.

Zanella S. *et al.* (1999) Analysis of the impact of intra-die variance on clock skew, *1999 4th International Workshop on Statistical Metrology*, pp. 14–17.

Zhang Q. and Liou J. J. (2001) SPICE modeling and quick estimation of MOSFET mismatch based on BSIM3 model and parametric tests, *IEEE J. Solid-State Circuits*, **36**(10), 1592–1595.

12

Quality Assurance of MOSFET Models

12.1 INTRODUCTION

The present MOSFET models for analog circuit simulation (such as BSIM3, BSIM4, and Philips MOS Model 9) are all based on the one-dimensional (1D) theory initially developed for long-channel MOSFETs. To keep pace with technology, phenomenological modifications of the device models are necessary, resulting in a steady erosion of the physical basis and a plethora of model parameters of obscure origin. It is not uncommon today that the device models contain more than hundred parameters to take into account the many subtle mechanisms that govern the characteristics of deep submicron device structures. With this large number of parameters to play around with and if the model parameters are not extracted carefully and correctly, strange effects often surface in the modeled device characteristics, especially in the small-signal quantities such as transconductance and channel conductance. This has become a major problem for the analog circuit designers that rely on precise modeling of the devices to accurately predict the behavior of their designs to ensure first-time-right silicon and to reduce the time-to-market for a product.

An automated system has been implemented for quality assurance (QA) of MOSFET device models that was described for the first time in Risanger *et al.* (2000). Typically, the device models are supplied from the foundries. Upon arrival at the design house, the device models are run through a set of benchmark tests and qualified before circuit designers are allowed to use them.

In this chapter we describe the various aspects of this QA system. We start out by giving a thorough motivation for the importance of such a system. Then we discuss some of the benchmark tests that are included in the QA system. Finally, we describe how the tasks of performing the tests are automated.

12.2 MOTIVATION

As mentioned above, the device models used today may contain more than hundred parameters. Without careful parameter extraction and good methodology, the model parameter

Device Modeling for Analog and RF CMOS Circuit Design. T. Ytterdal, Y. Cheng and T. A. Fjeldly
© 2003 John Wiley & Sons, Ltd ISBN: 0-471-49869-6

sets produced by the automated characterization systems at the foundries may cause the models to behave strangely. This has become a major obstacle for the analog circuit designers that rely on reliable models to simulate their designs accurately and efficiently. In this section we will discuss the most severe problems encountered by design engineers that motivated the introduction of a QA system.

One problem discovered is related to BSIM3v3.1 and its approach to binning. In Figure 12.1, we show the plot of modeled channel conductance g_{ds} versus gate length for a commercially available 0.5-µm CMOS process. The characterization of this particular process utilizes the built-in binning feature of the BSIM3 model. Binning is a model extraction approach in which the geometrical space is divided into regions, each having a separate model parameter set. A common problem with this approach is discontinuities of electrical characteristics across bin boundaries as shown in Figure 12.1. Note that this model even predicts a higher output conductance at a gate length of 2.1 µm compared to a length of 1.9 µm, which is obviously not the case for the real device.

Another problem we have observed in the modeling of g_{ds} is shown in Figure 12.2. In this case, the model predicts a monotonically increasing output conductance with gate length for devices with gate lengths larger than 10 µm. This example is also from a commercially available 0.5-µm CMOS process, but the model used here was Philips MOS Model 9.

Without proper model parameter extraction, the intrinsic gain (g_m/g_{ds}) of a MOSFET predicted by BSIM3v3.1 may contain a nonphysical bump for gate-source voltages V_{gs} close to the threshold voltage. In the example shown in Figure 12.3, we have plotted the intrinsic gain versus V_{gs} using a BSIM3v3.1 model parameter set for a commercially available 0.18-µm CMOS process. We notice a quite dramatic bump in the gain close to the threshold voltage. Such a behavior will certainly confuse a circuit design engineer who is looking for the optimum biasing for his/her amplifier design.

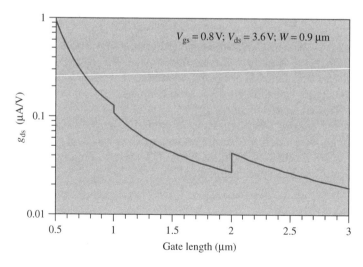

Figure 12.1   Discontinuous channel conductance across bins for BSIM3v3.1. Reproduced from Risanger J. S., Raaum J., and Ytterdal T. (2000) Quality assurance of MOSFET models for analog circuit design, *First Online Symposium for Electronics Engineers*, September 2000

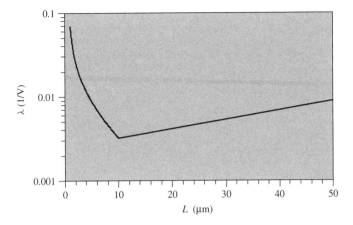

Figure 12.2 Nonphysical modeling of normalized output conductance λ. Reproduced from Risanger J. S., Raaum J., and Ytterdal T. (2000) Quality assurance of MOSFET models for analog circuit design, *First Online Symposium for Electronics Engineers*, September 2000

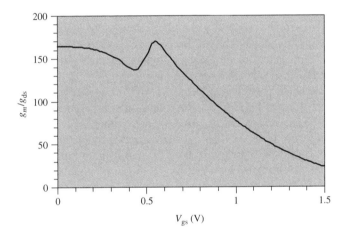

Figure 12.3 Nonphysical modeling of intrinsic gain. Reproduced from Risanger J. S., Raaum J., and Ytterdal T. (2000) Quality assurance of MOSFET models for analog circuit design, *First Online Symposium for Electronics Engineers*, September 2000

12.3 BENCHMARK CIRCUITS

Many of the tests implemented for MOSFET device models were originally described in Tsividis and Suyama (1994) and by the SEMATECH benchmark circuits described by the Compact MOSFET Council Web page at *http://www.eigroup.org/cmc/*. Since the introduction of the QA system, the number of tests implemented has grown to cover new features and new problems encountered in new technologies released for the commercial market. The following tests have been implemented:

- Drain current characteristics
- Transfer characteristics in weak and moderate inversion

- Transconductance to current ratio
- Channel conductance
- Temperature dependency of junction leakage currents
- Charge conservation
- Non-quasi-static operation
- Thermal noise
- Flicker noise
- Channel conductance versus channel length and width
- Intrinsic gain (g_m/g_{ds})
- Gate leakage current.

Selected tests are discussed in detail in the following sections.

12.3.1 Leakage Currents

With the continuous downscaling of the line widths in integrated circuit processes, the MOSFET leakage currents eventually dominate the current consumption of low power integrated circuits. Accurate modeling of the leakage currents at high and low temperatures are thus of vital importance for designing complex mixed-mode circuits having current consumption levels in the sub-microampere regime.

To reveal the leakage currents modeled in MOSFET device models, the test setup given in Figure 12.4 could be utilized. Here, an n-channel transistor is biased at a gate-source voltage of zero and a drain-source voltage equal to half of the power supply voltage. The drain current I_D is then calculated versus temperature. It is also possible to split the drain current into a drain-source I_{ds} and a drain-bulk I_{db} current to be able to study the two contributions individually. To illustrate qualitatively correct leakage current characteristics, we have simulated a state-of-the-art 0.13-μm nMOS transistor in a commercially available process. The width of the device was 0.8 μm. A qualitatively correct plot of the drain current is shown in Figure 12.5 and the corresponding plot of the individual contributions to the drain current is shown in Figure 12.6.

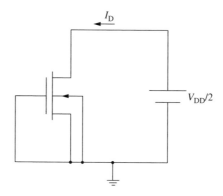

Figure 12.4   Test setup for calculating the leakage currents

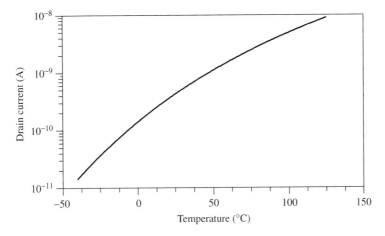

Figure 12.5 Qualitatively correct drain current of an n-channel MOS transistor having a gate-source voltage of 0 V. The threshold voltage of this process is about 0.4 V

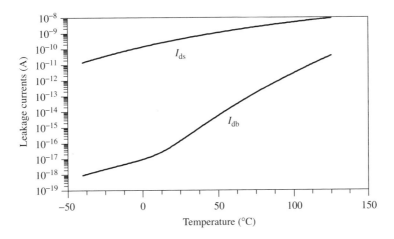

Figure 12.6 Drain-source I_{ds} and drain-bulk I_{db} leakage currents versus temperature

We note from Figure 12.5 that the temperature dependency of the drain current is close to exponential in the entire temperature range. To reproduce a similar plot in your circuit simulator, the option GMIN must be set to a very low value, much lower than its default value of 10^{-12} S. To check that the leakage of the bulk diodes is properly modeled versus temperature, a plot of I_{db} as the one shown in Figure 12.6 should be produced by the model.

12.3.2 Transfer Characteristics in Weak and Moderate Inversion

As the power supply voltages are scaled down from one technology generation to the next, the gate voltage overdrives used in analog CMOS circuits keep shrinking. Thus, it becomes increasingly important to accurately model the moderate and weak inversion regions. This test focuses on the three following issues: inspection of the moderate inversion region, check for proper inclusion of the drain-induced barrier lowering (DIBL) effect (see Section 1.5.2.2), and assure correct modeling of the body effect.

The DIBL is recognized as one of the short-channel effects. Thus, all models describing typical short-channel MOSFETs should be tested. DIBL clearly influences both the leakage current and the operating current of the MOSFET device. The test setup is shown in Figure 12.7. This test should be run twice, once for investigating the DIBL and once for investigating the body effect.

In the simulation of the DIBL effect, the body and the source terminals are shorted. In Figure 12.8 a qualitatively correct plot is shown. The drain current clearly increases with the drain voltage owing to the induced shift in the threshold voltage.

12.3.3 Gate Leakage Current

As the CMOS technologies are scaled down into the very deep submicron regime, the gate oxide thickness becomes thinner and thinner. As a consequence, the gate current is

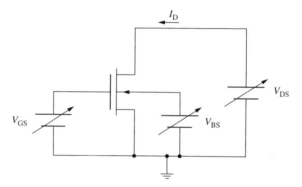

Figure 12.7 Test setup for simulating weak and moderate inversion characteristics

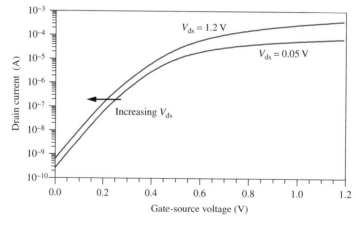

Figure 12.8 Simulated transfer characteristics for a deep submicrometer nMOS with a gate length of $0.13\,\mu m$

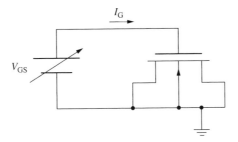

Figure 12.9   Test setup for simulating gate leakage current

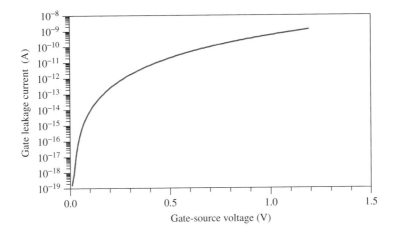

Figure 12.10   Simulated gate current versus gate-source voltage. The oxide thickness was 3 nm

increased considerably. At the 0.13-μm technology node, the static gate current is on the order of nanoamperes. Thus, in the design of low-power circuits, this current component can no longer be neglected and proper modeling of the gate current characteristics becomes important. On the basis of this, we have recently defined a gate leakage current test. The setup of this test is shown in Figure 12.9.

As shown in Figure 12.9, all terminals of the transistor except the gate are connected to ground. The gate-source voltage is swept from zero to the maximum value allowed for the process. An example of the gate voltage dependency on the gate current is shown in Figure 12.10. The simulated results shown in the figure were performed at room temperature. Since the gate current is dominated by tunneling, the temperature dependency is very weak.

12.4 AUTOMATION OF THE TESTS

The benchmark tests have been implemented at Nordic VLSI (www.nvlsi.com) using the command language called SimPilot and are performed automatically every time a new MOSFET model parameter set is received from a foundry. A schematic overview of the system implementation is given in Figure 12.11. SimPilot connects directly to the

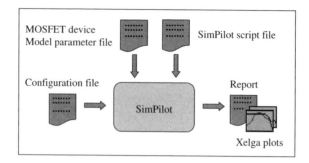

Figure 12.11   Overview of the implementation of the automated system. Reproduced from Risanger J. S., Raaum J., and Ytterdal T. (2000) Quality assurance of MOSFET models for analog circuit design, *First Online Symposium for Electronics Engineers*, September 2000

commercial circuit simulator Eldo and produces output files adapted for the graphical post-processor Xelga[1]. In addition to the command script file, SimPilot must be loaded with the actual MOSFET device model parameter file and a configuration file holding information about device geometries, temperatures, voltages, and so on. The command script file has a syntax almost identical to that used in UNIX scripts with the addition of numerous postprocessing commands specific to SimPilot. The command script may be executed in batch mode for efficient computer resource utilization. Although the script is considered static, it is easy to expand it with new functional tests if necessary. However, the user settings given in the configuration file have to follow a predefined setup.

The graphical output files created by SimPilot are given names in accordance to the present benchmark test and are observed visually to detect possible shortcomings in the MOSFET device model under test. Finally, a report file containing the test results and important notes is generated and published online on the Intranet to make the results available to the circuit designers.

REFERENCES

Risanger J. S., Raaum J., and Ytterdal T. (2000) Quality assurance of MOSFET models for analog circuit design, *First Online Symposium for Electronics Engineers*, September 2000, see *http://www.techonline.com/osee/*.

Tsividis Y. P. and Suyama K. (1994) MOSFET modeling for analog circuit CAD: problems and prospects, *IEEE J. Solid-State Circuits*, **29**(3), 210–216.

[1] SimPilot, Eldo, and Xelga are products of Mentor Graphics.

Index

Device Modeling for Analog and RF CMOS Circuit Design. T. Ytterdal, Y. Cheng and T. A. Fjeldly
© 2003 John Wiley & Sons, Ltd ISBN: 0-471-49869-6